Lecture Notes in Statistics 191

Edited by P. Bickel, P. Diggle, S. Fienberg, U. Gather,
I. Olkin, S. Zeger

For other titles published in this series, go to
www.springer.com/series/694

Vlad Stefan Barbu
Nikolaos Limnios

Semi-Markov Chains and Hidden Semi-Markov Models toward Applications

Their Use in Reliability and DNA Analysis

 Springer

Vlad Stefan Barbu
Université de Rouen, Rouen
France
vladstefan.barbu@univ-rouen.fr

Nikolaos Limnios
Université de Technologie de Compiègne
Compiègne
France
nikolaos.limnios@utc.fr

ISBN: 978-0-387-73171-1   e-ISBN: 978-0-387-73173-5
DOI: 10.1007/978-0-387-73173-5

Library of Congress Control Number: 2008930930

Printed on acid-free paper

9 8 7 6 5 4 3 2 1

springer.com

To our parents
Cristina and Vasile
Myrsini and Stratis

Preface

Semi-Markov processes are a generalization of Markov and of renewal processes. They were independently introduced in 1954 by Lévy (1954), Smith (1955) and Takacs (1954), who essentially proposed the same type of process. The basic theory was given by Pyke in two articles (1961a,b). The theory was further developed by Pyke and Schaufele (1964), Çinlar (1969, 1975), Koroliuk and his collaborators, and many other researchers around the world.

Nowadays, semi-Markov processes have become increasingly important in probability and statistical modeling. Applications concern queuing theory, reliability and maintenance, survival analysis, performance evaluation, biology, DNA analysis, risk processes, insurance and finance, earthquake modeling, etc.

This theory is developed mainly in a continuous-time setting. Very few works address the discrete-time case (see, e.g., Anselone, 1960; Howard, 1971; Mode and Pickens, 1998; Vassiliou and Papadopoulou, 1992; Barbu et al., 2004; Girardin and Limnios, 2004; Janssen and Manca, 2006). The present book aims at developing further the semi-Markov theory in the discrete-time case, oriented toward applications.

This book presents the estimation of discrete-time finite state space semi-Markov chains under two aspects. The first one concerns an observable semi-Markov chain Z, and the second one an unobservable semi-Markov chain Z with a companion observable chain Y depending on Z. This last setting, described by a coupled chain (Z, Y), is called a hidden semi-Markov model (HSMM).

In the first case, we observe a single truncated sample path of Z and then we estimate the semi-Markov kernel $\mathbf{q}$, which governs the random evolution of the chain. Having an estimator of $\mathbf{q}$, we obtain plug-in-type estimators for other functions related to the chain. More exactly, we obtain estimators of reliability, availability, failure rates, and mean times to failure and we present their asymptotic properties (consistency and asymptotic normality) as the length of the sample path tends to infinity. Compared to the common use of

Markov processes in reliability studies, semi-Markov processes offer a much more general framework.

In the second case, starting from a truncated sample path of chain Y, we estimate the characteristics of the underlying semi-Markov chain as well as the conditional distribution of Y. This type of approach is particularly useful in various applications in biology, speech and text recognition, and image processing. A lot of work using hidden Markov models (HMMs) has been conducted thus far in these fields. Combining the flexibility of the semi-Markov chains with the advantages of HMMs, we obtain hidden semi-Markov models, which are suitable application tools and offer a rich statistical framework.

The aim of this book is threefold:

- To give the basic theory of finite state space semi-Markov processes in discrete time;
- To perform a reliability analysis of semi-Markov systems, modeling and estimating the reliability indicators;
- To obtain estimation results for hidden semi-Markov models.

The book is organized as follows.

In Chapter 1 we present an overview of the book.

Chapter 2 is an introduction to the standard renewal theory in discrete time. We establish the basic renewal results that will be needed subsequently.

In Chapter 3 we define the Markov renewal chain, the semi-Markov chain, and the associated processes and notions. We investigate the Markov renewal theory for a discrete-time model. This probabilistic chapter is an essential step in understanding the rest of the book. We also show on an example how to practically compute the characteristics of such a model.

In Chapter 4 we construct nonparametric estimators for the main characteristics of a discrete-time semi-Markov system (kernel, sojourn time distributions, transition probabilities, etc.). We also study the asymptotic properties of the estimators. We continue the example of the previous chapter in order to numerically illustrate the qualities of the obtained estimators.

Chapter 5 is devoted to the reliability theory of discrete-time semi-Markov systems. First, we obtain explicit expressions for the reliability function of such systems and for its associated measures, like availability, maintainability, failure rates, and mean hitting times. Second, we propose estimators for these indicators and study their asymptotic properties. We illustrate these theoretical results for the model described in the example of Chapters 3 and 4, by computing and estimating reliability indicators.

In Chapter 6 we first introduce the hidden semi-Markov models (HSMMs), which are extensions of the well-known HMMs. We take into account two types of HSMMs. The first one is called SM-M0 and consists in an observed sequence of conditionally independent random variables and of a hidden (unobserved) semi-Markov chain. The second one is called SM-Mk and differs from the previous model in that the observations form a conditional Markov chain of order k. For the first type of model we investigate the asymptotic properties of the nonparametric maximum-likelihood estimator (MLE), namely, the consistency and the asymptotic normality. The second part of the chapter proposes an EM algorithm that allows one to find practically the MLE of a HSMM. We propose two different types of algorithms, one each for the SM-M0 and the SM-M1 models. As the MLE taken into account is nonparametric, the corresponding algorithms are very general and can also be adapted to obtain particular cases of parametric MLEs. We also apply this EM algorithm to a classical problem in DNA analysis, the CpG islands detection, which generally is treated by means of hidden Markov models.

Several exercises are proposed to the reader at the end of each chapter. Some appendices are provided at the end of the book, in order to render it as self contained as possible. Appendix A presents some results on semi-Markov chains that are necessary for the asymptotic normality of the estimators proposed in Chapters 4 and 5. Appendix B includes results on the conditional independence of hidden semi-Markov chains that will be used for deriving an EM algorithm (Chapter 6). Two additional complete proofs are given in Appendix C. In Appendix D some basic definitions and results on finite-state Markov chains are presented, while Appendix E contains several classic probabilistic and statistical results used throughout the book (dominated convergence theorem in discrete time, asymptotic results of martingales, Delta method, etc.).

A few words about the title of the book. We chose the expression "toward applications" so as to make it clear from the outset that throughout the book we develop the theoretical material in order to offer tools and techniques useful for various fields of application. Nevertheless, because we speak only about reliability and DNA analysis, we wanted to specify these two areas in the subtitle. In other words, this book is not only theoretical, but it is also application-oriented.

The book is mainly intended for applied probabilists and statisticians interested in reliability and DNA analysis and for theoretically oriented reliability and bioinformatics engineers; it can also serve, however, as a support for a six-month Master or PhD research-oriented course on semi-Markov chains and their applications in reliability and biology.

The prerequisites are a background in probability theory and finite state space Markov chains. Only a few proofs throughout the book require elements of measure theory. Some alternative proofs of asymptotic properties of estimators require a basic knowledge of martingale theory, including the central limit theorem.

The authors express their gratitude to Mei-Ling Ting Lee (Ohio State University) for having drawn their attention to hidden semi-Markov models for DNA analysis. They are also grateful to their colleagues, in particular N. Balakrishnan, G. Celeux, V. Girardin, C. Huber, M. Iosifescu, V.S. Koroliuk, M. Nikulin, G. Oprisan, B. Ouhbi, A. Sadek, and N. Singpurwalla for numerous discussions and/or comments. Our thanks go to M. Boussemart, for his initial participation in this research project, and to the members of the statistical work group of the Laboratory of Mathematics Raphaël Salem (University of Rouen), for helpful discussions on various topics of the book. The authors also wish to thank J. Chiquet and M. Karaliopoulou who read the manuscript and made valuable comments. Particularly, we are indebted to S. Trevezas for having tracked mistakes in the manuscript and for our discussions on this subject.

The authors would equally like to thank Springer editor John Kimmel for his patience, availability, advice, and comments, as well as the anonymous referees who helped improve the presentation of this book by way of useful comments and suggestions.

It is worth mentioning that this book owes much, though indirectly, to the "European seminar" (http://www.dma.utc.fr/~nlimnios/SEMINAIRE/).

Rouen, France   *Vlad Stefan BARBU*
Compiègne, France   *Nikolaos LIMNIOS*
March, 2008

Contents

1

Introduction

This introductory chapter has a twofold purpose: to answer the basic question "Why do we think that the topic presented in the book is interesting and worth studying?" and to give an overall presentation of the main features discussed in the book.

1.1 Object of the Study

First, we want to look at the main reasons which motivated us to study the topic of this book. Basically, here we answer two questions: "Why semi-Markov?" and "Why work in discrete time?"

1.1.1 The Underlying Idea in Semi-Markov Models

Much work has been carried out in the field of Markov processes, and a huge amount of Markov process applications can be found in the literature of the last 50 years. One of the reasons for applying Markov process theory in various fields is that the Markovian hypothesis is very intuitive and convenient when dealing with applications and the underlying computations are quite simple. One can formulate this hypothesis as follows: if the past and the present of a system are known, then the future evolution of the system is determined only by its present state, or equivalently, the past and the future are conditionally independent given the present (state). Thus the past history of a system plays no role in its future evolution, which is usually known as the "memoryless property of a Markov process." But the Markovian hypothesis imposes restrictions on the distribution of the sojourn time in a state, which should be exponentially distributed (continuous case) or geometrically distributed (discrete case). This is the main drawback when applying Markov processes in real applications.

V.S. Barbu, N. Limnios, *Semi-Markov Chains and Hidden Semi-Markov Models toward Applications*, DOI: 10.1007/978-0-387-73173-5_1,

What came naturally was to relax the underlying Markov assumption in order to:

- Allow arbitrarily distributed sojourn times in any state;
- Still have the Markovian hypothesis, but in a more flexible manner.

A process that has these two properties will be called a *semi-Markov process*.

To be more specific, let us consider a random system with finite state space $E = \{1, \ldots, s\}$, whose evolution in time is governed by a stochastic process $Z = (Z_k)_{k \in \mathbb{N}}$. We note here that all stochastic processes taken into account throughout this book are considered to evolve in discrete time. We will use the term *chain* for a discrete-time stochastic process.

Let us also denote by $S = (S_n)_{n \in \mathbb{N}}$ the successive time points when state changes in $(Z_n)_{n \in \mathbb{N}}$ occur and by $J = (J_n)_{n \in \mathbb{N}}$ the chain which records the visited states at these time points. Let $X = (X_n)_{n \in \mathbb{N}}$ be the successive sojourn times in the visited states. Thus, $X_n = S_n - S_{n-1}$, $n \in \mathbb{N}^*$, and, by convention, we set $X_0 = S_0 = 0$.

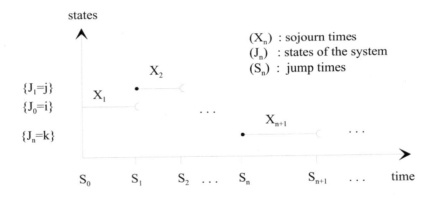

Fig. 1.1. Sample path of a semi-Markov chain

The relation between process Z and process J of the successively visited states is given by

$$Z_k = J_{N(k)}, \text{ or, equivalently, } J_n = Z_{S_n}, n, k \in \mathbb{N},$$

where $N(k) := \max\{n \in \mathbb{N} \mid S_n \leq k\}$ is the discrete-time counting process of the number of jumps in $[1, k] \subset \mathbb{N}$. Thus Z_k gives the system's state at time k.

Suppose the following conditional independence relation holds true almost surely:

$$\mathbb{P}(J_{n+1} = j, S_{n+1} - S_n = k \mid J_0, \ldots, J_n; S_0, \ldots, S_n)$$
$$= \mathbb{P}(J_{n+1} = j, S_{n+1} - S_n = k \mid J_n). \qquad (1.1)$$

This equation means: if we know the past visited states and jump times of the system, as well as its present state, the future visited state and the sojourn time in the present state depend only on the present state. In other words, we basically have a Markovian-type hypothesis, with the difference that the memoryless property does not act on the calendar time $(0, 1, \ldots, k, k+1, \ldots)$ but on a time governed by the jump time process J, $(J_0, J_1, \ldots, J_n, J_{n+1}, \ldots)$. This is what we called before a more flexible Markovian hypothesis.

If Equation (1.1) holds true, then $Z = (Z_n)_{n \in \mathbb{N}}$ is called a *semi-Markov chain* (SMC) and the couple $(J, S) = (J_n, S_n)_{n \in \mathbb{N}}$ is called a *Markov renewal chain* (MRC). Moreover, if the right-hand-side term of Relation (1.1) is independent of n, then Z and (J, S) are said to be (time) homogeneous and we define the *discrete-time semi-Markov kernel* $\mathbf{q} = (q_{ij}(k); \ i, j \in E, k \in \mathbb{N})$ by

$$q_{ij}(k) := \mathbb{P}(J_{n+1} = j, X_{n+1} = k \mid J_n = i).$$

The semi-Markov kernel $\mathbf{q}$ is the essential quantity which defines a semi-Markov chain (together with an initial distribution $\boldsymbol{\alpha}$, $\alpha_i := \mathbb{P}(J_0 = i), i \in E$). All the work in this book will be carried out for homogeneous Markov renewal/semi-Markov chains.

A few remarks are in order at this point.

First, Equation (1.1), together with the hypothesis of time homogeneity, shows that the visited-state chain $(J_n)_{n \in \mathbb{N}}$ is a homogeneous Markov chain (MC), called the embedded Markov chain (EMC). We denote by $\mathbf{p} = (p_{ij})_{i,j \in E}$ its transition matrix,

$$p_{ij} = \mathbb{P}(J_{n+1} = j \mid J_n = i), \ i, j \in E, \ n \in \mathbb{N}.$$

We do not allow transitions to the same state, i.e., we set $p_{ii} = 0$ for any $i \in E$.

Second, let us set $\mathbf{f} = (f_{ij}(k); \ i, j \in E, k \in \mathbb{N})$ for the sojourn time distributions, conditioned by the next state to be visited, $f_{ij}(k) = \mathbb{P}(X_{n+1} = k \mid J_n = i, J_{n+1} = j)$. Obviously, for any states $i, j \in E$ and nonnegative integer k we have

$$q_{ij}(k) = p_{ij} f_{ij}(k).$$

We want the chain to spend at least one time unit in a state, that is, $f_{ij}(0) = q_{ij}(0) = 0$, for any states i, j.

Consequently, the evolution of a sample path of a semi-Markov chain can be described as follows: the first state i_0 is chosen according to the initial distribution $\boldsymbol{\alpha}$; then, the next visited state i_1 is determined according to the transition matrix $\mathbf{p}$ and the chain stays in state i_0 for a time k determined by the sojourn time distribution in state i_0 before going to state i_1,

$(f_{i_0 i_1}(k); \ k \in \mathbb{N})$. Note that the sojourn time distributions $(f_{ij}(k); \ k \in \mathbb{N})$, $i, j \in E$ can be any discrete distribution (or continuous distribution in the continuous case), as opposed to the sojourn time constraints in the Markov case. This is why the semi-Markov processes are much more general and better adapted to applications than the Markov ones.

We note here that a Markov chain is a particular case of a semi-Markov chain (Example 3.2).

1.1.2 Discrete Time

As mentioned above, all the work presented in this book is carried out in discrete time. Here we simply want to explain why we think that discrete time is important.

Since the introduction of semi-Markov processes in the mid-1950s, their probabilistic and statistical properties have been widely studied in the continuous-time case. By contrast, discrete-time semi-Markov processes are almost absent from the literature (see Chapter 3 for references). In the authors' opinion, there are at least two reasons why discrete-time semi-Markov processes are interesting and worth studying.

The first reason for our interest in discrete time comes from specific semi-Markov applications where the time scale is intrinsically discrete. For instance, in reliability theory, one could be interested in the number of cycles done by a system or in the number of times (hours, days, etc.) that a specific event occurs. Note also that in DNA analysis, any natural (hidden) semi-Markov approach is based on discrete time because we are dealing with discrete sequences of, say, the four bases A, C, G, T.

The second reason relies on the simplicity of modeling and calculus in discrete time. A discrete-time semi-Markov process makes only a bounded number of jumps in a finite time interval (a fortiori, it does not explode). For this reason, any quantity of interest in a discrete-time semi-Markov model can be expressed as a finite series of semi-Markov kernel convolution products instead of an infinite series as in the continuous case.

Let us briefly explain this phenomenon. As will be seen in Chapter 3, the functionals of the semi-Markov kernel that we are interested in can be expressed as finite combinations of $\sum_{n=0}^{\infty} \mathbf{q}^{(n)}(k)$, $k \in \mathbb{N}$, with $\mathbf{q}^{(n)}$ the n-fold convolution of the semi-Markov kernel. In Proposition 3.1 we will see that for any states $i, j \in E$, the (i, j) element of the nth kernel convolution power has the probabilistic expression

$$q_{ij}^{(n)}(k) = \mathbb{P}(J_n = j, S_n = k \mid J_0 = i),$$

which means: $q_{ij}^{(n)}(k)$ is the probability that, starting from state i at initial time, the semi-Markov chain will do the nth jump at time k to state j. But the

SMC stays at least one unit of time in a state, so we obviously have $\mathbf{q}^{(n)}(k) = 0$ for any $n \geq k+1$ (Lemma 3.2). Consequently, the infinite series $\sum_{n=0}^{\infty} \mathbf{q}^{(n)}(k)$ becomes the finite series $\sum_{n=0}^{k} \mathbf{q}^{(n)}(k)$, and, as previously mentioned, all the quantities of interest can be written as finite series.

For these reasons, we think that it is interesting and legitimate to have probabilistic and statistical tools for (hidden) semi-Markov models (SMMs), adapted to the discrete case. This is also the case in Markov process studies, where we have a discrete-time theory and a continuous-time theory.

1.2 Discrete-Time Semi-Markov Framework

Our main objective concerning the semi-Markov models is to provide estimators for the main characteristics and to investigate their asymptotic properties. In order to achieve this, we first need to present briefly the classical theory of discrete-time renewal processes and then give the basic definitions and properties of a semi-Markov chain. We also need some basic definitions and results on Markov chains, which can be found in Appendix D.

1.2.1 Discrete-Time Renewal Processes

Roughly speaking, a *renewal process* (RP) $(S_n)_{n \in \mathbb{N}}$ represents the successive instants when a specific (fixed but random) event occurs. The term renewal comes from the assumption that when this event occurs, the process starts anew (this is a regeneration point of time). For this reason, that specific event will be called a *renewal* and $(S_n)_{n \in \mathbb{N}}$ will be called a *renewal process*. Since we will be concerned only with discrete-time renewal processes, we will generally use the term *renewal chain* (RC).

For a renewal process we have

$$S_n := X_0 + X_1 + \ldots + X_n,$$

where X_n, $n \in \mathbb{N}$, called *waiting times*, represent the times between two successive renewals. Note that the fundamental fact that the chain starts anew each time a renewal occurs means that $(X_n)_{n \in \mathbb{N}^*}$ is a sequence of i.i.d. random variables (in the simplest case, we suppose $X_0 = S_0 = 0$).

In Figure 1.2 we present the *counting process of the number of renewals in the time interval* $[1, n]$, denoted by $N(n)$.

The need for studying renewal chains before investigating the probabilistic behavior of semi-Markov chains comes from the fact that, as will be seen in Chapters 3 and 4, a semi-Markov chain can be analyzed through the so-called embedded renewal chains. That means that by taking into account only some particular aspects of the evolution of a SMC (successive visits of a specific

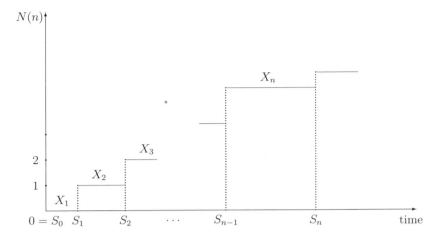

Fig. 1.2. Sample path of a renewal chain

state, for instance), we obtain a renewal chain. Due to this property, results on RCs will be of great help when investigating the behavior of SMCs.

For these reasons, Chapter 2 is devoted to renewal chains. First, we give some definitions and notation on RCs and introduce the basic notion of *renewal equation in discrete time*, given by

$$g_n = b_n + \sum_{k=0}^{n} f_k g_{n-k}, \quad n \in \mathbb{N}, \tag{1.2}$$

where $b = (b_n)_{n \in \mathbb{N}}$ is a known sequence, $g = (g_n)_{n \in \mathbb{N}}$ is an unknown sequence, and $f_k := \mathbb{P}(X_1 = k)$ is the *waiting time distribution of the renewal chain* (i.e., the distribution of the time elapsed between two successive renewals). We also suppose that the chain stays at least one time unit in a state, i.e., $f_0 = 0$.

Under some conditions, the solution of the renewal equation is given by (Theorem 2.2)

$$g_n = (u * b)_n, \quad n \geq 1, \tag{1.3}$$

where u_n represents the probability that a renewal will occur at time n and $*$ is the usual sequence convolution product (Definition 2.2).

Second, we study the asymptotic properties of quantities related to a renewal chain. Two cases – one periodic and the other aperiodic – are considered (Definitions 2.4 and 2.5). In this introductory part we consider only the aperiodic case. The main results can be summarized as follows.

- We give the strong law of large numbers (SLLN) and the central limit theorem (CLT) for the counting chain $(N(n))_{n \in \mathbb{N}^*}$ of the number of renewals in the time interval $[1, n]$.
- Another result consists in the asymptotic behavior of the expected number of renewals up to time n, denoted by $\Psi(n)$ and called *renewal function*. This is the so-called *elementary renewal theorem*, which states that

$$\lim_{n \to \infty} \frac{\Psi(n)}{n} = \frac{1}{\mu},$$

  where $\mu := \mathbb{E}(X_1)$ is the expected waiting time between two successive renewals.
- Similarly, the limit probability that a renewal will ever occur is provided by the *renewal theorem*:

$$\lim_{n \to \infty} u_n = \frac{1}{\mu}.$$

- A crucial result is the *key renewal theorem*, which gives the limit of the solution of renewal equation (1.2):

$$\lim_{n \to \infty} \sum_{k=0}^{n} b_k u_{n-k} = \frac{1}{\mu} \sum_{n=0}^{\infty} b_n.$$

The importance of this theorem stems from the fact that quantities related to a RC verify associated renewal equations. After solving these equations (via Theorem 2.2), the key renewal theorem immediately provides the limit of those quantities.

In the same chapter, we also introduce delayed renewal chains, stationary renewal chains, and alternating renewal chains and we give some useful related results.

1.2.2 Semi-Markov Chains

In order to undertake the estimation problems on semi-Markov chains, it is important to investigate them first from a probabilistic point of view.

Generally speaking, the probabilistic study of semi-Markov chains follows approximatively the same steps as for renewal chains, but with a different degree of complexity. Obviously, this complexity comes from the structural difference between the renewal framework and the semi-Markov one. We pass from an i.i.d. context to a semi-Markov one, introducing the dependence on a randomly chosen state.

As was done for renewal chains, the first part of this study developed in Chapter 3 is centered around a type of equation similar to the renewal equation. For a discrete-time semi-Markov framework, the variables in such an equation are not real sequences, like for renewal chains, but matrix-valued functions defined on $\mathbb{N}$ and denoted by $\mathbf{A} = (\mathbf{A}(k);\ k \in \mathbb{N})$, where $\mathbf{A}(k) = (A_{ij}(k);\ i, j \in E)$, $k \in \mathbb{N}$, are matrices on $E \times E$. Thus we define the *discrete-time Markov renewal equation* (DTMRE or MRE) by

$$\mathbf{L}(k) = \mathbf{G}(k) + \mathbf{q} * \mathbf{L}(k),\ k \in \mathbb{N}, \tag{1.4}$$

where $\mathbf{q}$ is the semi-Markov kernel, $\mathbf{L} = (L_{ij}(k);\ i, j \in E, k \in \mathbb{N})$ is an unknown matrix-valued function, $\mathbf{G} = (G_{ij}(k);\ i, j \in E, k \in \mathbb{N})$ is a known matrix-valued function, and $*$ denotes the discrete-time matrix convolution product (Definition 3.5).

First, the solution of such an equation is given in Theorem 3.1 by

$$\mathbf{L}(k) = \left(\sum_{n=0}^{k} \mathbf{q}^{(n)}(k) \right) * \mathbf{G}(k), \tag{1.5}$$

where $\mathbf{q}^{(n)}$ represents the n-fold convolution of the semi-Markov kernel $\mathbf{q}$. We see here what we already stated above: considering a quantity of interest of a discrete-time semi-Markov process, after solving the corresponding Markov renewal equation, we are able to express it as a series involving only a finite number of terms.

Second, let us denote by $\boldsymbol{\psi}$ this finite series of matrix convolution powers, $\boldsymbol{\psi}(k) := \sum_{n=0}^{k} \mathbf{q}^{(n)}(k)$. As will be seen in Chapter 3, the (i, j) element of $\mathbf{q}^{(n)}(k)$ has a simple probabilistic meaning: $q_{ij}^{(n)}(k)$ is the probability that, starting from state i at time zero, the semi-Markov chain will do the nth jump at time k to state j. Thus $\psi_{ij}(k)$ represents the probability that, starting from state i at time zero, the semi-Markov chain will do a jump at time k to state j. Again, recalling that in a renewal context we denoted by u_n the probability that a renewal will occur at time n, we see that we have a perfect correspondence between the solution of a renewal equation ($g_n = (u * b)_n)$) and the solution of a Markov renewal equation.

As an application of the Markov renewal theory, we obtain the explicit form of an important quantity of a semi-Markov chain, the semi-Markov transition function (matrix) $\mathbf{P} = (P_{ij}(k);\ i, j \in E, k \in \mathbb{N})$, defined by

$$P_{ij}(k) := \mathbb{P}(Z_k = j \mid Z_0 = i),\ i, j \in E,\ k \in \mathbb{N}.$$

First, we obtain its associated Markov renewal equation (Proposition 3.2)

$$\mathbf{P} = \mathbf{I} - \mathbf{H} + \mathbf{q} * \mathbf{P},$$

where:

$\mathbf{I} := (\mathbf{I}(k);\ k \in \mathbb{N})$, with $\mathbf{I}(k)$ the identity matrix for any $k \in \mathbb{N}$;
$\mathbf{H} := (\mathbf{H}(k);\ k \in \mathbb{N})$, $\mathbf{H}(k) := diag(H_i(k);\ i \in E)$, with $H_i(k)$ the cumulative sojourn time distribution in state i of the SMC (Definition 3.4).

Second, solving this Markov renewal equation we obtain the explicit expression of the semi-Markov transition function

$$\mathbf{P}(k) = \boldsymbol{\psi} * (\mathbf{I} - \mathbf{H})(k), \ k \in \mathbb{N}. \tag{1.6}$$

The second part of the probabilistic study of semi-Markov chains presented in Chapter 3 consists in limit results.

- We first have the analogy of the renewal theorem, called *Markov renewal theorem* (Theorem 3.2), which states that, under some conditions, for any states i and j we have

$$\lim_{k \to \infty} \psi_{ij}(k) = \frac{1}{\mu_{jj}},$$

  where μ_{jj} is the mean recurrence time of state j for the semi-Markov chain.
- Similarly, we have the *key Markov renewal theorem* (Theorem 3.3). Let us associate to each state $j \in E$ a real-valued function $v_j(n)$, defined on $\mathbb{N}$, with $\sum_{n \geq 0} | v_j(n) | < \infty$. Then, for any states i and j we have

$$\lim_{k \to \infty} \psi_{ij} * v_j(k) = \frac{1}{\mu_{jj}} \sum_{n \geq 0} v_j(n).$$

- As an application of the key Markov renewal theorem, we obtain the limit distribution of a semi-Markov chain (Proposition 3.9). Under some mild conditions, for any states i and j we have

$$\lim_{k \to \infty} P_{ij}(k) = \frac{\nu(j)m_j}{\sum_{l \in E} \nu(l)m_l},$$

  where $(\nu(1), \ldots, \nu(s))$ is the stationary distribution of the EMC and m_j is the mean sojourn time of the SMC in state j.
- For each $M \in \mathbb{N}$, we consider the following functional of the semi-Markov chain

$$W_f(M) := \sum_{n=1}^{N(M)} f(J_{n-1}, J_n, X_n),$$

  where f is a real measurable function defined on $E \times E \times \mathbb{N}^*$. For the functional $W_f(M)$ we have the SLLN and the CLT (Theorems 3.4 and 3.5, respectively). This CLT will be extensively used in Chapters 4 and 5 for proving the asymptotic normality of different estimators.

In Section 3.5 we give a Monte Carlo algorithm for obtaining a trajectory of a given SMC in the time interval $[0, M]$. Another Monte Carlo algorithm is proposed in Exercise 3.6.

1.2.3 Semi-Markov Chain Estimation

Let us consider a sample path of a semi-Markov chain censored at fixed arbitrary time $M \in \mathbb{N}^*$, that is, a sequence of successively visited states and sojourn times

$$\mathcal{H}(M) := (J_0, X_1, \ldots, J_{N(M)-1}, X_{N(M)}, J_{N(M)}, u_M), \qquad (1.7)$$

where $u_M := M - S_{N(M)}$ represents the censored sojourn time in the last visited state $J_{N(M)}$ (recall that $N(M)$ is the discrete-time counting process of the number of jumps in $[1, M]$).

Our objective was, first, to obtain estimators for any quantity of a semi-Markov model and, second, to investigate the asymptotic properties of these estimators.

The likelihood function corresponding to the sample path $\mathcal{H}(M)$ is

$$L(M) = \alpha_{J_0} \prod_{k=1}^{N(M)} p_{J_{k-1} J_k} f_{J_{k-1} J_k}(X_k) \overline{H}_{J_{N(M)}}(u_M),$$

where $\overline{H}_i(\cdot) := \mathbb{P}(X_1 > \cdot \mid J_0 = i)$ is the survival function in state i and α_i is the initial distribution of state i.

By Lemma 4.1 we will show that $u_M/M \xrightarrow[M \to \infty]{a.s.} 0$. Consequently, the term $\overline{H}_{J_{N(M)}}(u_M)$ corresponding to u_M has no contribution to the likelihood when M tends to infinity, and for this reason it can be neglected. On the other side, the information on α_{J_0} contained in the sample path $\mathcal{H}(M)$ does not increase with M, because $\mathcal{H}(M)$ contains only one observation of the initial distribution $\boldsymbol{\alpha}$ of $(J_n)_{n \in \mathbb{N}}$. As we are interested in large-sample theory of semi-Markov chains, the term α_{J_0} will be also neglected in the expression of the likelihood function. For this reasons, instead of maximizing $L(M)$ we will maximize the *approached likelihood function* defined by

$$L_1(M) = \prod_{k=1}^{N(M)} p_{J_{k-1} J_k} f_{J_{k-1} J_k}(X_k)$$

and we will call the obtained estimators "approached maximum-likelihood estimators."

Starting from a semi-Markov sample path $\mathcal{H}(M)$, we define for any states i, j, and positive integer k, $1 \leq k \leq M$:

- $N_i(M)$: the number of visits to state i of the EMC, up to time M;
- $N_{ij}(M)$: the number of transitions of the EMC from i to j, up to time M;
- $N_{ij}(k, M)$: the number of transitions of the EMC from i to j, up to time M, with sojourn time in state i equal to k.

For any states i and j, the estimators of the (i, j) element of the transition matrix $\mathbf{p}$, of the conditional sojourn time distributions $\mathbf{f}$, and of the discrete-time semi-Markov kernel $\mathbf{q}$ are given by (Proposition 4.1 and Relations (4.1)–(4.3))

$$\widehat{p}_{ij}(M) = N_{ij}(M)/N_i(M),$$
$$\widehat{f}_{ij}(k, M) = N_{ij}(k, M)/N_{ij}(M),$$
$$\widehat{q}_{ij}(k, M) = N_{ij}(k, M)/N_i(M).$$

Once the estimator of the kernel is obtained, any quantity of the semi-Markov chain can be estimated, after being expressed as a function of the semi-Markov kernel.

The second step in the semi-Markov estimation is to derive the asymptotic properties of the proposed estimators.

First, we need asymptotic results on $N_i(M), N_{ij}(M)$ and $N(M)$, as M tends to infinity. Using the fact that $(J_n)_{n\in\mathbb{N}}$ is a Markov chain, some results are directly obtained from Markov chain theory (Convergences (3.34) and (3.35)), whereas others are specific to the semi-Markov context (Convergences (3.36)–(3.38)).

Second, we look at the strong convergence of the estimators $\widehat{p}_{ij}(M)$, $\widehat{f}_{ij}(k, M), \widehat{q}_{ij}(k, M), \widehat{P}_{ij}(k, M)$, etc. Results of this type are given in Corollary 4.1, Proposition 4.2, and Theorems 4.1, 4.3, 4.4, and 4.6.

Third, we are interested in the asymptotic normality of the estimators. Two different proofs of this kind of result can be given. The first one is based on the CLT for Markov renewal chains (Theorem 3.5) and on Lemmas A.1–A.3. The second uses the Lindeberg–Lévy CLT for martingales (Theorem E.4). When proving the asymptotic normality of the kernel estimator (Theorem 4.2), we give complete proofs based on both methods. In the other cases (Theorems 4.5 and 4.7), we only present the main steps of the proof based on the martingale approach. Note equally that the asymptotic normality allows us also to construct asymptotic confidence intervals for the estimated quantities.

1.3 Reliability Theory of Discrete-Time Semi-Markov Systems

Much work has been conducted in recent decades on probabilistic and statistical methods in reliability. Most of the existing mathematical models for system reliability suppose that a system's evolution is in continuous time. As mentioned before, there are particular applications where it is natural to consider that the time is discrete. For example, we think of systems working on demand, like electric gadgets, in which case we are interested in the number of times the system functioned up to failure. Other examples are those systems whose lifetimes are expressed as the number of cycles/days/months up to failure. For these kinds of problems we think that it is important to have discrete-time mathematical models of reliability. But even when the initial problem is in continuous time, we can discretize it and handle it in discrete time. Since in computer implementation we always need to discretize, discrete-time semi-Markov systems can be useful as schemes of discretization of continuous-time systems.

Among examples of existing works on discrete-time reliability, we cite Nakagawa and Osaki (1975), Roy and Gupta (1992), Xie et al. (2002), Bracquemond and Gaudoin (2003) (in a general i.i.d. context), Balakrishnan et al. (2001), Platis et al. (1998) (reliability modeling via homogeneous and non-homogeneous Markov chains), Sadek and Limnios (2002) (reliability metric estimation of discrete-time Markov systems), Csenki (2002), Barbu et al. (2004) (reliability modeling via semi-Markov chains), and Barbu and Limnios (2006a,b) (reliability metric estimation of discrete-time semi-Markov systems).

In the work presented in Chapter 5 we consider a system $\mathcal{S}$ whose possible states during its evolution in time are $E = \{1, \ldots, s\}$. We denote by $U = \{1, \ldots, s_1\}$ the subset of working states of the system (the up states) and by $D = \{s_1 + 1, \ldots, s\}$ the subset of failure states (the down states), with $0 < s_1 < s$. We suppose that $E = U \cup D$ and $U \cap D = \emptyset$, $U \neq \emptyset$, $D \neq \emptyset$.

The first part of the chapter is concerned with the modeling of reliability for discrete-time semi-Markov systems. We obtain explicit forms for reliability indicators: reliability, availability, maintainability, failure rates, and mean hitting times (mean time to repair, mean time to failure, mean up time, mean down time, mean time between failures) (Propositions 5.1–5.6).

NB: Generally, there are two different ways of obtaining these explicit forms. The first one is the straight Markov renewal theoretic way: first we find the Markov renewal equation associated to the respective quantity; then we solve this equation (via Theorem 3.1) and get the desired result. The second way is a case-adapted one, constructed for each reliability indicator. As an example, see the two proofs of Proposition 5.3, where we use both methods in order to find the explicit form of reliability.

From a theoretical point of view, the method based on Markov renewal theory is more attractive due to its generality. But the direct way can be more intuitive and can tell us information that is kept "hidden" by the general method. An example justifying this assertion can be seen in the two different ways of obtaining the reliability (Proposition 5.3).

The second part of the chapter looks at reliability indicator estimation. Just like for the estimation of the SM model, we start with a semi-Markov path censored at fixed arbitrary time $M \in \mathbb{N}^*$, as defined in (1.7). As the reliability indicators have explicit formulas in terms of the basic quantities of the semi-Markov chain, the estimators of these quantities obtained in Chapter 4 allow us to immediately derive plug-in estimators of reliability, availability, failure rates, and mean times. For instance, the reliability of a discrete-time semi-Markov system at time $k \in \mathbb{N}$ is given by

$$R(k) = \boldsymbol{\alpha}_1 \, \mathbf{P}_{11}(k) \, \mathbf{1}_{s_1},$$

where $\boldsymbol{\alpha}_1$ and $\mathbf{P}_{11}$ are (roughly speaking) partitions of the initial distribution vector $\boldsymbol{\alpha}$ and of the semi-Markov transition matrix $\mathbf{P}(k)$ according to the state partition $E = U \cup D$, and $\mathbf{1}_{s_1}$ denotes the s_1-column vector whose elements are all 1. Using this expression of reliability, we immediately have its estimator

$$\widehat{R}(k, M) = \boldsymbol{\alpha}_1 \cdot \widehat{\mathbf{P}}_{11}(k, M) \cdot \mathbf{1}_{s_1}.$$

We also investigate the consistency and the asymptotic normality of these estimators as the sample size M tends to infinity (Theorems 5.1–5.4 and Corollary 5.1). The techniques are the same as those previously used for the asymptotic properties of the estimators of quantities associated to semi-Markov chains. The only difference is the greater complexity of computations. Anyway, due to the particularities of discrete-time semi-Markov systems (finite-series expressions), the numerical computations of complicated quantities are fast.

1.4 Hidden Semi-Markov Models

The basic idea of a hidden model is the following: we observe the evolution in time of a certain phenomenon (observed process), but we are interested in the evolution of another phenomenon, which we are not able to observe (hidden process). The two processes are related in the sense that the state occupied by the observed process depends on the state that the hidden process is in.

To get one of the most intuitive and general examples of a hidden model, one can think the observed process as a received signal and the hidden process as the emitted signal.

If we add to this elementary hidden model the assumption that the hidden process (denoted by Z) is a Markov process of a certain order, we obtain a hidden Markov model (HMM). The observed process (denoted by Y) is either a conditional Markov chain of order k or a sequence of conditionally independent random variables. A schematic representation of this kind of model is given in Figure 1.3. The arrows in Y or Z lines denote the dependence on the past (k arrows if Markov of order k).

Since being introduced by Baum and Petrie (1966), hidden Markov models have become very popular, being applied in a wide range of areas. The reason is that many types of concrete applications can be investigated using this kind of model. Some examples: Y is the GPS position of a car, while Z represents the real (and unknown) position of the car; in reliability and maintenance, Y

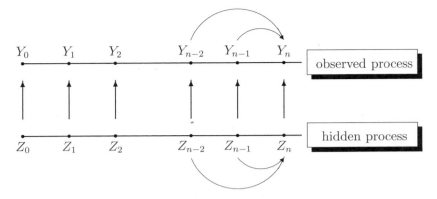

Fig. 1.3. A hidden (semi-)Markov model

can be any indicator of the state of an engine (temperature, pressure, noise), while Z is the real state of the engine; Y is an observed DNA sequence and Z is an unknown chain related to a DNA coding mechanism (for example, Z can be the sequence of indicators for CpG islands, as in Example 6.1).

The main drawback of hidden Markov models comes from the Markov property, which requires that the sojourn time in a state be geometrically distributed. This makes the hidden Markov models too restrictive from a practical point of view. In order to solve this kind of problem in the field of speech recognition, Ferguson (1980) proposed a model that allows arbitrary sojourn time distributions for the hidden process. This is called a hidden semi-Markov model, represented by a coupled process $(Z, Y) = (Z_n, Y_n)_{n \in \mathbb{N}}$, where Z is a semi-Markov process and Y is the observed process, depending on Z. Thus we have a model that combines the flexibility of semi-Markov processes with the modeling capacity of hidden-Markov models.

An important example of practical interest of hidden semi-Markov models (also known as explicit/variable state duration hidden Markov models), is GENSCAN, a program for gene identification (see Burge and Karlin, 1997) developed by Chris Burge at Stanford University. The underlying mathematical model is based on hidden semi-Markov chains.

Chapter 6 focuses on hidden semi-Markov model estimation. In the first part, we introduce two types of hidden semi-Markov models. First, we consider that the observed process Y, with finite state space $A := \{1, \ldots, d\}$, is a sequence of conditionally independent random variables, given a sample path of the hidden semi-Markov process Z, i.e., for all $a \in A, j \in E, n \in \mathbb{N}^*$, we have

$$\mathbb{P}(Y_n = a \mid Y_{n-1} = \cdot, \dots, Y_0 = \cdot, Z_n = i, Z_{n-1} = \cdot, \dots, Z_0 = \cdot)$$
$$= \mathbb{P}(Y_n = a \mid Z_n = i) =: R_{i;a}. \tag{1.8}$$

The process $(Z, Y) = (Z_n, Y_n)_{n \in \mathbb{N}}$ is called a *hidden semi-Markov chain of type SM-M0*, where the index 0 stands for the order of Y regarded as a conditional Markov chain.

Second, we suppose that Y is a homogeneous Markov chain of order $k, k \geq 1$, conditioned on Z, i.e., Relation (6.3) holds. In this case, the process (Z, Y) is called a *hidden semi-Markov chain of type SM-Mk*, where the index k stands for the order of the conditional Markov chain Y.

The second part of the chapter is devoted to nonparametric maximum-likelihood estimation for a hidden semi-Markov chain (Z, Y) of type SM-M0. We present the strong consistency and the asymptotic normality of the MLEs.

The basic idea is to associate a hidden Markov model to the initial model in order to be able to apply the classical estimation results of hidden Markov models. To this purpose, we consider $U = (U_n)_{n \in \mathbb{N}}$ the backward-recurrence times of the semi-Markov chain $(Z_n)_{n \in \mathbb{N}}$ defined by $U_n := n - S_{N(n)}$.

It is known that the couple (Z, U) is a Markov chain (Proposition 6.1), so we have a hidden Markov chain $((Z, U), Y)$ associated to the initial hidden semi-Markov chain. The results are obtained under the assumption that all the conditional sojourn time distributions $f_{ij}(\cdot), i, j \in E$, have bounded support. From a practical point of view, this assumption does not represent an important drawback because in applications we always take into account only a finite support.

Then, applying the classical results of asymptotic properties of the MLEs for a hidden Markov model (Baum and Petrie, 1966; Bickel et al., 1998), we obtain the strong consistency of the model's estimators (Theorems 6.1–6.3) and the asymptotic normality (Theorems 6.4–6.6).

When performing maximum-likelihood estimation on a hidden (semi-) Markov model, we cannot obtain explicit expressions for the MLEs of the model parameter, $\boldsymbol{\theta} = ((q_{ij}(k))_{i,j;k}, (R_{i;a})_{i;a})$, since we do not observe the hidden process. For this reason, in the last part of Chapter 6 we propose an EM algorithm for obtaining the nonparametric MLEs of a hidden semi-Markov chain (Z, Y) of type SM-M0 or SM-M1. We present here the main features of the SM-M0 case.

Our interest in developing an EM algorithm comes, first, from the fact that the existing literature on the EM algorithms for hidden semi-Markov models (Ferguson, 1980; Levinson, 1986; Guédon and Cocozza-Thivent, 1990; Sansom and Thomson, 2001; Guédon, 2003) considers, generally, parametric families for sojourn time distributions. Second, the semi-Markov chain we consider is general (in the sense that the kernel has the form $q_{ij}(k) = p_{ij} f_{ij}(k)$), while most of the hidden semi-Markov models in the literature assume particular cases of semi-Markov processes (see Remark 3.3 for more details).

Let us consider a sample path of the hidden semi-Markov chain, censored at fixed arbitrary time $M \in \mathbb{N}^*$, $(Z_0^M, Y_0^M) := (Z_0, \ldots, Z_M, Y_0, \ldots, Y_M)$. Set also $\mathbf{y}_0^M := \{Y_0^M = y_0^M\}$. The likelihood function of the complete data for a hidden system SM-M0 is given by

$$f_M(Y_0^M, Z_0^M \mid \boldsymbol{\theta}) = \mu(J_0, Y_0) \prod_{k=1}^{N(M)} \left[p_{J_{k-1}J_k} f_{J_{k-1}J_k}(X_k) \prod_{l=S_{k-1}}^{S_k-1} R_{J_{k-1},Y_l} \right]$$

$$\times \overline{H}_{J_{N(M)}}(U_M) \prod_{l=S_{N(M)}}^{M} R_{J_{N(M)},Y_l},$$

where $\mu(i, a)$ is the initial distribution of the the state $(i, a) \in E \times A$. Let us also denote by $g_M(Y_0^M \mid \boldsymbol{\theta})$ the likelihood function of the incomplete data (the known data) Y_0^M,

$$g_M(Y_0^M \mid \boldsymbol{\theta}) = \sum_{(Z_0, \ldots, Z_M) \in E^{M+1}} f_M(Y_0^M, Z_0^M \mid \boldsymbol{\theta}).$$

The basic idea of the EM algorithm (cf. Dempster et al., 1977; Baum et al., 1970) is to start with a given value $\boldsymbol{\theta}^{(m)}$ of the parameter and to maximize with respect to $\boldsymbol{\theta}$ the expectation of the log-likelihood function of the complete data, conditioned by $\mathbf{y}_0^M$,

$$\mathbb{E}_{\boldsymbol{\theta}^{(m)}} \left[\log(f_M(Y_0^M, Z_0^M \mid \boldsymbol{\theta})) \mid \mathbf{y}_0^M \right],$$

instead of maximizing the likelihood function of observed data $g_M(Y_0^M \mid \boldsymbol{\theta})$. The maximum, denoted by $\boldsymbol{\theta}^{(m+1)}$, is obtained by an iterative scheme, alternating two steps, called *expectation* and *maximization*.

So, we start with an initial value of the parameter, denoted by $\boldsymbol{\theta}^{(0)}$. After the first iteration of the algorithm we obtain an update $\boldsymbol{\theta}^{(1)}$ of the parameter and we continue recurrently up to the moment when a particular stopping condition is fulfilled. The sequence $\left(\boldsymbol{\theta}^{(m)} \right)_{m \in \mathbb{N}}$ of successively obtained parameters satisfies

$$g_M(Y_0^M \mid \boldsymbol{\theta}^{(m+1)}) \geq g_M(Y_0^M \mid \boldsymbol{\theta}^{(m)}), \ m \in \mathbb{N},$$

so the EM algorithm provides updates of the parameter that increase the likelihood function of the incomplete data.

Other algorithmic approaches, not discussed in this book, could be developed using Markov Chain Monte Carlo (MCMC) methods (see, e.g., Muri-Majoube, 1997, for the use of MCMC methods in a hidden Markov framework).

2

Discrete-Time Renewal Processes

The purpose of this chapter is to provide an introduction to the theory of discrete-time renewal processes. We define renewal chains, delayed renewal chains, and associated quantities. Basic results are presented and asymptotic behavior is investigated. This is a preliminary chapter useful for the study of semi-Markov chains.

Renewal processes (RPs) provide a theoretical framework for investigating the occurrence of patterns in repeated independent trials. Roughly speaking, the reason for using the term "renewal" comes from the basic assumption that when the pattern of interest occurs for the first time, the process starts anew, in the sense that the initial situation is reestablished. This means that, starting from this "renewal instant," the waiting time for the second occurrence of the pattern has the same distribution as the time needed for the first occurrence. The process continues like this indefinitely.

Renewal processes have known a huge success in the last 60 years and much work is still being carried out in the field. The reasons for this phenomenon are manifold and we give some of them below.

- First, there are many real situations where a renewal modeling is appropriate and gives good results. It is worth mentioning that renewal theory was developed in the first place for studying system reliability, namely, for solving problems related to the failure and replacement of components.
- Second, it has become clear that fundamental renewal results are of intrinsic importance for probability theory and statistics of stochastic processes.
- Third, there are some more complex stochastic processes (called regenerative processes) in which one or more renewal processes are embedded. When this is the case, the analysis of the limit behavior of the phenomena modeled by this kind of process can be performed using the corresponding results for the embedded renewal process.
- For Markov and semi-Markov processes, under some regularity conditions, the successive times of entering a fixed state form a renewal process.

V.S. Barbu, N. Limnios, *Semi-Markov Chains and Hidden Semi-Markov Models toward Applications*, DOI: 10.1007/978-0-387-73173-5_2,

In other words, renewal processes are very simple but nonetheless general enough. This generality is provided by the renewal phenomenon, which can be encountered in very different types of problems. As these problems can be very complex, by using their renewal feature we are at least able to answer some specific questions.

As mentioned before, a semi-Markov process has embedded renewal processes, so the results presented in this chapter will be useful for the understanding of the renewal mechanism in a semi-Markov setting and for obtaining the results derived from this mechanism. Since in the rest of the book we will be concerned only with discrete-time processes, we present here elements of a renewal theory in discrete time.

For the present chapter we have mainly used the paper of Smith (1958) and the monographs of Feller (1993, 1971), Port (1994), Karlin and Taylor (1975, 1981), Durrett (1991), and Cox (1962).

In the first part of the chapter, the (simple) renewal chain (RC) is defined, together with some related quantities of interest, and some basic results are given. We also stress a number of specific differences between renewal theory in a discrete-time and in a continuous-time framework. The second part of the chapter is devoted to the asymptotic analysis of renewal chains. Several fundamental results are provided, together with their proofs. We also introduce delayed renewal chains, which model the same type of phenomenon as do the simple renewal chains, with the only difference that there is a "delay" in the onset of observations. Associated quantities are defined and basic results are provided. We end the chapter by presenting alternating renewal chains, particularly useful in reliability analysis. Some examples spread throughout the chapter give different situations where a renewal mechanism occurs.

2.1 Renewal Chains

Let $(X_n)_{n\in\mathbb{N}}$ be a sequence of extended positive integer-valued random variables, $X_n > 0$ a.s. for $n \geq 1$ (we use the term "extended" in the sense that X_n can take the value ∞). Denote by $(S_n)_{n\in\mathbb{N}}$ the associated sequence of partial sums:

$$S_n := X_0 + X_1 + \ldots + X_n. \tag{2.1}$$

If for a certain $n \in \mathbb{N}$ we have $X_n = \infty$, then $S_n = S_{n+1} = \ldots = \infty$. Note that we have $S_0 \leq S_1 \leq \ldots$, where equality holds only for infinite S_n.

The sequence $(X_n)_{n\in\mathbb{N}^*}$ is called a *waiting time sequence* and X_n is the *nth waiting time*. The sequence $(S_n)_{n\in\mathbb{N}}$ is called an *arrival time sequence* and S_n is the *nth arrival time*.

Intuitively, $(S_n)_{n\in\mathbb{N}}$ can be seen as the successive instants when a specific event occurs, while $(X_n)_{n\in\mathbb{N}^*}$ represent the interarrival times, i.e., the times elapsed between successive occurrences of the event. See Figure 2.1, page 25, for a graphical representation of these chains.

Definition 2.1 (renewal chain).
    An arrival time sequence $(S_n)_{n\in\mathbb{N}}$, for which the waiting times $(X_n)_{n\in\mathbb{N}^*}$ form an i.i.d. sequence and $S_0 = X_0 = 0$, is called a (simple) renewal chain (RC), and every S_n is called a renewal time.

Later on we will also consider the case $S_0 > 0$.

There are many real situations which could be modeled by renewal chains.

Example 2.1. Suppose, for instance, that we have a single-bulb lamp and we have replacement lightbulbs with i.i.d. lifetimes $(X_n)_{n\in\mathbb{N}^*}$. Suppose the lamp is always on as follows: at time 0 the first bulb with lifetime X_1 is placed in the lamp; when the bulb burns out at time $S_1 = X_1$, it is replaced with a second bulb with lifetime X_2, which will burn out at time $S_2 = X_1 + X_2$, and so on. In this example, the sequence $(S_n)_{n\in\mathbb{N}}$ of bulb replacements is a renewal chain.

Example 2.2. Consider the following DNA sequence of HEV (hepatitis E virus):

AGGCAGACCACATATGTGGTCGATGCCATGGAGGCCCATCAGTTTATTA
AGGCTCCTGGCATCACTACTGCTATTGAGCAGGCTGCTCTAGCAGCGGC
CATCCGTCTGGACACCAGCTACGGTACCTCCGGGTAGTCAAATAATTCC
GAGGACCGTAGTGATGACGATAACTCGTCCGACGAGATCGTCGCCGGT

Suppose that the bases $\{A, C, G, T\}$ are independent of each other and have the same probability of appearing in a location, which is equal to $1/4$. Thus the occurrences of one of them, say, C, form a renewal chain. The interarrival distribution is geometric with parameter (i.e., probability of success) equal to $1/4$.

The common distribution of $(X_n)_{n\in\mathbb{N}^*}$ is called the *waiting time distribution of the renewal chain*. Denote it by $f = (f_n)_{n\in\mathbb{N}}$, $f_n := \mathbb{P}(X_1 = n)$, with $f_0 := 0$, and denote by F the cumulative distribution function of the waiting time, $F(n) := \mathbb{P}(X_1 \leq n)$. Set $\overline{f} := \sum_{n\geq 0} f_n \leq 1 = \mathbb{P}(X_1 < \infty)$ for the probability that a renewal will ever occur.
    A renewal chain is called *recurrent* if $\overline{f} = \mathbb{P}(X_1 < \infty) = 1$ and *transient* if $\overline{f} = \mathbb{P}(X_1 < \infty) < 1$. Note that if the renewal chain is transient, then X_n are improper (defective) random variables. Set $\mu := \mathbb{E}(X_1)$ for the common expected value of $X_n, n \in \mathbb{N}^*$. A recurrent renewal chain is called *positive recurrent* (resp. *null recurrent*) if $\mu < \infty$ (resp. $\mu = \infty$).

Remark 2.1. It is worth stressing here an important phenomenon which is specific to discrete-time renewal processes. As $f_0 = \mathbb{P}(X_1 = 0) = 0$, the waiting time between two successive occurrences of a renewal is at least one unit of time. Consequently, in a finite interval of time, of length, say, n, we can have at most n renewals. As will be seen in the sequel, this is the reason why many variables of interest can be expressed as finite series of basic quantities, whereas in a continuous-time setting the corresponding series are infinite. This is one advantage of a renewal theory carried out in a discrete-time setting.

Let us consider $(Z_n)_{n \in \mathbb{N}}$ the sequence of indicator variables of the events {a renewal occurs at instant n}, i.e.,

$$Z_n := \begin{cases} 1, & \text{if } n = S_m \text{ for some } m \geq 0 \\ 0, & \text{otherwise} \end{cases} = \sum_{m=0}^{n} \mathbf{1}_{\{S_m = n\}}.$$

Note that the previous series is finite for the reasons given in Remark 2.1. As $S_0 = 0$ by definition (time 0 is considered to be a renewal time), we have $Z_0 = 1$ a.s. Let u_n be the probability that a renewal occurs at instant n, i.e., $u_n = \mathbb{P}(Z_n = 1)$, with $u_0 = 1$. Obviously, we can have $\overline{u} := \sum_{n \geq 0} u_n = \infty$. As we will see in Theorem 2.1, this will always be the case for a recurrent renewal chain.

We can express $u_n, n \in \mathbb{N}$, in terms of convolution powers of the waiting time distribution $f = (f_n)_{n \in \mathbb{N}}$. To this end, let us first recall the definition of the convolution product and of the n-fold convolution.

Definition 2.2. *Let $f, g : \mathbb{N} \to \mathbb{R}$. The* discrete-time convolution product *of f and g is the function $f * g : \mathbb{N} \to \mathbb{R}$ defined by*

$$f * g(n) := \sum_{k=0}^{n} f(n-k)g(k), \ n \in \mathbb{N}.$$

For notational convenience, we use both $f * g(n)$ and $(f * g)_n$ for denoting the sequence convolution.

Recall that if X and Y are two independent positive-integer-valued random variables, with $f = (f_n)_{n \in \mathbb{N}}$ and $g = (g_n)_{n \in \mathbb{N}}$ the corresponding distributions, then $f * g = ((f * g)_n)_{n \in \mathbb{N}}$ is the distribution of $X + Y$.

Note that the discrete-time convolution is associative, commutative, and has the *identity element* $\delta : \mathbb{N} \to \mathbb{R}$ defined by

$$\delta(k) := \begin{cases} 1, & \text{if } k = 0, \\ 0, & \text{elsewhere.} \end{cases}$$

The power in the sense of convolution is defined by induction.

Definition 2.3. *Let $f : \mathbb{N} \to \mathbb{R}$. The n-fold convolution of f is the function $f^{(n)} : \mathbb{N} \to \mathbb{R}$, defined recursively by*

$$f^{(0)}(k) := \delta(k), \quad f^{(1)}(k) := f(k),$$

and

$$f^{(n)}(k) := \underbrace{(f * f * \ldots * f)}_{n-times}(k), \quad n \geq 2.$$

Note that we have $f^{(n+m)} = f^{(n)} * f^{(m)}$ for all $n, m \in \mathbb{N}$.

Recall that, if $X_1, \ldots, X_n$ are n i.i.d. positive-integer-valued random variables, with $f = (f_n)_{n \in \mathbb{N}}$ the common distribution, then $f^{(n)}$ is the distribution of $X_1 + \ldots + X_n$. In our case, for $f = (f_n)_{n \in \mathbb{N}}$ the waiting time distribution of a renewal chain, $f^{(n)}$ is the distribution of S_n, i.e., $f^{(n)}(k)$ is the probability that the $(n+1)$th occurrence of a renewal takes place at instant k, $f^{(n)}(k) = \mathbb{P}(S_n = k)$. Using this fact, we can obtain the desired expression of u_n in terms of convolution powers of f. Indeed, we have

$$u_n = \mathbb{P}(Z_n = 1) = \mathbb{E}(Z_n) = \sum_{m=0}^{n} \mathbb{P}(S_m = n) = \sum_{m=0}^{n} f_n^{(m)}, i.e.,$$

$$u_n = \sum_{m=0}^{n} f_n^{(m)}. \tag{2.2}$$

Note that the last term in (2.2) can be expressed as a finite series of convolution powers of f due to the fact that

$$f_n^{(m)} = \mathbb{P}(S_m = n) = 0, \quad m > n, \tag{2.3}$$

for the reasons described in Remark 2.1.

Let us now investigate further the relation between $(f_n)_{n \in \mathbb{N}}$ and $(u_n)_{n \in \mathbb{N}}$. Note that, for any $n \in \mathbb{N}^*$, we have

$$u_n = \mathbb{P}(Z_n = 1) = \mathbb{P}(S_1 = n) + \sum_{k=1}^{n-1} \mathbb{P}(S_1 = k)\mathbb{P}(Z_n = 1 \mid S_1 = k)$$

$$= \mathbb{P}(X_1 = n) + \sum_{k=1}^{n-1} \mathbb{P}(X_1 = k)\mathbb{P}(Z_{n-k} = 1)$$

$$= f_n + \sum_{k=1}^{n-1} f_k u_{n-k}.$$

Thus, we have obtained

$$u_n = \sum_{k=1}^{n} f_k u_{n-k}, \quad n \in \mathbb{N}^*, \tag{2.4}$$

or, in convolution notation,

$$u_n = (f * u)_n, \quad n \in \mathbb{N}^*. \tag{2.5}$$

A tool frequently used in the analysis of renewal chains is the generating function. When applied to equations like (2.5) it simplifies the computations because it transforms convolutions into ordinary products. As the sequences $(f_n)_{n \in \mathbb{N}}$ and $(u_n)_{n \in \mathbb{N}}$ are fundamental quantities associated to a renewal chain, let us denote their generating functions by

$$\Phi(s) := \sum_{n=0}^{\infty} f_n s^n, \quad U(s) := \sum_{n=0}^{\infty} u_n s^n.$$

The series are convergent at least for $|s| < 1$, while $\Phi(s)$ is convergent also for $s = 1$. Note that in the case of $\Phi(s)$, we may start the summation at index 1, since $f_0 = 0$.

Proposition 2.1. *The generating functions of $(f_n)_{n \in \mathbb{N}}$ and $(u_n)_{n \in \mathbb{N}}$ are related by*

$$\Phi(s) = \frac{U(s) - 1}{U(s)}, \quad U(s) = \frac{1}{1 - \Phi(s)}. \tag{2.6}$$

Proof. Multiplying Equation (2.5) by s^n and summing over $n = 1, 2, \ldots$ we obtain $U(s) - 1 = U(s)\Phi(s)$, where we have used the fact that the generating function of $f * u$ is the product of the corresponding generating functions, i.e., $U(s)\Phi(s)$ (see, e.g. Port, 1994). □

By means of the generating functions $U(s)$ and $\Phi(s)$ we can easily prove a necessary and sufficient condition for a renewal chain to be transient.

Theorem 2.1. *A renewal chain is transient iff $\overline{u} := \sum_{n=0}^{\infty} u_n < \infty$. If this is the case, the probability that a renewal will ever occur is given by $\overline{f} = (\overline{u} - 1)/\overline{u}$.*

Proof. As $U(s)$ is an increasing function of s, $\lim_{\substack{s \to 1 \\ s < 1}} U(s)$ exists, but it can be infinite. For any $N \in \mathbb{N}$ we have

$$\sum_{n=0}^{N} u_n \leq \lim_{\substack{s \to 1 \\ s < 1}} U(s) \leq \overline{u} = \sum_{n=0}^{\infty} u_n.$$

As this is valid for any $N \in \mathbb{N}$, we obtain that $\lim_{\substack{s \to 1 \\ s < 1}} U(s) = \overline{u}$.

By definition, the chain is transient iff $\overline{f} < 1$. In this case, applying Proposition 2.1 and the fact that $\Phi(1) = \overline{f}$ we obtain that $\lim_{\substack{s \to 1 \\ s < 1}} U(s) = 1/(1 - \overline{f})$, thus $\overline{u} = 1/(1 - \overline{f})$. □

Note that the u_n, $n \in \mathbb{N}$, verify the equation

$$u_n = \delta(n) + (f * u)_n, \quad n \in \mathbb{N}, \tag{2.7}$$

where $\delta(0) = 1$ and $\delta(n) = 0$ for $n \in \mathbb{N}^*$. Indeed, this is exactly Equation (2.5) for $n \in \mathbb{N}^*$, whereas for $n = 0$ the equality obviously holds true, because $u_0 = 1$.

Equation (2.7) is a special case of what is called a *renewal equation in discrete time* and has the following form:

$$g_n = b_n + \sum_{k=0}^{n} f_k g_{n-k}, \quad n \in \mathbb{N}, \tag{2.8}$$

where $b = (b_n)_{n \in \mathbb{N}}$ is a known sequence and $g = (g_n)_{n \in \mathbb{N}}$ is an unknown sequence.

The following result proves that the renewal equation has a unique solution. We give two different proofs of the theorem, the first one is straightforward, whereas the second one is based on generating function technique.

Theorem 2.2 (solution of a discrete-time renewal equation).
If $b_n \geq 0, n \in \mathbb{N}$ and $\sum_{n=0}^{\infty} b_n < \infty$, then the discrete-time renewal equation (2.8) has the unique solution

$$g_n = (u * b)_n, \quad n \in \mathbb{N}. \tag{2.9}$$

Proof (1). First, it can be immediately checked that $g_n = (u * b)_n, n \in \mathbb{N}$, is a solution of the renewal equation (2.8). Indeed, for all $n \in \mathbb{N}$, the right-hand side of (2.8) becomes

$$b_n + (f * g)_n = b_n + (f * u * b)_n = (b * (\delta + f * u))_n = (b * u)_n,$$

where we have used Equation (2.7).

Second, for $g_n', n \in \mathbb{N}$, another solution of the renewal equation (2.8) we obtain

$$(g - g')_n = (f * (g - g'))_n = (f^{(m)} * (g - g'))_n = \sum_{k=0}^{n} f_k^{(m)} * (g - g')_{n-k},$$

for any $m \in \mathbb{N}$. Taking $m > n$ and using Equation (2.3), we get $f_k^{(m)} = 0$ for any $k = 0, 1, \ldots, n$, so $g_n = g_n'$ for any $n \in \mathbb{N}$. □

Proof (2). Let us set $\bar{b} := \sum_{n=0}^{\infty} b_n$. From Equation (2.8) we have

$$g_n \leq \bar{b} + \max(g_0, \ldots, g_{n-1}),$$

and by induction on n we get $g_n \leq (n+1)\bar{b}$. Denote by $G(s)$ and $B(s)$ the generating functions of $(g_n)_{n \in \mathbb{N}}$ and $(b_n)_{n \in \mathbb{N}}$, respectively.

For $0 \leq s < 1$ we have

$$G(s) = \sum_{n=0}^{\infty} g_n s^n \leq \bar{b} \sum_{n=0}^{\infty} (n+1)s^n = \bar{b} \, \frac{d \left(\sum_{n=0}^{\infty} s^n \right)}{ds} = \frac{\bar{b}}{(1-s)^2} < \infty.$$

Thus, the generating function of $(g_n)_{n \in \mathbb{N}}$ exists and is finite, for $0 \leq s < 1$. Multiplying Equation (2.8) by s^n and summing over all $n = 0, 1, \ldots$, we obtain

$$G(s) = B(s) + \Phi(s)G(s).$$

Using Proposition 2.1 we get

$$G(s) = \frac{B(s)}{1 - \Phi(s)} = B(s)U(s),$$

and equalizing the coefficients of s^n we obtain the desired result. $\square$

There are some stochastic processes, related to a renewal chain, whose study is important for understanding the asymptotic behavior of the renewal chain. First, for all $n \in \mathbb{N}^*$, we define the counting process of the number of renewals in the time interval $[1, n]$ by

$$N(n) := \max\{k \mid S_k \leq n\} = \sum_{k=1}^{n} Z_k = \sum_{k=1}^{n} \mathbf{1}_{[0,n]}(S_k) = \sum_{k=1}^{n} \mathbf{1}_{\{S_k \leq n\}}. \quad (2.10)$$

By convention, we set $N(0) := 0$. The definition of $N(n)$ shows that $N(n) \geq k$ iff $S_k \leq n$, so the equality

$$\mathbb{P}(N(n) \geq k) = \mathbb{P}(S_k \leq n) \quad (2.11)$$

holds true for any $k \in \mathbb{N}$. As $N(n) \leq n$, the previous relation is trivially fulfilled for $k > n$, i.e., $\mathbb{P}(N(n) \geq k) = \mathbb{P}(S_k \leq n) = 0$.

Second, let us introduce the following stochastic processes:

- $U_n := n - S_{N(n)}, n \in \mathbb{N}$, called the *backward recurrence time* on time n (also called the *current lifetime* or *age*);
- $V_n := S_{N(n)+1} - n, n \in \mathbb{N}$, called the *forward recurrence time* on time n (also called the *residual* or *excess lifetime*);
- $L_n := S_{N(n)+1} - S_{N(n)} = U_n + V_n, n \in \mathbb{N}$, called the *total lifetime* on time n.

Figure 2.1 presents the counting process of the number of renewals in a renewal chain.

It can be proved that $(U_n)_{n \in \mathbb{N}}$ and $(V_n)_{n \in \mathbb{N}}$ are homogeneous Markov chains (see the following example and Exercises 2.6 and 2.7).

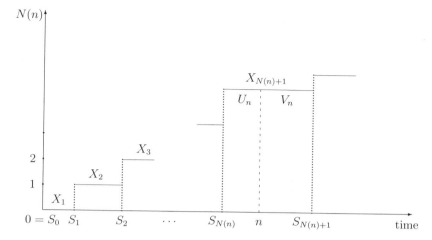

Fig. 2.1. Renewal chain

Example 2.3. Consider a recurrent renewal chain $(S_n)_{n \in \mathbb{N}}$ with waiting time distribution $(f_n)_{n \in \mathbb{N}}$. Set $m := \sup\{n \mid f_n > 0\}$ and note $E := \{1, \dots, m\}$, if $m < \infty$, and $E := \{1, 2, \dots\} = \mathbb{N}^*$, if $m = \infty$. Define the transition probabilities $(p_{ij})_{i,j \in E}$ by $p_{i\,i-1} := 1$, if $i > 1$, $p_{1i} := f_i$ for any $i \in E$ and $p_{ij} := 0$ for all other $i, j \in E$. It can be easily checked that $(V_n)_{n \in \mathbb{N}}$ is an irreducible and recurrent Markov chain having $(p_{ij})_{i,j \in E}$ as transition probabilities.

On the other hand, suppose that we have an irreducible and recurrent Markov chain with state space $E := \{1, \dots, m\}$, $m < \infty$, or $E = \mathbb{N}^*$. Suppose also that the Markov chain starts in state 0 with probability 1. Then, the successive returns to state 0 form a renewal chain.

This example shows that there is a correspondence in both senses between renewal chains and Markov chains. For this reason, techniques and results developed for Markov chains can be applied to renewal chains and vice versa.

A function of great importance in renewal theory is the *renewal function* $\Psi(n), n \in \mathbb{N}$, defined as the expected number of renewals up to time n:

$$\Psi(n) := \mathbb{E}[N(n) + 1] = \sum_{k=0}^{n} \mathbb{P}(S_k \leq n) = \sum_{k=0}^{n} \sum_{l=0}^{n} f_l^{(k)}, \quad n \in \mathbb{N}. \quad (2.12)$$

As $N(0) = 0$, we get $\Psi(0) = 1$.

Remark 2.2. Some authors define the renewal function without taking into account the renewal that occurs at the origin; if this were the case, we would have $\Psi(n) = \mathbb{E}(N(n))$. There is only a question of convenience for technical reasons, and we have chosen here to include the renewal at the origin.

The renewal function can be expressed in terms of $(u_n)_{n\in\mathbb{N}}$. Indeed,

$$\Psi(n) = \mathbb{E}(N(n) + 1) = \mathbb{E}\left(\sum_{k=0}^{n} Z_k\right) = \sum_{k=0}^{n} u_k, \quad n \in \mathbb{N}. \qquad (2.13)$$

2.2 Limit Theorems

In this section we present the basic results on the asymptotic behavior of a renewal chain. The limit theorems concern different quantities related to a renewal chain: the counting process of the number of renewals $(N(n))_{n\in\mathbb{N}}$ (Theorems 2.3 and 2.4), the expected value of the number of renewals, i.e., the renewal function $\Psi(n)$ (Theorem 2.5 and 2.6), the sequence $u = (u_n)_{n\in\mathbb{N}}$ of the probabilities that renewals will occur at instants $n, n \in \mathbb{N}$ (Theorem 2.6), and the solution of the general renewal equation given in Equation (2.8) (Theorem 2.7). In the rest of the chapter we will be using the convention $1/\infty = 0$.

Lemma 2.1.

1. *For a renewal chain we have* $\lim_{n\to\infty} S_n = \infty$ *a.s.*
2. *For a recurrent renewal chain we have* $\lim_{n\to\infty} N(n) = \infty$ *a.s.*

Proof.
1. Since $S_n \geq n$, the conclusion is straightforward.
2. The result is a direct consequence of 1. and of Relation (2.11) between $(N(n))_{n\in\mathbb{N}}$ and $(S_k)_{k\in\mathbb{N}}$. Indeed, for any fixed $n \in \mathbb{N}^*$ we have

$$\mathbb{P}(\lim_{k\to\infty} N(k) \leq n) = \lim_{k\to\infty} \mathbb{P}(N(k) \leq n) = \lim_{k\to\infty} \mathbb{P}(S_n \geq k),$$

where the above permutation of probability and limit is obtained applying the continuity from above of the probability (Theorem E.1) to the nonincreasing sequence of events $\{N(k) \leq n\}_{k\in\mathbb{N}}$. As the renewal chain is recurrent, $\mathbb{P}(X_n < \infty) = 1$ for any $n \in \mathbb{N}$, so $S_n = X_0 + \ldots + X_n < \infty$ a.s. for any $n \in \mathbb{N}$. Finally, we get

$$\mathbb{P}(\lim_{k\to\infty} N(k) \leq n) = \lim_{k\to\infty} \mathbb{P}(S_n \geq k) = 0,$$

which proves the result. □

The following two results investigate the asymptotic behavior of the counting chain of the number of renewals $(N(n))_{n\in\mathbb{N}}$.

Theorem 2.3 (strong law of large numbers for $N(n)$).
For a recurrent renewal chain $(S_n)_{n\in\mathbb{N}}$ we have

$$\lim_{n\to\infty} \frac{N(n)}{n} = \frac{1}{\mu} \quad a.s.$$

Proof. We give the proof only for a positive recurrent renewal chain, i.e., $\mu = \mathbb{E}(X_1) < \infty$. Note that for all $n \in \mathbb{N}$ we have

$$S_{N(n)} \leq n < S_{N(n)+1}.$$

Using the definition of S_n given in Equation (2.1) and dividing the previous inequalities by $N(n)$ we get

$$\frac{1}{N(n)} \sum_{k=1}^{N(n)} X_k \leq \frac{n}{N(n)} \leq \frac{1}{N(n)} \sum_{k=1}^{N(n)+1} X_k,$$

or, equivalently,

$$\frac{1}{N(n)} \sum_{k=1}^{N(n)} X_k \leq \frac{n}{N(n)} \leq \frac{N(n)+1}{N(n)} \frac{1}{N(n)+1} \sum_{k=1}^{N(n)+1} X_k. \qquad (2.14)$$

As $N(n) \xrightarrow[n\to\infty]{a.s.} \infty$, we have $(N(n)+1)/N(n) \xrightarrow[n\to\infty]{a.s.} 1$. Applying the SLLN to the sequence of i.i.d. random variables $(X_k)_{k\in\mathbb{N}}$ and Theorem E.5, we obtain from (2.14) that $n/N(n)$ tends to $\mathbb{E}(X_1) = \mu$, as n tends to infinity. $\qquad\square$

Theorem 2.4 (central limit theorem for $N(n)$).
Consider a positive recurrent renewal chain $(S_n)_{n\in\mathbb{N}}$, with $\mu = \mathbb{E}(X_1) < \infty$ and $0 < \sigma^2 := Var(X_1) < \infty$. Then

$$\frac{N(n) - n/\mu}{\sqrt{n\sigma^2/\mu^3}} \xrightarrow[n\to\infty]{\mathcal{D}} \mathcal{N}(0, 1).$$

Proof. We mainly follow the proof of Feller (1993) and Karlin and Taylor (1975). The main idea is to use the CLT for $S_n = X_1 + \ldots + X_n$ ($X_n, n \in \mathbb{N}^*$ i.i.d.) and Relation (2.11) between $(S_n)_{n\in\mathbb{N}}$ and $(N(n))_{n\in\mathbb{N}}$.
Set $\xi_n := \frac{\sqrt{n}}{\sqrt{\sigma^2/\mu^3}} \left[\frac{N(n)}{n} - \frac{1}{\mu} \right]$. Then we can write

$$\mathbb{P}(\xi_n \leq k) = \mathbb{P}\left(N_n \leq \frac{n}{\mu} + k\sqrt{\frac{n\sigma^2}{\mu}} \right).$$

Set $\rho_n := \lfloor \frac{n}{\mu} + k\sqrt{\frac{n\sigma^2}{\mu^3}} \rfloor$, where $\lfloor \ \ \rfloor$ denotes the integer part. Note that, as $n \to \infty$, $\rho_n \sim \frac{n}{\mu} + k\sqrt{\frac{n\sigma^2}{\mu^3}}$, so $(n - \rho_n\mu) \sim -k\sqrt{\frac{n\sigma^2}{\mu^3}}$. Further, as $n \to \infty$, we have $\sigma\sqrt{\rho_n} \sim \sigma\sqrt{\frac{n}{\mu}}$, and consequently we get

$$\frac{n - \rho_n\mu}{\sigma\sqrt{\rho_n}} \sim -k, \ \text{ as } n \to \infty.$$

Applying the CLT to S_n, we obtain

$$\mathbb{P}(\xi_n \leq k) = \mathbb{P}(N(n) \leq \rho_n) = \mathbb{P}(S_{\rho_n} \geq n) = \mathbb{P}\left(\frac{S_{\rho_n} - \rho_n \mu}{\sigma \sqrt{\rho_n}} \geq \frac{n - \rho_n \mu}{\sigma \sqrt{\rho_n}}\right).$$

Finally, taking the limit as n tends to infinity, we get

$$\mathbb{P}(\xi_n \leq k) \xrightarrow[n \to \infty]{} 1 - \Phi(-k) = \Phi(k),$$

where $\Phi(k)$ is the cumulative distribution function of $\mathcal{N}(0,1)$.   □

The standard asymptotic results in renewal theory consist in the asymptotic behavior of the renewal function. The following three theorems investigate this topic.

Theorem 2.5 (elementary renewal theorem).
For a recurrent renewal chain $(S_n)_{n \in \mathbb{N}}$ we have

$$\lim_{n \to \infty} \frac{\Psi(n)}{n} = \frac{1}{\mu}.$$

Proof. Consider the case where the renewal chain is positive recurrent, that is, $\mu < \infty$.

As $\mathbb{E}[N(n)] < \infty$, $n \leq S_{N(n)+1}$, and $N(n) + 1$ is a stopping time for $(X_n)_{n \in \mathbb{N}}$, applying Wald's lemma (Theorem E.8) we have

$$n \leq \mathbb{E}[S_{N(n)+1}] = \mu \, \Psi(n)$$

and we obtain

$$\liminf_{n \to \infty} \frac{\Psi(n)}{n} \geq \frac{1}{\mu}. \tag{2.15}$$

Let $c \in \mathbb{N}^*$ be an arbitrary fixed integer and define $X_n^c := \min(X_n, c)$. Consider $(S_n^c)_{n \in \mathbb{N}}$ a truncated renewal chain having the interarrival times $(X_n^c)_{n \in \mathbb{N}}$. Denote by $(N^c(n))_{n \in \mathbb{N}}$ the corresponding counting process of the number of renewals in $[1, n]$. Since $X_n^c \leq c$, we obtain $n + c \geq S_{N^c(n)+1}^c$, which implies

$$n + c \geq \mathbb{E}[S_{N^c(n)+1}^c] = \mu^c \, \Psi^c(n), \tag{2.16}$$

where $\Psi^c(n) := \mathbb{E}[N^c(n) + 1]$ and $\mu^c := \mathbb{E}[X_1^c]$.

Since $X_n^c \leq X_n$, we obviously have $N^c(n) \geq N(n)$ and, consequently, $\Psi^c(n) \geq \Psi(n)$. From this relation and (2.16) we immediately obtain

$$\limsup_{n \to \infty} \frac{\Psi(n)}{n} \leq \frac{1}{\mu^c}. \tag{2.17}$$

As $X_n^c \xrightarrow[c \to \infty]{a.s.} X_n$ and the sequence $(X_n^c(\omega))_{c \in \mathbb{N}^*}$ is a nondecreasing sequence in c, for any $n \in \mathbb{N}^*$ and $\omega \in \Omega$, by monotone convergence theorem (Theorem E.2) we obtain that $\lim_{c \to \infty} \mu^c = \mu$. Consequently, convergence (2.17) yields

$$\limsup_{n\to\infty} \frac{\Psi(n)}{n} \leq \frac{1}{\mu},\tag{2.18}$$

and from Relations (2.15) and (2.18) we get the desired result.

When the renewal chain is null recurrent, i.e., $\mu = \infty$, the result is obtained by the same method, in which case we set $1/\infty = 0$. □

The asymptotic behavior of certain quantities of a renewal chain is strictly related to the notion of periodicity of the chain. The general notion of periodicity of a distribution is recalled in the following definition, while the periodicity of a renewal chain is introduced thereafter.

Definition 2.4. *A distribution* $g = (g_n)_{n\in\mathbb{N}}$ *on* $\mathbb{N}$ *is said to be* periodic *if there exists an integer* $d > 1$ *such that* $g_n \neq 0$ *only when* $n = d, 2d, \ldots$. *The greatest* d *with this property is called the* period *of* g. *If* $d = 1$, *the distribution* g *is said to be* aperiodic.

Definition 2.5. *A renewal chain* $(S_n)_{n\in\mathbb{N}}$ *is said to be* periodic of period d, $d \in \mathbb{N}^*, d > 1$, *if its waiting times distribution* $f = (f_n)_{n\in\mathbb{N}}$ *is periodic of period* d. *If* f *is aperiodic, then the renewal chain is called* aperiodic.

Theorem 2.6 (renewal theorem).

1. For a recurrent aperiodic renewal chain $(S_n)_{n\in\mathbb{N}}$ *we have*

$$\lim_{n\to\infty} u_n = \frac{1}{\mu}.\tag{2.19}$$

2. For a periodic recurrent renewal chain $(S_n)_{n\in\mathbb{N}}$ *of period* $d > 1$ *we have*

$$\lim_{n\to\infty} u_{nd} = \frac{d}{\mu}\tag{2.20}$$

and $u_k = 0$ *for all* k *not multiple of* d.

The proof of the renewal theorem will be provided at the end of next section.

Remark 2.3. The renewal theorem in a continuous-time setting states that for a continuous-time recurrent aperiodic renewal process, for any $h > 0$, we have

$$\lim_{n\to\infty} [\Psi(n+h) - \Psi(n)] = \frac{h}{\mu}.\tag{2.21}$$

The same result holds true for a continuous-time recurrent periodic renewal process of period d, provided that h is a multiple of d.

It is easy to see that in the discrete-time case Convergence (2.21) is immediately obtained from Theorem 2.6. Indeed, recall that for $n \in \mathbb{N}$ we have $\Psi(n) = \sum_{k=0}^{n} u_k$ (cf. Equation (2.13)). Consequently, for any $h \in \mathbb{N}^*$, we have

$$\Psi(n+h) - \Psi(n) = \sum_{k=n+1}^{n+h} u_k.$$

Since this sum is finite, Convergence (2.21) follows from Theorem 2.6 for both periodic and aperiodic recurrent renewal chains. As in discrete time Convergence (2.21) is not more general than Convergences (2.19) and (2.20), the form of the discrete-time version of the renewal theorem usually met in the literature is the one we gave in Theorem 2.6.

The following result is an equivalent form of the renewal theorem.

Theorem 2.7 (key renewal theorem).
  Consider a recurrent renewal chain $(S_n)_{n \in \mathbb{N}}$ and a real sequence $(b_n)_{n \in \mathbb{N}}$.

1. If the chain is aperiodic and $\sum_{n=0}^{\infty} | b_n | < \infty$, then

$$\lim_{n \to \infty} \sum_{k=0}^{n} b_k u_{n-k} = \frac{1}{\mu} \sum_{n=0}^{\infty} b_n. \tag{2.22}$$

2. If the chain is periodic of period $d > 1$ and if for a certain positive integer l, $0 \leq l < d$, we have $\sum_{n=0}^{\infty} | b_{l+nd} | < \infty$, then

$$\lim_{n \to \infty} \sum_{k=0}^{l+nd} b_k u_{l+nd-k} = \frac{d}{\mu} \sum_{n=0}^{\infty} b_{l+nd}. \tag{2.23}$$

Remark 2.4. Recall that the solution of a discrete-time renewal equation (Equation 2.8) was obtained in Theorem 2.2 as $g = b * u$. Thus, the key renewal theorem describes the asymptotic behavior of this solution g, in both periodic and aperiodic cases.

Proof.
1. Note that $\lim_{n \to \infty} u_{n-k} = 1/\mu$ (cf. Theorem 2.6 (1), with the convention $1/\infty = 0$), $| b_k u_{n-k} | \leq | b_k |$, and $\sum_{k=0}^{\infty} | b_k | < \infty$. Thus, we are under the hypotheses of Proposition E.1 and we obtain the desired result.

2. The proof is similar to that of the aperiodic case, using the second assertion of Theorem 2.6. As the chain is periodic of period $d > 1$, all the u_m with m not multiple of d are zero. Consequently, setting $k = l + id$, $i = 0, 1, \ldots, n$, the series from the left member of (2.23) becomes

$$\sum_{k=0}^{l+nd} b_k u_{l+nd-k} = \sum_{i=0}^{n} b_{l+id} u_{(n-i)d}.$$

From Theorem 2.6(2) we have $\lim_{n \to \infty} u_{(n-i)d} = d/\mu$, with the convention $1/\infty = 0$. As $\sum_{i=0}^{\infty} | b_{l+id} | < \infty$ and $| b_{l+id} u_{(n-1)d} | \leq | b_{l+id} |$, the desired result is obtained from Proposition E.1. $\square$

Remark 2.5. In the case where $b_0 = 1$ and $b_n = 0, n \geq 1$, Convergence (2.22) becomes (2.19). Note also that we have used (2.19) in order to prove (2.22). So, the key renewal theorem and renewal theorem are equivalent in the aperiodic case, and the same remark holds true for the periodic case.

2.3 Delayed Renewal Chains

Delayed renewal chains are used for modeling the same type of phenomenon as renewal chains do, with the only difference that we do not consider the origin as the occurrence of the first renewal, that is, $S_0 = X_0 > 0$. In other words, we want to observe a normal renewal chain, but we missed the beginning and we denote by S_0 the time when we observe the first renewal, which is not identically 0 but follows a certain distribution.

Definition 2.6 (delayed renewal chain).
An arrival time sequence $(S_n)_{n \in \mathbb{N}}$ *for which the waiting times* $(X_n)_{n \in \mathbb{N}^*}$ *form an i.i.d. sequence and* X_0 *is independent of* $(X_n)_{n \in \mathbb{N}^*}$ *is called a* delayed renewal chain *and every* S_n *is called a* renewal time.

The chain $(S_n - S_0)_{n \in \mathbb{N}}$ is an ordinary renewal chain, called the *associated renewal chain*. Note that $(S_n)_{n \in \mathbb{N}}$ is a renewal chain iff $S_0 = 0$ a.s.

Figure 2.3 presents the counting process of the number of renewals in a delayed renewal chain.

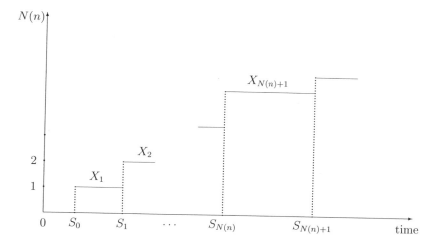

Fig. 2.2. Delayed renewal chain

Example 2.4. Consider again the Example 2.1 of the light lamp with exchange lightbulbs of lifetimes $(X_n)_{n \in \mathbb{N}}$. Suppose this time that the bulb which is on at time 0 started to function sometimes in the past, but all the others are new, with the same lifetime distribution. In other words, let $(X_n)_{n \in \mathbb{N}}$ be independent and $(X_n)_{n \in \mathbb{N}^*}$ have the same distribution. Then, the sequence $(S_n)_{n \in \mathbb{N}}$ defined by $S_n := X_0 + \ldots + X_n$, $n \in \mathbb{N}$, is an example of a delayed renewal chain.

Example 2.5. A typical example of a delayed renewal chain can be obtained as follows: consider an irreducible, finite state space Markov chain and let x be a certain state. It is easy to check that the successive returns to state x, when starting at $y \neq x$, represent a delayed renewal chain.

Example 2.6. Consider a sequence of i.i.d. Bernoulli trials. We can prove that the times of occurrence of different finite patterns form renewal chains, generally delayed. For instance, suppose we have the following sequence issued from repeated i.i.d. Bernoulli trials

$$S\ S\ S\ S\ F\ S\ F\ F\ S\ F\ S\ F\ S\ S\ S \cdots$$
$$\uparrow\ \uparrow\ \uparrow\ \uparrow\ \uparrow\ \uparrow\ \uparrow\ \uparrow\ \uparrow\ \uparrow\ \uparrow\ \uparrow\ \uparrow\ \uparrow\ \uparrow$$
$$1\ 2\ 3\ 4\ 5\ 6\ 7\ 8\ 9\ 10\ 11\ 12\ 13\ 14\ 15 \cdots$$

and suppose that we are interested in the occurrence of the pattern SFS in this sample. Two different counting procedures can be considered.

First, suppose that the counting starts anew when the searched pattern occurs, that is, we do not allow overlapping. Note that SFS occurs in the 6th and 11th trials. Note also that SFS occurs in the 13th trial, but we do not count this occurrence, because we do not allow overlapping, so we have started anew the counting at the 12th trial (that is, after the occurrence of SFS in the 11th trial).

Second, suppose that we allow overlapping. In this case, the pattern SFS occurs in the 6th, 11th and 13th trial.

In both situations the occurrence times of the pattern SFS is a delayed renewal chain. See also Example 2.7 for more details.

Let us set out some definitions and notation for a delayed renewal chain $(S_n)_{n \in \mathbb{N}}$. The distribution $b = (b_n)_{n \in \mathbb{N}}$ of S_0 will be called the *initial distribution* (or the *delayed distribution*) of the delayed renewal chain, $b_n := \mathbb{P}(S_0 = n)$. Its generating function will be denoted by $B(s)$. Set $v_n := \sum_{k=0}^{\infty} \mathbb{P}(S_k = n)$ for the probability that a renewal will occur at instant n in the delayed renewal chain and $V(s)$ for the corresponding generating function. By u_n we denote the same probability of occurrence of a renewal at time n, but in the associated renewal chain, and we set $U(s)$ for the generating function.

As was done before, denote by $f = (f_n)_{n \in \mathbb{N}}$, with $f_0 := 0$, the common distribution of the waiting times $(X_n)_{n \in \mathbb{N}^*}$ in the delayed renewal chain. Also, let F be the cumulative distribution function of the waiting times. Since $X_n := S_n - S_{n-1} = (S_n - S_0) - (S_{n-1} - S_0)$, we have the same distribution for the waiting times considered in the delayed renewal chain or in the associated renewal chain.

A delayed renewal chain is called *aperiodic* (resp. *periodic of period $d > 1$*) if the associated renewal chain is aperiodic (resp. periodic of period $d > 1$). Similarly, a delayed renewal chain is said to be *transient* or *positive (null) recurrent* if the associated renewal chain is a such.

We can easily derive a recurrence formula for $(v_n)_{n \in \mathbb{N}}$, as we already did for $(u_n)_{n \in \mathbb{N}}$ in an ordinary renewal chain (Equation (2.4)), and obtain an expression for the generating function $V(s)$ (as we did in Proposition 2.1 for $U(s)$). Indeed, we have

$$v_n = \mathbb{P}(S_0 = n) + \sum_{k=0}^{n-1} \mathbb{P}(S_0 = k) \sum_{r=1}^{n-k} \mathbb{P}(S_r - S_0 = n - k) = b_n + \sum_{k=0}^{n-1} b_k u_{n-k}.$$

Thus we have obtained

$$v_n = \sum_{k=0}^{n} b_k u_{n-k}, \quad i.e., \quad v_n = (b * u)_n, \quad n \in \mathbb{N}. \tag{2.24}$$

Multiplying Equation (2.24) by s^n and summing over $n = 0, 1, 2, \ldots$ we get

$$V(s) = B(s)U(s) = \frac{B(s)}{1 - \Phi(s)}. \tag{2.25}$$

The following theorem describes the asymptotic behavior of $(v_n)_{n \in \mathbb{N}}$, the sequence of probabilities that a renewal will occur at time $n, n \in \mathbb{N}$.

Theorem 2.8 (renewal theorem for delayed RCs).
Consider a delayed recurrent renewal chain $(S_n)_{n \in \mathbb{N}}$ with initial distribution $b = (b_n)_{n \in \mathbb{N}}$.

1. If the chain is aperiodic, then

$$\lim_{n \to \infty} v_n = \frac{1}{\mu} \sum_{n=0}^{\infty} b_n. \tag{2.26}$$

2. If the chain is periodic of period $d > 1$, then for any positive integer l, $0 \le l < d$,

$$\lim_{n \to \infty} v_{l+nd} = \frac{d}{\mu} \sum_{n=0}^{\infty} b_{l+nd}. \tag{2.27}$$

Note that $\sum_{n=0}^{\infty} b_n = \mathbb{P}(S_0 < \infty) < 1$ means that $(b_n)_{n \in \mathbb{N}}$ is an improper distribution. Thus, the limit in (2.26) is $1/\mu$ for $(b_n)_{n \in \mathbb{N}}$ a proper distribution. The same type of remark is true for the limit in (2.27).

Proof. As $v = (b * u)$ (cf. Equation (2.24)), the proof is a direct application of the key renewal theorem (Theorem 2.7) for nondelayed renewal chains. □

Example 2.7. Let us continue Example 2.6 by considering a sequence of i.i.d. Bernoulli trials, with probability of success $\mathbb{P}(S) = p$ and probability of failure $\mathbb{P}(F) = q = 1 - p, 0 < p < 1$. We are interested in the occurrence of the pattern SFS in the case when we allow overlapping. More specifically, we want to:

1. Prove that the successive occurrence times of the pattern form an aperiodic delayed recurrent renewal chain;

2. Compute the main characteristics of the chain: the initial distribution $(b_n)_{n \in \mathbb{N}}$, the probability v_n that a renewal occurs at instant n, the common distribution $(f_n)_{n \in \mathbb{N}}$ of the waiting times $(X_n)_{n \in \mathbb{N}^*}$ and $\mu = \mathbb{E}(X_1)$, the mean waiting time between two successive occurrences of the pattern.

To answer the first question, note first that the independence of the Bernoulli trials implies the independence of the waiting times. Second, let us compute the distribution of X_m for $m \in \mathbb{N}^*$.

$f_0 = \mathbb{P}(X_m = 0) = 0$ (by definition);

$f_1 = \mathbb{P}(X_m = 1) = \mathbb{P}(\text{renewal at } (S_{m-1} + 1)) = p^2 q;$

$f_2 = \mathbb{P}(X_m = 2) = \mathbb{P}(\text{renewal at } (S_{m-1} + 2), \text{not a renewal at } (S_{m-1} + 1))$
$\quad = p^2 q (1 - p^2 q);$

$f_n = \mathbb{P}(X_m = n) = p^2 q (1 - p^2 q)^{n-1}, n \geq 2.$

Thus, we see that the distribution of X_m does not dependent on $m \in \mathbb{N}^*$, so we have a renewal chain. Moreover, as $f_1 = \mathbb{P}(X_m = 1) = p^2 q \neq 0$, we get that the chain is aperiodic. In order to see that the renewal chain is delayed, we have to check that the distribution of X_0 is different than the distribution of $X_m, m \in \mathbb{N}^*$. Indeed, we have $b_0 = \mathbb{P}(X_0 = 0) = 0$, $b_1 = \mathbb{P}(X_0 = 1) = 0$ (because the pattern SFS cannot occur in the first two trials), and $b_n = \mathbb{P}(X_0 = n) = p^2 q (1 - p^2 q)^{n-2}, n \geq 2$. In conclusion, the successive occurrence times of SFS form an aperiodic delayed renewal chain. As $\sum_{n \geq 0} f_n = 1$, we see that the chain is recurrent.

Concerning the characteristics of the chain, we have already computed $(f_n)_{n \in \mathbb{N}}$ and $(b_n)_{n \in \mathbb{N}}$. Similarly, we obtain $v_0 = v_1 = 0$ (a renewal cannot occur in the first two trials) and for $n \geq 2$ we have

$v_n = \mathbb{P}(Z_n = 1)$
$\quad = \mathbb{P}(S \text{ at the } n\text{th trial}, F \text{ at the } (n-1)\text{th trial}, S \text{ at the } (n-2)\text{th trial})$
$\quad = p^2 q.$

We can immediately see that the mean waiting time between two successive occurrences is $\mu := \mathbb{E}(X_1) = 1/p^2 q$. One can also check that the relation $\lim_{n \to \infty} v_n = \frac{1}{\mu} \sum_{n=0}^{\infty} b_n$ proved in the previous theorem holds true.

The Stationary Renewal Chain

Our objective is to construct a particular case of delayed renewal chain which has important stationary or time-invariant properties. The question we want to answer is the following: For what kind of delayed renewal chain is v_n constant with respect to n ?

Let us consider $(S_n)_{n \in \mathbb{N}}$ a delayed renewal chain such that $\mathbb{P}(S_0 < \infty) = 1$, i.e., $(b_n)_{n \in \mathbb{N}}$ is a proper distribution. Suppose that $v_n = \Delta$ for all $n \in \mathbb{N}$. Consequently, the generating function $V(s)$ is given by

$$V(s) = \sum_{n=0}^{\infty} \Delta s^n = \frac{\Delta}{1-s}.$$

Using Equation (2.25) we obtain

$$B(s) = \frac{\Delta}{1-s}(1 - \Phi(s)),$$

$$\sum_{n=0}^{\infty} b_n s^n = \Delta \left(\sum_{n=0}^{\infty} s^n \right) \left(1 - \sum_{n=1}^{\infty} f_n s^n \right).$$

Equalizing the coefficients of s^n in the left-hand and right-hand side, for $n \in \mathbb{N}$, we get

$$b_0 = \Delta,$$

$$b_n = \Delta \left(1 - \sum_{k=1}^{n} f_k \right) = \Delta \mathbb{P}(X_1 > n), \quad n \in \mathbb{N}^*.$$

Taking into account the fact that $\sum_{n=0}^{\infty} b_n = 1$ and using $\sum_{n=0}^{\infty} P(X_1 > n) = \mu$ we obtain

$$\Delta = 1/\mu,$$

provided that the delayed renewal chain is positive recurrent. Thus, we have shown that if a positive recurrent delayed renewal chain satisfies $v_n = \Delta$ for all $n \in \mathbb{N}$, then $\Delta = 1/\mu$ and $b_n = \mathbb{P}(X_1 > n)/\mu$.

Now, starting with a positive recurrent delayed renewal chain $(S_n)_{n \in \mathbb{N}}$ such that $b_n = \mathbb{P}(X_1 > n)/\mu$ for all $n \in \mathbb{N}$, we want to prove that $v_n = 1/\mu$ for all $n \in \mathbb{N}$. For $0 \le s < 1$, the generating function of the first occurrence of a renewal is

$$B(s) = \frac{1}{\mu} \sum_{n=0}^{\infty} \mathbb{P}(X_1 > n)s^n = \frac{1}{\mu} \sum_{n=0}^{\infty} \sum_{k=n+1}^{\infty} \mathbb{P}(X_1 = k)s^n$$

$$= \frac{1}{\mu} \sum_{k=1}^{\infty} \sum_{n=0}^{k-1} s^n f_k = \frac{1}{\mu} \frac{1}{1-s} \left(\sum_{k=1}^{\infty} f_k - \sum_{k=1}^{\infty} s^k f_k \right)$$

$$= \frac{1}{\mu} \frac{1 - \Phi(s)}{1-s}.$$

From Equation (2.25) we obtain

$$V(s) = \frac{1}{\mu} \frac{1}{1-s} = \sum_{n=0}^{\infty} \frac{1}{\mu} s^n,$$

so $v_n = 1/\mu$ for all $n \in \mathbb{N}$.

This entire discussion can be summarized in the following result.

Proposition 2.2. *Let* $(S_n)_{n \in \mathbb{N}}$ *be a positive recurrent delayed renewal chain with waiting times* $(X_n)_{n \in \mathbb{N}}$ *and* $\mu := \mathbb{E}(X_1) < \infty$. *Then,* $\mathbb{P}(S_0 = n) := \mathbb{P}(X_1 > n)/\mu$ *is the unique choice for the initial distribution of the delayed renewal chain such that* $v_n \equiv$ *constant for all* $n \in \mathbb{N}$. *Moreover, this common constant value is* $1/\mu$.

The delayed renewal chain with $v_n = 1/\mu$ for all $n \in \mathbb{N}$ is called a *stationary renewal chain* and its initial distribution defined by $\mathbb{P}(S_0 = n) := \mathbb{P}(X_1 > n)/\mu$ for all $n \in \mathbb{N}$ is called the *stationary distribution of the delayed renewal chain* $(S_n)_{n \in \mathbb{N}}$.

Remark 2.6. It can be shown that for $(S_n)_{n \in \mathbb{N}}$ a simple renewal chain and m a fixed integer, $m \in \mathbb{N}$, we have the following limiting distribution of the current lifetime $U_m = m - S_{N(m)}$ and of the residual lifetime $V_m = S_{N(m)+1} - m$:

$$\lim_{m \to \infty} \mathbb{P}(U_m = n) = \mathbb{P}(X_1 > n)/\mu = \lim_{m \to \infty} \mathbb{P}(V_m = n + 1).$$

Consequently, a stationary renewal chain can be seen as a simple renewal chain which started indefinitely far in the past, such that the distribution $(b_n)_{n \in \mathbb{N}}$ of the first renewal we observe starting from time 0 is the same as the limit distribution of the current and residual lifetime in the simple renewal chain. This phenomenon explains intuitively the time-invariant property of a stationary renewal chain given in Proposition 2.2.

We end this section by proving Theorem 2.6.

Proof (of renewal theorem–Theorem 2.6).
(1) Consider $(S_n)_{n \in \mathbb{N}}$ a recurrent aperiodic renewal chain (not delayed, i.e., $S_0 = X_0 = 0$ a.s.), with $(X_n)_{n \in \mathbb{N}^*}$ the interrenewal times. We will prove the result for the positive recurrent case, i.e., for $\mu := \mathbb{E}(X_1) < \infty$, following the proof of Karlin and Taylor (1981) (proof of Theorem 1.1, pages 93–95) based on the technique of coupling random processes.

Let $(T_n)_{n \in \mathbb{N}}$ be the stationary renewal chain associated to the renewal chain $(S_n)_{n \in \mathbb{N}}$ (cf. Proposition 2.2). More specifically, $T_n := Y_0 + \ldots + Y_n$, where $(Y_n)_{n \in \mathbb{N}^*}$ i.i.d. such that Y_n has the same distribution as X_n for $n \geq 1$, and

$$\mathbb{P}(T_0 = n) = \mathbb{P}(Y_0 = n) := \mathbb{P}(X_1 > n)/\mu, \quad n \in \mathbb{N}.$$

Let us define the chain $(U_n)_{n \in \mathbb{N}}$ by induction as follows:

$$U_0 := X_0 - Y_0 = -Y_0,$$
$$U_n := U_{n-1} + (X_n - Y_n) = \ldots = (X_1 + \ldots + X_n) - (Y_0 + \ldots + Y_n), n \geq 1.$$

Denote by N the first instant when the same number of renewals takes place at the same time in the chains $(S_n)_{n \in \mathbb{N}}$ and in $(T_n)_{n \in \mathbb{N}}$, i.e., $N := \min\{n \in \mathbb{N} \mid U_n = 0\}$. As $\mathbb{E}(X_n - Y_n) = 0$ for $n \geq 1$, applying Theorem D.1, we obtain that the Markov chain $(U_n - U_0)_{n \in \mathbb{N}}$ is recurrent. Consequently,

$\mathbb{P}(N < \infty) = 1$. Thus, for $n \geq N$ we have that S_n and T_n have the same distribution and we obtain

$$
\begin{aligned}
u_n &= \mathbb{P}(S_k = n \text{ for some } k \in \mathbb{N}) \\
&= \mathbb{P}(T_k = n \text{ for some } k \geq N) + \mathbb{P}(S_k = n \text{ for some } k < N) \\
&= \mathbb{P}(T_k = n \text{ for some } k \in \mathbb{N}) - \mathbb{P}(T_k = n \text{ for some } k < N) \\
&\quad + \mathbb{P}(S_k = n \text{ for some } k < N).
\end{aligned}
\tag{2.28}
$$

First, note that $\mathbb{P}(T_k = n \text{ for some } k \in \mathbb{N}) = 1/\mu$ (Proposition 2.2). Second, as $\{T_k = n \text{ for some } k < N\} \subset \{T_N > n\}$ for $k < N$, we have

$$
\lim_{n \to \infty} \mathbb{P}(T_k = n \text{ for some } k < N) \leq \lim_{n \to \infty} \mathbb{P}(T_N > n) = \mathbb{P}(T_N = \infty),
$$

where the last equality is obtained using the continuity from above of the probability (Theorem E.1) applied to the nonincreasing sequence $\{T_N > n\}_{n \in \mathbb{N}}$. Since $\mathbb{P}(N < \infty) = 1$ and $(T_n)_{n \in \mathbb{N}}$ is a recurrent renewal chain (because $(S_n)_{n \in \mathbb{N}}$ is so), we get $\mathbb{P}(T_N = \infty) = 0$, so

$$
\lim_{n \to \infty} \mathbb{P}(T_k = n \text{ for some } k < N) = 0.
$$

In the same way, we get $\lim_{n \to \infty} \mathbb{P}(S_k = n \text{ for some } k < N) = 0$ and, from Equation (2.28), we obtain $\lim_{n \to \infty} u_n = 1/\mu$.

(2) The case when the recurrent renewal chain $(S_n)_{n \in \mathbb{N}}$ is periodic of period $d > 1$ can be easily reduced to the aperiodic case. Let $S'_n := S_{dn}$. The renewal chain $(S'_n)_{n \in \mathbb{N}}$ is aperiodic and we denote by μ_d its mean waiting time. Note that we have $\mu = d\mu_d$, where μ is the mean waiting time of the original renewal chain $(S_n)_{n \in \mathbb{N}}$. Using (1) we obtain

$$
\lim_{n \to \infty} u_{nd} = 1/\mu_d = d/\mu,
$$

which accomplishes the proof. $\qquad\qquad\qquad\qquad\qquad\qquad\qquad\qquad\qquad\quad\square$

2.4 Alternating Renewal Chain

We present here a particular case of a renewal chain that is important in reliability theory due to its simplicity and intuitive interpretation.

Definition 2.7 (alternating renewal chain). *Let $(X_n)_{n \in \mathbb{N}^*}$ be a sequence of i.i.d. random variables, with common distribution $h = (h_n)_{n \in \mathbb{N}}$, $h_0 := 0$. Similarly, let $(Y_n)_{n \in \mathbb{N}^*}$ be a sequence of i.i.d. random variables, with common distribution $g = (g_n)_{n \in \mathbb{N}}$, $g_0 := 0$. We also suppose that the sequences $(X_n)_{n \in \mathbb{N}^*}$ and $(Y_n)_{n \in \mathbb{N}^*}$ are independent between them. Define $V_n := X_n + Y_n, n \in \mathbb{N}^*$, and $S_n := \sum_{i=1}^{n} V_i, n \in \mathbb{N}^*$, $S_0 := 0$. The sequence $(S_n)_{n \in \mathbb{N}}$ is called an* alternating renewal chain, *with up-time distribution h and down-time distribution g.*

One can easily check that an alternating renewal chain $(S_n)_{n\in\mathbb{N}}$ is an ordinary renewal chain with waiting times V_n, $n \in \mathbb{N}^*$ and waiting time distribution $f := h * g$.

We give now a reliability example where the use of an alternating renewal chain arises naturally.

Example 2.8. Consider a component of a system whose evolution in time is as follows: at time 0, a new component starts to work for a random time X_1, when it fails and is repaired during a random time Y_1 (or replaced with a new, identical component); then, it works again for a random time X_2, when it is again repaired (or replaced) for a time Y_2, and so on. Suppose that $X := (X_n)_{n\in\mathbb{N}^*}$ is a sequence of i.i.d. random variables, that is, the repair process is perfect (or the replacement components are identical to the used ones). Suppose also that $Y := (Y_n)_{n\in\mathbb{N}^*}$ is a sequence of i.i.d. random variables, i.e., all the repairing (or replacement) conditions are identical. Suppose also that sequences X and Y are independent between them. The sequence $S_n := \sum_{i=1}^{n}(X_i + Y_i)$, $n \in \mathbb{N}^*$, $S_0 := 0$, forms an alternating renewal chain. See also Exercise 2.2.

Let us consider the system given in the previous example. Thus, we suppose that $X_n, n \in \mathbb{N}^*$, is the nth up time (working time) of the system, while $Y_n, n \in \mathbb{N}^*$, is the nth down time (repair time) of the system. First, we want to obtain the probability that the system will be working at time k. Second, we want to see what this probability will be for large k (when k tends to infinity). To answer this questions, we introduce:

- The *reliability of the system at time* $k \in \mathbb{N}$ – the probability that the system has functioned without failure in the period $[0, k]$,
- The *availability of the system at time* $k \in \mathbb{N}$ – the probability that the system will be working at time $k \in \mathbb{N}$.

In our context, the reliability at time $k \in \mathbb{N}$ is

$$R(k) = \mathbb{P}(X_1 > k) = \sum_{m \geq 1} f_{k+m}.$$

We want to obtain the availability of the system as the solution of a renewal equation. For any $k \in \mathbb{N}^*$, we can write

$$A(k) = \sum_{n \geq 1} \mathbb{P}(S_{n-1} \leq k < S_{n-1} + X_n)$$

$$= \mathbb{P}(X_1 > k) + \sum_{n \geq 2} \mathbb{P}(S_{n-1} \leq k < S_{n-1} + X_n)$$

$$= R(k) + \sum_{n \geq 2} \sum_{m=1}^{k} \mathbb{P}(S_{n-1} \leq k < S_{n-1} + X_n, \ S_1 = m)$$

$$= R(k) + \sum_{m=1}^{k} \sum_{n \geq 2} \mathbb{P}(S_{n-2} \leq k - m < S_{n-2} + X_{n-1})\mathbb{P}(S_1 = m)$$

$$= R(k) + \sum_{m=1}^{k} A(k-m)f_m,$$

and we obtain the renewal equation associated to the availability:

$$A(k) = R(k) + f * A(k), k \in \mathbb{N}.$$

Although we proved this equality only for $k \in \mathbb{N}^*$, it is obviously satisfied for $k = 0$, because $f_0 := 0$ and the system is working at time 0, i.e., $A(0) = R(0) = 1$.

Solving this renewal equation (Theorem 2.2), we get the probability that the system will be working at time k in terms of the reliability

$$A(k) = u * R(k), \ k \in \mathbb{N},$$

where u_n, the probability that a renewal will occur at time n, represents in our case the probability that the system is just starting to function again after a repairing period.

We are interested now in the probability that the system will be working at time k, for large k. Thus, we want to obtain the limit of $A(k)$, the availability at time k, as k tends to infinity. This is called *steady-state* (or *limit*) *availability*. From the key renewal theorem (Theorem 2.7) we get

$$\lim_{k \to \infty} A(k) = \lim_{k \to \infty} u * R(k) = \frac{1}{\mu} \sum_{m=0}^{\infty} R(m),$$

where $\mu = \mathbb{E}(X_1 + Y_1)$ is the mean waiting time in the renewal chain $(S_n)_{n \in \mathbb{N}}$. Noting that

$$\sum_{m=0}^{\infty} R(m) = \sum_{m=0}^{\infty} \mathbb{P}(X_1 > m) = \mathbb{E}(X_1),$$

we get the steady-state availability

$$\lim_{k \to \infty} A(k) = \frac{\mu_X}{\mu_X + \mu_Y},$$

where we set $\mu_X := \mathbb{E}(X_1)$ for the mean lifetime and $\mu_Y := \mathbb{E}(Y_1)$ for the mean repair time.

Remark 2.7. It is worth noticing that for the results concerning availability we do not need that the sequences of random variables $(X_n)_{n \in \mathbb{N}^*}$ and $(Y_n)_{n \in \mathbb{N}^*}$ to be independent between them, but only that $(X_n + Y_n)_{n \in \mathbb{N}^*}$ is an i.i.d. sequence.

Exercises

Exercise 2.1. Show that the alternating renewal chain introduced by Definition 2.7 is an ordinary renewal chain (i.e., nondelayed) with waiting times $V_n = X_n + Y_n$, $n \in \mathbb{N}^*$, and waiting time distribution $f := h * g$.

Exercise 2.2 (binary component). Consider the binary component (or system) given in Example 2.8. The component starts to work at time $n = 0$. Consider that the lifetimes $(X_n)_{n \in \mathbb{N}^*}$ of the component have a common geometric distribution on $\mathbb{N}^*$, of parameter $p, 0 < p < 1$, denoted by $(h_n)_{n \in \mathbb{N}}$, $h_0 := 0$, $h_n := p(1-p)^{n-1}, n \geq 1$. Consider also that the repair times $(Y_n)_{n \in \mathbb{N}^*}$ have a common geometric distribution on $\mathbb{N}^*$, of parameter $q, 0 < q < 1$, denoted by $(g_n)_{n \in \mathbb{N}}$, $g_0 := 0$, $g_n := q(1-q)^{n-1}, n \geq 1$.

1. Show that the sequence $S_n := \sum_{i=1}^{n} (X_i + Y_i)$, $n \in \mathbb{N}*$, $S_0 := 0$ forms a positive recurrent renewal chain and compute the characteristics of the chain: the waiting time distribution $f = (f_n)_{n \in \mathbb{N}}$, the corresponding generating function, and the sequence $(u_n)_{n \in \mathbb{N}}$ of probabilities that a renewal occurs at time n.

2. For these types of components and for a large time n, compute approximatively the number of repairs (replacements) needed during the time interval $[0; n]$.
   Numerical application: take $n = 2000, p = 0.01$, and $q = 0.1$.

Exercise 2.3 (binary component: continuation). Consider the binary component given in Exercise 2.2. Denote by 0 the working state and by 1 the failure state. Let T_n be a random variable of state space $E = \{0, 1\}$, defined as follows: $T_n = 0$ if the component is in the working state at time n and $T_n = 1$ if the component is in the failure state at time n.

1. Show that $(T_n)_{n \in \mathbb{N}}$ is a Markov chain with state space $E = \{0, 1\}$, initial distribution $\alpha = (1, 0)$, and transition matrix

$$\mathbf{p} = \begin{pmatrix} 1 - p & p \\ q & 1 - q \end{pmatrix}.$$

2. Show that the n step transition matrix is

$$\mathbf{p}^n = \frac{1}{p+q} \begin{pmatrix} q & p \\ q & p \end{pmatrix} + \frac{(1 - p - q)^n}{p + q} \begin{pmatrix} p & -p \\ -q & q \end{pmatrix}.$$

3. Give the state probability vector $P(n) = (P_1(n), P_2(n))$.

Exercise 2.4 (transient or stopped discrete-time renewal processes). Let us consider a discrete-time renewal process with waiting time distribution $(f_n)_{n \in \mathbb{N}}$ such that $\sum_{n \geq 0} f_n = p$, with $0 < p < 1$. We define the lifetime of the chain by

$$T := S_N = X_1 + \ldots + X_N,$$

where $S_{N+1} = S_{N+2} = \ldots = \infty$.

1. Compute the distribution of N.
2. Show that $\mathbb{P}(T = n) = p\mathbf{1}_{\{n=0\}} + \sum_{k=1}^{n} \mathbb{P}(T = n - k)f(k), n \geq 0$.
3. Give the distribution of T.

Exercise 2.5. Let $F_n(x) = \frac{1}{n} \sum_{k=1}^{n} \mathbf{1}_{\{X_k \leq x\}}$ be the empirical distribution function of the i.i.d. random sample $X_1, \ldots, X_n$. Denote by F the common distribution function of $X_i, i = 1, \ldots, n$. The Glivenko–Cantelli theorem tells us that

$$\sup_{x \in \mathbb{R}} | F_n(x) - F(x) | \xrightarrow[n \to \infty]{a.s.} 0.$$

Suppose that $N(m), m \geq 0$, is a family of positive integer-valued random variables such that $N(m) \xrightarrow[m \to \infty]{a.s.} \infty$. Define another empirical distribution as follows:

$$\widetilde{F}_m(x) = \frac{1}{N(m)} \sum_{k=1}^{N(m)} \mathbf{1}_{\{X_k \leq x\}}.$$

Prove that

$$\sup_{x \in \mathbb{R}} | \widetilde{F}_m(x) - F(x) | \xrightarrow[m \to \infty]{a.s.} 0.$$

Exercise 2.6 (forward recurrence times). Consider the settings of Example 2.3, i.e., start with $(S_n)_{n \in \mathbb{N}}$, a recurrent renewal chain, with waiting time distribution $(f_n)_{n \in \mathbb{N}}$, take $E := \{1, \ldots, m\}$, if $m < \infty$, or $E := \{1, 2, \ldots\} = \mathbb{N}^*$, if $m = \infty$, where $m := \sup\{n \mid f_n > 0\}$.

1. Show that the forward recurrence times $(V_n)_{n \in \mathbb{N}}$ form an irreducible recurrent Markov chain.
2. Show that $(V_n)_{n \in \mathbb{N}}$ has the transition matrix $(p_{ij})_{i,j \in E}$ defined by $p_{1i} := f_i$ for any $i \in E$, $p_{i\,i-1} := 1$, if $i > 1$, and $p_{ij} := 0$ for all other $i, j \in E$.

Exercise 2.7 (backward recurrence times). Let $(S_n)_{n \in \mathbb{N}}$ be a recurrent renewal chain, with waiting time distribution $(f_n)_{n \in \mathbb{N}}$.

1. Show that the backward recurrence times $(U_n)_{n \in \mathbb{N}}$ form an irreducible recurrent Markov chain.
2. As in Exercise 2.6 for the forward recurrence times, find the transition probabilities of the backward recurrence times in terms of the waiting time distribution $(f_n)_{n \in \mathbb{N}}$.

3

Semi-Markov Chains

In this chapter we define the discrete-time semi-Markov processes and we obtain their basic probabilistic properties. We present elements of Markov renewal theory in the discrete case and we propose a simple forward algorithm for solving the discrete-time Markov renewal equation.

Compared to the attention given to the continuous-time semi-Markov processes, the discrete-time semi-Markov processes (semi-Markov chains) are almost absent in the literature. The main works on the topic are those of Anselone (1960), Howard (1971), Gerontidis (1994), Mode and Pickens (1998), Mode and Sleeman (2000), Vassiliou and Papadopoulou (1992, 1994) (the last two references are on nonhomogeneous semi-Markov chains), Barbu, Boussemart and Limnios (2004), and Girardin and Limnios (2004, 2006). Recent works on discrete-time hidden semi-Markov models will be presented in Chapter 6.

The discrete-time framework offers, especially for applications, some important advantages compared to the continuous-time case. We stress the simplicity of semi-Markov chains and, furthermore, the fact that a discrete-time semi-Markov process cannot explode, which means that in a bounded interval of time the process can visit only a finite number of states. For this reason, the Markov renewal function is expressed as a finite series of semi-Markov kernel convolution products instead of an infinite series in the continuous case. Consequently, all the numerical computations for a functional of a discrete semi-Markov kernel are much faster and more accurate than for a continuous one. From this point of view, the discrete-time Markov renewal processes represent a good support for the numerical calculus of the continuous-time Markov renewal processes, after their discretization.

We begin this chapter by defining the Markov renewal chains, the semi-Markov chains, and associated notions, in analogy with the corresponding notions of continuous-time semi-Markov processes (using, basically, Pyke, 1961a,b; Çinlar, 1969, 1975; Limnios and Oprişan, 2001). Afterward, we present the elements of a Markov renewal theory for a discrete-time model,

V.S. Barbu, N. Limnios, *Semi-Markov Chains and Hidden Semi-Markov Models toward Applications*, DOI: 10.1007/978-0-387-73173-5_3,
© Springer Science+Business Media, LLC 2008

we propose a simple forward algorithm for solving the discrete-time Markov renewal equation and we derive asymptotic results for Markov renewal chains. An example (Example 3.4), for which all the computational details are given, illustrates how the characteristics of a discrete-time semi-Markov model are obtained.

The results presented in this chapter are based on Barbu, Boussemart and Limnios (2004).

3.1 Markov Renewal Chains and Semi-Markov Chains

Consider a random system with finite state space $E = \{1, \ldots, s\}$. We denote by $\mathcal{M}_E$ the set of real matrices on $E \times E$ and by $\mathcal{M}_E(\mathbb{N})$ the set of matrix-valued functions defined on $\mathbb{N}$, with values in $\mathcal{M}_E$. For $\mathbf{A} \in \mathcal{M}_E(\mathbb{N})$, we write $\mathbf{A} = (\mathbf{A}(k); \ k \in \mathbb{N})$, where, for $k \in \mathbb{N}$ fixed, $\mathbf{A}(k) = (A_{ij}(k); \ i, j \in E) \in \mathcal{M}_E$. Set $\mathbf{I}_E \in \mathcal{M}_E$ for the identity matrix and $\mathbf{0}_E \in \mathcal{M}_E$ for the null matrix. When space E is clear from the context, we will write $\mathbf{I}$ and $\mathbf{0}$ instead of $\mathbf{I}_E$ and $\mathbf{0}_E$.

Let us also define $\mathbf{I} := (\mathbf{I}(k); \ k \in \mathbb{N})$ as the constant matrix-valued function whose value for any nonnegative integer k is the identity matrix, that is, $\mathbf{I}(k) := \mathbf{I}$ for any $k \in \mathbb{N}$. Similarly, we set $\mathbf{0} := (\mathbf{0}(k); \ k \in \mathbb{N})$, with $\mathbf{0}(k) := \mathbf{0}$ for any $k \in \mathbb{N}$.

We suppose that the evolution in time of the system is described by the following chains:

- The chain $J = (J_n)_{n \in \mathbb{N}}$ with state space E, where J_n is the system state at the nth jump time;
- The chain $S = (S_n)_{n \in \mathbb{N}}$ with state space $\mathbb{N}$, where S_n is the nth jump time. We suppose that $S_0 = 0$ and $0 < S_1 < S_2 < \ldots < S_n < S_{n+1} < \ldots$;
- The chain $X = (X_n)_{n \in \mathbb{N}}$ with state space $\mathbb{N}$, $X_n := S_n - S_{n-1}$ for all $n \in \mathbb{N}^*$ and $X_0 := 0$ a.s. Thus for all $n \in \mathbb{N}^*$, X_n is the sojourn time in state J_{n-1}, before the nth jump.

Figure 1.1, page 2, gives a representation of the evolution of the system. One fundamental notion for our work is that of semi-Markov kernel in discrete time.

Definition 3.1 (discrete-time semi-Markov kernel).
   A matrix-valued function $\mathbf{q} = (q_{ij}(k)) \in \mathcal{M}_E(\mathbb{N})$ *is said to be a* discrete-time semi-Markov kernel *if it satisfies the following three properties:*

1. $0 \leq q_{ij}(k), \ i, j \in E, \ k \in \mathbb{N}$,
2. $q_{ij}(0) = 0, \ i, j \in E$,
3. $\displaystyle\sum_{k=0}^{\infty} \sum_{j \in E} q_{ij}(k) = 1, \ i \in E$.

Remark 3.1. Instead of considering semi-Markov kernels as defined above, one can be interested in nonnegative matrix-valued functions $\mathbf{q} = (q_{ij}(k)) \in \mathcal{M}_E(\mathbb{N})$ that satisfy

$$\sum_{k=0}^{\infty} \sum_{j \in E} q_{ij}(k) \le 1$$

for all $i \in E$. Such a matrix-valued function is called a *semi-Markov subkernel*. With only one exception (second proof of Proposition 5.1), we will not consider subkernels along this book.

Definition 3.2 (Markov renewal chain).

The chain $(J, S) = (J_n, S_n)_{n \in \mathbb{N}}$ *is said to be a* Markov renewal chain (MRC) *if for all* $n \in \mathbb{N}$, *for all* $i, j \in E$, *and for all* $k \in \mathbb{N}$ *it satisfies almost surely*

$$\mathbb{P}(J_{n+1} = j, S_{n+1} - S_n = k \mid J_0, \ldots, J_n; S_0, \ldots, S_n)$$
$$= \mathbb{P}(J_{n+1} = j, S_{n+1} - S_n = k \mid J_n). \tag{3.1}$$

Moreover, if Equation (3.1) is independent of n, *then* (J, S) *is said to be* homogeneous *and the discrete-time semi-Markov kernel* $\mathbf{q}$ *is defined by*

$$q_{ij}(k) := \mathbb{P}(J_{n+1} = j, X_{n+1} = k \mid J_n = i).$$

Throughout this book, we consider homogeneous MRCs only.

Note that, if (J, S) is a (homogeneous) Markov renewal chain, we can easily see that $(J_n)_{n \in \mathbb{N}}$ is a (homogeneous) Markov chain, called the *embedded Markov chain* (EMC) associated to the MRC (J, S). We denote by $\mathbf{p} = (p_{ij})_{i,j \in E} \in \mathcal{M}_E$ the transition matrix of (J_n), defined by

$$p_{ij} := \mathbb{P}(J_{n+1} = j \mid J_n = i), \ i, j \in E, \ n \in \mathbb{N}.$$

Note also that, for any $i, j \in E$, p_{ij} can be expressed in terms of the semi-Markov kernel by $p_{ij} = \sum_{k=0}^{\infty} q_{ij}(k)$.

Remark 3.2.

1. Note that we have imposed $S_0 = X_0 = 0$ and we get what we can call, by analogy with renewal chains, a simple (i.e., nondelayed) Markov renewal chain. All the Markov renewal models throughout this book will follow this assumption. Anyway, the reader interested in delayed Markov renewal models, defined by taking $S_0 = X_0 \ne 0$, can easily adapt the results presented here for the nondelayed case (as for simple and delayed renewal chains).

2. Note that we do not allow instantaneous transitions, i.e., $q_{ij}(0) = 0, i, j \in E$. This is a direct consequence of the fact that the chain $(S_n)_{n \in \mathbb{N}}$ is supposed to be increasing $(0 = S_0 < S_1 < S_2 < \ldots)$ or, equivalently, the random variables $X_n, n \in \mathbb{N}^*$, are positive. Nor do we allow transitions to the same state, i.e., $p_{ii} = 0, i \in E$.

3. If we have a MRC with $p_{jj} \neq 0$ for some $j \in E$, then we can transform it into a MRC as defined above, i.e., with transition matrix of the embedded Markov chain $p'_{ii} = 0$ for any $i \in E$, using the following transformation:

$$p'_{ij} := p_{ij} / \sum_{r \neq i} p_{ir}, \quad \text{for } i \neq j,$$

$$p'_{ii} := 0.$$

Let us also introduce the *cumulated semi-Markov kernel* $\mathbf{Q} = (\mathbf{Q}(k); k \in \mathbb{N}) \in \mathcal{M}_E(\mathbb{N})$ defined, for all $i, j \in E$ and $k \in \mathbb{N}$, by

$$Q_{ij}(k) := \mathbb{P}(J_{n+1} = j, X_{n+1} \leq k \mid J_n = i) = \sum_{l=0}^{k} q_{ij}(l). \tag{3.2}$$

When investigating the evolution of a Markov renewal chain we are interested in two types of holding time distributions: the sojourn time distributions in a given state and the conditional distributions depending on the next state to be visited.

Definition 3.3 (conditional distributions of sojourn times).
For all $i, j \in E$, let us define:

1. $f_{ij}(\cdot)$, the conditional distribution of X_{n+1}, $n \in \mathbb{N}$:

$$f_{ij}(k) := \mathbb{P}(X_{n+1} = k \mid J_n = i, J_{n+1} = j), \ k \in \mathbb{N}. \tag{3.3}$$

2. $F_{ij}(\cdot)$, the conditional cumulative distribution of X_{n+1}, $n \in \mathbb{N}$:

$$F_{ij}(k) := \mathbb{P}(X_{n+1} \leq k \mid J_n = i, J_{n+1} = j) = \sum_{l=0}^{k} f_{ij}(l), \ k \in \mathbb{N}. \tag{3.4}$$

Obviously, for all $i, j \in E$ and for all $k \in \mathbb{N}$, we have

$$f_{ij}(k) = q_{ij}(k)/p_{ij} \text{ if } p_{ij} \neq 0 \tag{3.5}$$

and, by convention, we put $f_{ij}(k) = \mathbf{1}_{\{k=\infty\}}$ if $p_{ij} = 0$.

Definition 3.4 (sojourn time distributions in a given state).
For all $i \in E$, let us denote by:

1. $h_i(\cdot)$ the sojourn time distribution in state i:

$$h_i(k) := \mathbb{P}(X_{n+1} = k \mid J_n = i) = \sum_{j \in E} q_{ij}(k), \ k \in \mathbb{N}.$$

2. $H_i(\cdot)$ *the sojourn time cumulative distribution in state i:*

$$H_i(k) := \mathbb{P}(X_{n+1} \le k \mid J_n = i) = \sum_{l=1}^{k} h_i(l), \ k \in \mathbb{N}.$$

For G the cumulative distribution of a certain r.v. X, we denote the *survival function* by $\overline{G}(n) := 1 - G(n) = \mathbb{P}(X > n)$, $n \in \mathbb{N}$. Thus for all states $i, j \in E$ we establish $\overline{F}_{ij}$ and $\overline{H}_i$ as the corresponding survival functions.

Remark 3.3. As we saw in Equation (3.5), the semi-Markov kernel introduced in Definition 3.2 verifies the relation

$$q_{ij}(k) = p_{ij} f_{ij}(k), \text{ for all } i, j \in E \text{ and } k \in \mathbb{N} \text{ such that } p_{ij} \ne 0.$$

We can also define two particular semi-Markov kernels for which $f_{ij}(k)$ does not depend on i or j, by setting the semi-Markov kernels of the form $q_{ij}(k) = p_{ij} f_j(k)$ or $q_{ij}(k) = p_{ij} f_i(k)$. Note that this $f_i(\cdot)$ is simply $h_i(\cdot)$, the sojourn time distribution in state i, as defined above. These particular types of Markov renewal chains could be adapted for some applications, where practical arguments justify that the sojourn times in a state depend only on the next visited state ($f_{ij}(k) := f_j(k)$) or only on the present visited state $f_{ij}(k) := f_i(k)$). Note that these particular Markov renewal chains can be obtained by transforming the general Markov renewal chain (i.e., with the kernel $q_{ij}(k) = p_{ij} f_{ij}(k)$). Only this type of general kernel will be considered throughout this book.

Example 3.1. Let $(J_n, S_n)_{n \in \mathbb{N}}$ be a Markov renewal chain, with the semi-Markov kernel $q_{ij}(k) = p_{ij} f_{ij}(k)$ and sojourn time distribution in state i $h_i(k) = \sum_{j \in E} q_{ij}(k)$. Note that the process $(J_n, S_{n+1})_{n \in \mathbb{N}}$ is a Markov renewal chain with semi-Markov kernel $\widetilde{q}_{ij}(k) = p_{ij} h_j(k)$. Similarly, the process $(J_{n+1}, S_n)_{n \in \mathbb{N}}$ is a Markov renewal chain with kernel $\breve{q}_{ij}(k) = p_{ij} h_i(k)$.

Example 3.2 (Markov chain). A Markov chain with the transition matrix $(p_{ij})_{i,j \in E}$ (Appendix D) is a particular case of a MRC with semi-Markov kernel (see Proposition D.2)

$$q_{ij}(k) = \begin{cases} p_{ij} (p_{ii})^{k-1}, & \text{if } i \ne j \text{ and } k \in \mathbb{N}^*, \\ 0, & \text{elsewhere.} \end{cases}$$

Example 3.3 (alternating renewal chain). A discrete-time alternating renewal process (Section 2.4), with up-time distribution f and down-time distribution g, is a MRC with two states and with semi-Markov kernel

$$\mathbf{q}(k) = \begin{pmatrix} 0 & f(k) \\ g(k) & 0 \end{pmatrix}, \ k \in \mathbb{N}.$$

The operation that will be commonly used when working on the space $\mathcal{M}_E(\mathbb{N})$ of matrix-valued functions will be the discrete-time matrix convolution product. In the sequel we recall its definition, we see that there exists an identity element and we define recursively the n-fold convolution.

Definition 3.5 (discrete-time matrix convolution product).
   Let $\mathbf{A}, \mathbf{B} \in \mathcal{M}_E(\mathbb{N})$ *be two matrix-valued functions. The matrix convolution product* $\mathbf{A} * \mathbf{B}$ *is the matrix-valued function* $\mathbf{C} \in \mathcal{M}_E(\mathbb{N})$ *defined by*

$$C_{ij}(k) := \sum_{r \in E} \sum_{l=0}^{k} A_{ir}(k-l) \, B_{rj}(l), \quad i, j \in E, \quad k \in \mathbb{N},$$

or, in matrix form,

$$\mathbf{C}(k) := \sum_{l=0}^{k} \mathbf{A}(k-l) \, \mathbf{B}(l), \quad k \in \mathbb{N}.$$

The following result concerns the existence of the identity element for the matrix convolution product in discrete time.

Lemma 3.1. *Let* $\boldsymbol{\delta I} = (d_{ij}(k); \ i, j \in E) \in \mathcal{M}_E(\mathbb{N})$ *be the matrix-valued function defined by*

$$d_{ij}(k) := \begin{cases} 1, & \text{if } i = j \text{ and } k = 0, \\ 0, & \text{elsewhere}, \end{cases}$$

or, in matrix form,

$$\boldsymbol{\delta I}(k) := \begin{cases} \mathbf{I}, & \text{if } k = 0, \\ \mathbf{0}, & \text{elsewhere}. \end{cases}$$

   Then $\boldsymbol{\delta I}$ *satisfies*

$$\boldsymbol{\delta I} * \mathbf{A} = \mathbf{A} * \boldsymbol{\delta I} = \mathbf{A}, \quad \mathbf{A} \in \mathcal{M}_E(\mathbb{N}),$$

i.e., $\boldsymbol{\delta I}$ *is the identity element for the discrete-time matrix convolution product.*

The power in the sense of convolution is defined straightforwardly using Definition 3.5.

Definition 3.6 (discrete-time n-fold convolution).
   Let $\mathbf{A} \in \mathcal{M}_E(\mathbb{N})$ *be a matrix-valued function and* $n \in \mathbb{N}$. *The n-fold convolution* $\mathbf{A}^{(n)}$ *is the matrix-valued function defined recursively by:*

$$A_{ij}^{(0)}(k) := d_{ij}(k) = \begin{cases} 1, & \text{if } i = j \text{ and } k = 0, \\ 0, & \text{elsewhere}, \end{cases}$$

$$A_{ij}^{(1)}(k) := A_{ij}(k),$$

$$A_{ij}^{(n)}(k) := \sum_{r \in E} \sum_{l=0}^{k} A_{ir}(l) \, A_{rj}^{(n-1)}(k-l), \quad n \geq 2, k \in \mathbb{N},$$

that is,

$$\mathbf{A}^{(0)} := \delta \mathbf{I}, \ \mathbf{A}^{(1)} := \mathbf{A} \ and \ \mathbf{A}^{(n)} := \mathbf{A} * \mathbf{A}^{(n-1)}, \quad n \geq 2.$$

Some important quantities for investigating the evolution of a Markov renewal chain are the probabilities $\mathbb{P}(J_n = j, S_n = k \mid J_0 = i)$, $i, j \in E$, $n \in \mathbb{N}$. They are the analog of the n-step transition functions of a Markov chain (see Appendix D). Recall that, for a finite Markov chain $(X_n)_{n \in \mathbb{N}}$ of transition matrix $\mathbf{p} = (p_{ij})_{i,j \in E}$, the n-step transition function (Appendix D) can be written as

$$\mathbb{P}(X_n = j \mid X_0 = i) = p_{ij}^n, \quad \text{for any } n \in \mathbb{N},$$

where p_{ij}^n is the (i, j) element of the n-fold matrix product of $\mathbf{p}$. A similar result holds true for the probabilities $\mathbb{P}(J_n = j, S_n = k \mid J_0 = i)$ in a Markov renewal context:

Proposition 3.1. *For all $i, j \in E$, for all n and $k \in \mathbb{N}$, we have*

$$\mathbb{P}(J_n = j, S_n = k \mid J_0 = i) = q_{ij}^{(n)}(k). \tag{3.6}$$

Proof. We prove the result by induction. For $n = 0$, we have

$$\mathbb{P}(J_0 = j, S_0 = k \mid J_0 = i) = q_{ij}^{(0)}(k).$$

Obviously, for $k \neq 0$ or $i \neq j$, this probability is zero. On the other hand, if $i = j$ and $k = 0$, the probability is one, thus the result follows.

For $n = 1$, the result obviously holds true, using the definition of the semi-Markov kernel $\mathbf{q}$ and of $q_{ij}^{(1)}(k)$. For $n \geq 2$:

$$\mathbb{P}(J_n = j, S_n = k \mid J_0 = i)$$

$$= \sum_{r \in E} \sum_{l=1}^{k-1} \mathbb{P}(J_n = j, S_n = k, J_1 = r, S_1 = l \mid J_0 = i)$$

$$= \sum_{r \in E} \sum_{l=1}^{k-1} \mathbb{P}(J_n = j, S_n = k \mid J_1 = r, S_1 = l, J_0 = i) \, \mathbb{P}(J_1 = r, S_1 = l \mid J_0 = i)$$

$$= \sum_{r \in E} \sum_{l=1}^{k-1} \mathbb{P}(J_{n-1} = j, S_{n-1} = k - l \mid J_0 = r) \mathbb{P}(J_1 = r, X_1 = l \mid J_0 = i)$$

$$= \sum_{r \in E} \sum_{l=1}^{k-1} q_{rj}^{(n-1)}(k-l) \, q_{ir}(l) = q_{ij}^{(n)}(k),$$

and the result follows. $\square$

Remark 3.4. For $(J_n, S_n)_{n \in \mathbb{N}}$ a Markov renewal chain of state space E, one can check that it is also a Markov chain with state space $E \times \mathbb{N}$. Denoting by $\mathbf{p}^{(n)}$ the n-step transition function of this Markov chain, the previous proposition states that for all $i, j \in E$, for all n and $k \in \mathbb{N}$, we have

$$p_{(i,0),(j,k)}^{(n)} = q_{ij}^{(n)}(k).$$

As a direct application of the previous proposition, we have the following lemma, which will be seen to be a key result for Markov renewal chain computations.

Lemma 3.2. *Let $(J, S) = (J_n, S_n)_{n \in \mathbb{N}}$ be a Markov renewal chain and $\mathbf{q} \in \mathcal{M}_E(\mathbb{N})$ its associated semi-Markov kernel. Then, for all n, $k \in \mathbb{N}$ such that $n \geq k + 1$ we have $\mathbf{q}^{(n)}(k) = 0$.*

Proof. It is clear that the jump time process $(S_n)_{n \in \mathbb{N}}$ verifies the relation $S_n \geq n$, $n \in \mathbb{N}$. Writing Equation (3.6) for n and $k \in \mathbb{N}$ such that $n \geq k + 1$, we obtain the desired result. $\square$

As we will see in the sequel, the property of the discrete-time semi-Markov kernel stated above is essential for the simplicity and the numerical accuracy of the results obtained in discrete time. We need to stress the fact that this property is intrinsic to the discrete-time process and it is no longer valid for a continuous-time Markov renewal process.

Let us now introduce the notion of semi-Markov chain, strictly related to that of the Markov renewal chain.

Definition 3.7 (semi-Markov chain).
Let (J, S) be a Markov renewal chain. The chain $Z = (Z_k)_{k \in \mathbb{N}}$ is said to be a semi-Markov chain *associated to the MRC (J, S) if*

$$Z_k := J_{N(k)}, k \in \mathbb{N},$$

where

$$N(k) := \max\{n \in \mathbb{N} \mid S_n \leq k\} \tag{3.7}$$

is the discrete-time counting process of the number of jumps in $[1, k] \subset \mathbb{N}$. Thus Z_k gives the system state at time k. We have also $J_n = Z_{S_n}$ and $S_n = \min\{k > S_{n-1} \mid Z_k \neq Z_{k-1}\}$, $n \in \mathbb{N}$.

We want to point out that it is only a matter of technical convenience that we have chosen to define $N(k)$ as the counting process of the number of jumps in $[1, k]$ instead of $[0, k]$.

Let the row vector $\boldsymbol{\alpha} = (\alpha_1, \ldots, \alpha_s)$ denote the *initial distribution* of the semi-Markov chain $Z = (Z_k)_{k \in \mathbb{N}}$, i.e., $\alpha_i := \mathbb{P}(Z_0 = i) = \mathbb{P}(J_0 = i)$, $i \in E$.

Definition 3.8. *The* transition function of the semi-Markov chain Z *is the matrix-valued function* $\mathbf{P} = (P_{ij}(k); \ i,j \in E, k \in \mathbb{N}) \in \mathcal{M}_E(\mathbb{N})$ *defined by*

$$P_{ij}(k) := \mathbb{P}(Z_k = j \mid Z_0 = i), \ i,j \in E, \ k \in \mathbb{N}.$$

The following result consists in a recursive formula for computing the transition function $\mathbf{P}$ of the semi-Markov chain Z, which is a first example of a Markov renewal equation, as we will see in the next section.

Proposition 3.2. *For all $i,j \in E$ and $k \in \mathbb{N}$, we have*

$$P_{ij}(k) = \delta_{ij}\left[1 - H_i(k)\right] + \sum_{r \in E}\sum_{l=0}^{k} q_{ir}(l)P_{rj}(k-l), \qquad (3.8)$$

where δ_{ij} is the Kronecker symbol, $\delta_{ij} := 1$ for $i = j$ and $\delta_{ij} := 0$ for $i \neq j$. For all $k \in \mathbb{N}$, let us define $\mathbf{H}(k) := diag(H_i(k); \ i \in E)$, $\mathbf{H} := (\mathbf{H}(k); \ k \in \mathbb{N})$, where $H_i(\cdot)$ is the sojourn time cumulative distribution function in state i (Definition 3.4). In matrix-valued function notation, Equation (3.8) becomes

$$\mathbf{P}(k) = (\mathbf{I} - \mathbf{H})(k) + \mathbf{q} * \mathbf{P}(k), k \in \mathbb{N}. \qquad (3.9)$$

Proof. For all $i,j \in E$ and for all $k \in \mathbb{N}$, we have

$$
\begin{aligned}
P_{ij}(k) &= \mathbb{P}(Z_k = j \mid Z_0 = i) \\
&= \mathbb{P}(Z_k = j, S_1 \leq k \mid Z_0 = i) + \mathbb{P}(Z_k = j, S_1 > k \mid Z_0 = i) \\
&= \sum_{r \in E}\sum_{l=0}^{k} \mathbb{P}(Z_k = j, Z_{S_1} = r, S_1 = l \mid Z_0 = i) + \delta_{ij}\left(1 - H_i(k)\right) \\
&= \sum_{r \in E}\sum_{l=0}^{k} \mathbb{P}(Z_k = j \mid Z_{S_1} = r, S_1 = l, Z_0 = i) \\
&\quad \times \mathbb{P}(J_1 = r, S_1 = l \mid J_0 = i) + \delta_{ij}\left(1 - H_i(k)\right) \\
&= \sum_{r \in E}\sum_{l=0}^{k} \mathbb{P}(Z_{k-l} = j \mid Z_0 = r)\mathbb{P}(J_1 = r, X_1 = l \mid J_0 = i) \\
&\quad + \delta_{ij}\left(1 - H_i(k)\right) \\
&= \delta_{ij}\left(1 - H_i(k)\right) + \sum_{r \in E}\sum_{l=0}^{k} P_{rj}(k-l)\,q_{ir}(l),
\end{aligned}
$$

and we obtain the desired result. □

3.2 Markov Renewal Theory

In this section, we will study equations like Equation (3.8) (or, equivalently, Equation (3.9)), given in the previous section, called Markov renewal equations (MRE). Our objective is to investigate the existence and the uniqueness of solutions for this type of equation. In Theorem 3.1 we will see that we can obtain an explicit solution, which will turn out to be unique. In particular, starting from Equation (3.9), we will find an explicit form of the transition function $\mathbf{P}$ of the semi-Markov chain Z, written in terms of the semi-Markov kernel $\mathbf{q}$.

Definition 3.9 (discrete-time Markov renewal equation).
Let $\mathbf{L} = (L_{ij}(k);\ i, j \in E, k \in \mathbb{N}) \in \mathcal{M}_E(\mathbb{N})$ be an unknown matrix-valued function and $\mathbf{G} = (G_{ij}(k);\ i, j \in E, k \in \mathbb{N}) \in \mathcal{M}_E(\mathbb{N})$ be a known one. The equation

$$\mathbf{L}(k) = \mathbf{G}(k) + \mathbf{q} * \mathbf{L}(k),\ k \in \mathbb{N},\tag{3.10}$$

is called a discrete-time Markov renewal equation (DTMRE).

Remark 3.5. Note that we can also consider the case of a MRE with $\mathbf{L}$ and $\mathbf{G}$ (column) vectors instead of matrices. This will be the case in the first proof of Proposition 5.1.

With the above definition, we see that Equation (3.9) is the Markov renewal equation of the transition matrix-valued function $\mathbf{P}(k)$.

Note that Equation (3.10) is equivalent to equation

$$(\boldsymbol{\delta}\mathbf{I} - \mathbf{q}) * \mathbf{L}(k) = \mathbf{G}(k),\ k \in \mathbb{N}.\tag{3.11}$$

In order to be able to solve this type of equation, let us define the inverse of a matrix-valued function in the convolution sense and see under what conditions this inverse exists.

Definition 3.10 (left inverse in the convolution sense).
Let $\mathbf{A} \in \mathcal{M}_E(\mathbb{N})$ be a matrix-valued function. If there exists $\mathbf{B} \in \mathcal{M}_E(\mathbb{N})$ such that

$$\mathbf{B} * \mathbf{A} = \boldsymbol{\delta}\mathbf{I},\tag{3.12}$$

then $\mathbf{B}$ is called the left inverse of $\mathbf{A}$ in the convolution sense and it is denoted by $\mathbf{A}^{(-1)}$.

The left inverse of A is not always defined. For example, for $k = 0$, we have to solve the equation

$$(\mathbf{B} * \mathbf{A})(0) = (\boldsymbol{\delta}\mathbf{I})(0) \Leftrightarrow \mathbf{B}(0)\,\mathbf{A}(0) = \mathbf{I}_E.$$

Obviously, taking for example $\mathbf{A}(0) = \mathbf{0}_E$, there is no solution.

The objective of the next proposition is to give a necessary and sufficient condition for the existence and uniqueness of the left inverse.

Proposition 3.3. *Let* $\mathbf{A} \in \mathcal{M}_E(\mathbb{N})$ *be a matrix-valued function. The left inverse* $\mathbf{B}$ *of* $\mathbf{A}$ *exists and is unique iff* $\det \mathbf{A}(0) \neq 0$. *Moreover,* $\mathbf{B} = \mathbf{A}^{(-1)} \in \mathcal{M}_E(\mathbb{N})$ *is given by the following recursive formula*

$$\mathbf{B}(n) = \begin{cases} [\mathbf{A}(0)]^{-1}, & \text{if } n = 0, \\ -\left(\sum_{l=0}^{n-1} \mathbf{B}(l)\mathbf{A}(n-l)\right)[\mathbf{A}(0)]^{-1}, & \text{if } n \geq 1. \end{cases} \tag{3.13}$$

Proof. We have to solve Equation (3.12), where $\mathbf{B} \in \mathcal{M}_E(\mathbb{N})$ is an unknown matrix-valued function. This equation can be written

$$\sum_{l=0}^{n} \mathbf{B}(n-l)\mathbf{A}(l) = \boldsymbol{\delta}\mathbf{I}(n), n \in \mathbb{N}.$$

First, $\mathbf{B}(0)\mathbf{A}(0) = \boldsymbol{\delta}\mathbf{I}(0) = \mathbf{I}$, which holds iff $\mathbf{A}(0)$ invertible. In this case, we get $\mathbf{B}(0) = [\mathbf{A}(0)]^{-1}$ and, for $n > 1$, we have

$$\sum_{l=0}^{n} \mathbf{B}(n-l)\mathbf{A}(l) = \boldsymbol{\delta}\mathbf{I}(n) = \mathbf{0},$$

which yields $\mathbf{B}(n) = \left(\sum_{l=0}^{n-1} \mathbf{B}(l)\mathbf{A}(n-l)\right)[\mathbf{A}(0)]^{-1}$. $\qquad\square$

It is worth noticing that we can solve Equation (3.10) if the left inverse of the matrix-valued function $(\boldsymbol{\delta}\mathbf{I} - \mathbf{q})$ exists. Using Proposition 3.3, it suffices to show that $(\boldsymbol{\delta}\mathbf{I} - \mathbf{q})(0)$ is invertible. Indeed, as $\mathbf{q}(0) = \mathbf{0}_E$, we have

$$(\boldsymbol{\delta}\mathbf{I} - \mathbf{q})(0) = \boldsymbol{\delta}\mathbf{I}(0) - \mathbf{q}(0) = \mathbf{I}_E.$$

As the inverse of $(\boldsymbol{\delta}\mathbf{I} - \mathbf{q})$ will be used when solving any Markov renewal equation, let us define the matrix-valued function $\boldsymbol{\psi} = (\boldsymbol{\psi}(k); \; k \in \mathbb{N}) \in \mathcal{M}_E(\mathbb{N})$ by

$$\boldsymbol{\psi}(k) := (\boldsymbol{\delta}\mathbf{I} - \mathbf{q})^{(-1)}(k). \tag{3.14}$$

The objective of the next result is to provide an algorithmic and explicit manner for computing $\boldsymbol{\psi}$.

Proposition 3.4. *The matrix-valued function* $\boldsymbol{\psi} = (\boldsymbol{\psi}(k); \; k \in \mathbb{N})$ *is given by*

$$\boldsymbol{\psi}(k) = \sum_{n=0}^{k} \mathbf{q}^{(n)}(k), \; k \in \mathbb{N} \tag{3.15}$$

and is computed using using the recursive formula

$$\boldsymbol{\psi}(k) = \begin{cases} \mathbf{I}_E, & \text{if } k = 0, \\ -\sum_{l=0}^{k-1} \boldsymbol{\psi}(l)\,(\boldsymbol{\delta}\mathbf{I} - \mathbf{q})(k-l), & \text{if } k > 0. \end{cases} \tag{3.16}$$

Proof. From the above discussion we know that $(\boldsymbol{\delta}\mathbf{I} - \mathbf{q})(0)$ is invertible. Applying Proposition 3.3, we obtain that the left inverse of the matrix-valued function $(\boldsymbol{\delta}\mathbf{I} - \mathbf{q})$ exists and is unique. Algorithm (3.16) is a direct application of the previous Algorithm (3.13) to the matrix-valued function $(\boldsymbol{\delta}\mathbf{I} - \mathbf{q})$.

For all $k \in \mathbb{N}$, we have

$$\left(\sum_{n=0}^{\infty} \mathbf{q}^{(n)} \right) * (\boldsymbol{\delta}\mathbf{I} - \mathbf{q})(k) = \left(\sum_{n=0}^{\infty} \mathbf{q}^{(n)} \right)(k) - \left(\sum_{n=0}^{\infty} \mathbf{q}^{(n)} \right) * \mathbf{q}(k)$$

$$= \sum_{l=0}^{\infty} \mathbf{q}^{(l)}(k) - \sum_{l=1}^{\infty} \mathbf{q}^{(l)}(k)$$

$$= \mathbf{q}^{(0)}(k) = \boldsymbol{\delta}\mathbf{I}(k).$$

As the left inverse of $(\boldsymbol{\delta}\mathbf{I} - \mathbf{q})$ is unique (Proposition 3.3), we obtain that

$$\boldsymbol{\psi}(k) := \sum_{n=0}^{\infty} \mathbf{q}^{(n)}(k). \tag{3.17}$$

Applying Lemma 3.2, we have $\mathbf{q}^{(n)}(k) = 0, n > k, \ n, k \in \mathbb{N}$, hence we obtain that $\boldsymbol{\psi}$ is given by

$$\boldsymbol{\psi}(k) = \sum_{n=0}^{k} \mathbf{q}^{(n)}(k). \tag{3.18}$$

which finishes the proof. $\qquad\qquad\qquad\qquad\qquad\qquad\qquad\qquad\qquad\qquad\square$

We point out that the passage from Equation (3.17) to Equation (3.18) in the previous proof is valid only in discrete time. In fact, Lemma 3.2 allows us to have a finite series of convolution powers of the kernel instead of an infinite series.

We would like to obtain now another expression of $\boldsymbol{\psi}$, more obviously related to the evolution of the semi-Markov system than those given in Equations (3.14) and (3.15).

First, for any states $i, j \in E$ (not necessary distinct) and any positive integer $k \in \mathbb{N}$, from Proposition 3.1 and Lemma 3.2 one immediately gets

$$\psi_{ij}(k) = \mathbb{P}\left(\bigcup_{n=0}^{k} \{J_n = j, S_n = k\} \mid J_0 = i \right) \le 1. \tag{3.19}$$

This expression of $\psi_{ij}(k)$ has a very simple and intuitive interpretation: $\psi_{ij}(k)$ represents the probability that starting at time 0 in state i, the semi-Markov chain will do a jump to state j at time k.

Second, we want to express $\psi_{ij}(k)$ in terms of renewal chains embedded in the semi-Markov chain. To this purpose, let $(S_n^j)_{n \in \mathbb{N}}$ be the successive passage times in a fixed state $j \in E$. More precisely, define them recursively as follows:

$$S_0^j = S_m, \text{ with } m = \min\{l \in \mathbb{N} \mid J_l = j\},$$
$$S_n^j = S_m, \text{ with } m = \min\{l \in \mathbb{N} \mid J_l = j, \ S_l > S_{n-1}^j\}, \ n \in \mathbb{N}^*.$$

One can check that $(S_n^j)_{n \in \mathbb{N}}$ is a renewal chain (i.e., nondelayed) on the event $\{J_0 = j\}$ and a delayed renewal chain otherwise. If $(S_n^j)_{n \in \mathbb{N}}$ is a delayed RC, note also that $(S_n^j - S_0^j)_{n \in \mathbb{N}}$ is the associated renewal chain (see Section 2.3 for the definition).

For any arbitrary states $i, j \in E$, $i \neq j$, we consider $g_{ij}(\cdot)$ the distribution of the first hitting time of state j, starting from state i :

$$g_{ij}(k) := \mathbb{P}_i(S_0^j = k), k \geq 0. \tag{3.20}$$

Note that $g_{ij}(0) = 0$. We set $G_{ij}(k) := \sum_{l=0}^{k} g_{ij}(l)$ for the corresponding cumulative distribution function and μ_{ij} for the *mean first passage time from state i to j for the semi-Markov chain* $(Z_k)_{k \in \mathbb{N}}$, i.e., $\mu_{ij} := \mathbb{E}_i(S_0^j) = \sum_{k \geq 0} g_{ij}(k)$.

Similarly, for any state $j \in E$, we consider $g_{jj}(\cdot)$ the distribution of the recurrence time in state j

$$g_{jj}(k) := \mathbb{P}_j(S_1^j = k), k \geq 0, \tag{3.21}$$

with $g_{jj}(0) = 0$. We also set $G_{jj}(k) := \sum_{l=0}^{k} g_{jj}(l)$ for the cumulative distribution function and μ_{jj} for the *mean recurrence time of state j for the semi-Markov chain* $(Z_k)_{k \in \mathbb{N}}$, $\mu_{jj} := \mathbb{E}_j(S_1^j) = \sum_{k \geq 0} g_{jj}(k)$.

For two states i and j we have

$$\psi_{jj}(k) = \sum_{r=0}^{k} g_{jj}^{(r)}(k) \tag{3.22}$$

and, for $i \neq j$,

$$\psi_{ij}(k) = \sum_{r=0}^{k} g_{ij} * g_{jj}^{(r)}(k) = g_{ij} * \psi_{jj}(k), \tag{3.23}$$

where $g_{jj}^{(r)}(\cdot)$ is the r-fold convolution of $g_{jj}(\cdot)$ (Definition 2.3).

To see that these two equalities hold true, note first that Equation (3.22) is exactly Equation (2.2), written for the RC $(S_n^j - S_0^j)_{n \in \mathbb{N}}$, associated to the (possibly) delayed RC $(S_n^j)_{n \in \mathbb{N}}$, with waiting time distribution $(g_{jj}(n))_{n \in \mathbb{N}}$. In an analogous manner, Equation (3.23) can be obtained, working directly on the delayed RC $(S_n^j)_{n \in \mathbb{N}}$, with the additional assumption that $\mathbb{P}(J_0 = i) = 1$ (or, in other words, taking $g_{ij}(\cdot)$ as the initial distribution of this delayed RC).

In conclusion, $\psi_{jj}(k)$ is the probability that a renewal will occur at time k in the RC $(S_n^j - S_0^j)_{n \in \mathbb{N}}$,

$$\psi_{jj}(k) = \sum_{n=0}^{k} \mathbb{P}_j(S_n^j = k) = \mathbb{P}_j\left(\bigcup_{n=0}^{k} \{S_n^j = k\}\right), \tag{3.24}$$

and, for two states $i \neq j$, $\psi_{ij}(k)$ has the same meaning in the delayed RC $(S_n^j)_{n \in \mathbb{N}}$, provided that $\mathbb{P}(J_0 = i) = 1$,

$$\psi_{ij}(k) = \sum_{n=0}^{k} \mathbb{P}_i(S_n^j = k) = \mathbb{P}_i \left(\bigcup_{n=0}^{k} \{S_n^j = k\} \right). \qquad (3.25)$$

We would like now to introduce some types of regularity of a SMC (MRC) using a description of the states. To this purpose, note first that $G_{ij}(\infty)$ is the probability that the SMC will go from i to j. Consequently, $G_{ij}(\infty) = 0$ if there is a zero probability that the SMC will arrive at state j, starting from state i. These remarks allow one to introduce a state classification for a SMC (MRC) and to describe the chain according to this state classification.

Definition 3.11. *Let* $(Z_k)_{k \in \mathbb{N}}$ *be a semi-Markov chain with state space* E *and* $(J_n, S_n)_{n \in \mathbb{N}}$ *the associated Markov renewal chain.*

1. *If* $G_{ij}(\infty)G_{ji}(\infty) > 0$, *we say that* i *and* j communicate *and we denote this by* $i \leftrightarrow j$. *One can check that the relation "$\leftrightarrow$" is an equivalence relation on* E. *The elements of the quotient space* $E/\leftrightarrow$ *are called* (communication) classes.
2. *The semi-Markov chain (Markov renewal chain) is said to be* irreducible *if there is only one class.*
3. *A state* i *is said to be* recurrent *if* $G_{ii}(\infty) = 1$ *and* transient *if* $G_{ii}(\infty) < 1$. *A recurrent state* i *is* positive recurrent *if* $\mu_{ii} < \infty$ *and* null recurrent *if* $\mu_{ii} = \infty$. *If all the states are (positive/null) recurrent, the SMC (MRC) is said to be (positive/null) recurrent.*
4. *A subset of states* C *is said to be a* final set *if* $\sum_{i \in C} P_{ij}(k) = 0$ *for any state* $j \in E \setminus C$ *and any integer* $k \in \mathbb{N}$. *An irreducible final set is a* closed (or essential) class.
5. *The SMC (MRC) is said to be* ergodic *if it is irreducible and positive recurrent.*
6. *Let* $d > 1$ *be a positive integer. A state* $i \in E$ *is said to be* d-periodic (aperiodic) *if the distribution* $g_{ii}(\cdot)$ *is* d-periodic (aperiodic).
7. *An irreducible SMC is* d-periodic, $d > 1$, *if all states are* d-periodic. *Otherwise, it is called* aperiodic.

Proposition 3.5.

1. *Two communicating states of a MRC are either both aperiodic or both periodic. In the later case they have the same period.*
2. *If the embedded Markov chain* J *is irreducible and the MRC is* d-periodic *then* $q_{ij}(k)$ *has the support* $\{a_{ij} + rd; r \in \mathbb{N}\}$, *where* a_{ij} *are nonnegative constants depending on states* i *and* j.

A proof of this result can be found in Çinlar (1975). See also Exercise 3.5.

Remark 3.6. From the first statement of this proposition we have that an irreducible MRC with one d-periodic state has all states d-periodic and, consequently, it is a d-periodic MRC.

The mean recurrence times μ_{jj} can be expressed in terms of $\boldsymbol{\nu}$, the stationary distribution of the embedded Markov chain $(J_n)_{n \in \mathbb{N}}$, and of the mean sojourn times $m_j, j \in E$.

Proposition 3.6. *For an aperiodic and ergodic MRC $(J_n, S_n)_{n \in \mathbb{N}}$, the mean recurrence time of an arbitrary state $j \in E$ is given by*

$$\mu_{jj} = \frac{\sum_{i \in E} \nu(i) m_i}{\nu(j)} = \frac{\overline{m}}{\nu(j)}, \tag{3.26}$$

where we set

$$\overline{m} := \sum_{i \in E} \nu(i) m_i \tag{3.27}$$

for the mean sojourn time of the MRC.

Proof. A proof based on generating functions can be found in Howard (1971). Here we give a direct proof. For i and $j \in E$ arbitrary states we have

$$\mu_{ij} = \mathbb{E}_i(S_0^j) = \mathbb{E}_i(S_0^j \mathbf{1}_{\{J_1 = j\}}) + \mathbb{E}_i(S_0^j \mathbf{1}_{\{J_1 \neq j\}})$$

$$= p_{ij} m_i + \sum_{k \neq j} \mathbb{E}_i(S_0^j \mid \mathbf{1}_{\{J_1 = k\}}) \mathbb{P}_i(J_1 = k)$$

$$= p_{ij} m_i + \sum_{k \neq j} p_{ik} \left[\mathbb{E}_k \left(S_0^j \right) + \mathbb{E}_i \left(S_1 \right) \right]$$

$$= p_{ij} m_i + \sum_{k \neq j} p_{ik} \mu_{kj} + (1 - p_{ij}) m_i.$$

Hence $\mu_{ij} = m_i + \sum_{k \neq j} p_{ik} \mu_{kj}$. Multiplying this equality by $\nu(i)$ and summing over $i \in E$ we get

$$\sum_{i \in E} \nu(i) \mu_{ij} = \sum_{i \in E} m_i \nu(i) + \sum_{i \in E} \nu(i) \sum_{k \neq j} p_{ik} \mu_{kj}$$

$$= \overline{m} + \sum_{k \neq j} \left(\sum_{i \in E} \nu(i) p_{ik} \right) \mu_{kj}$$

$$= \overline{m} + \sum_{k \in E} \nu(k) \mu_{kj} - \nu(j) \mu_{jj}.$$

Thus $\overline{m} = \nu(j) \mu_{jj}$.   □

Another quantity of interest, strictly related to ψ, is the Markov renewal function, defined as the expected number of visits to a certain state, up to a given time. More precisely, we have the following definition.

Definition 3.12 (Markov renewal function).

The Markov renewal function $\boldsymbol{\Psi} = (\boldsymbol{\Psi}(k); k \in \mathbb{N}) \in \mathcal{M}_E(\mathbb{N})$ *of the MRC is defined by*

$$\Psi_{ij}(k) = \mathbb{E}_i[\widetilde{N}_j(k)], \ i,j \in E, \ k \in \mathbb{N}, \tag{3.28}$$

where $\widetilde{N}_j(k)$ *is the number of visits to state* j *of the EMC, up to time* k,

$$\widetilde{N}_j(k) := \sum_{n=0}^{N(k)} \mathbf{1}_{\{J_n=i\}} = \sum_{n=0}^{k} \mathbf{1}_{\{J_n=j;S_n \leq k\}}. \tag{3.29}$$

It is easy to see that the Markov renewal function can be expressed as follows:

$$\Psi(k) = \sum_{l=0}^{k} \psi(l). \tag{3.30}$$

Indeed, we have

$$\Psi_{ij}(k) = \mathbb{E}_i \left[\sum_{n=0}^{k} \mathbf{1}_{\{J_n=j;S_n \leq k\}} \right] = \sum_{n=0}^{k} \mathbb{P}(J_n = j; S_n \leq k \mid J_0 = i)$$

$$= \sum_{n=0}^{k} \sum_{l=0}^{k} \mathbb{P}(J_n = j; S_n = l \mid J_0 = i) = \sum_{l=0}^{k} \sum_{n=0}^{k} q_{ij}^{(n)}(l).$$

From Lemma 3.2 we know that $q_{ij}^{(n)}(l) = 0$ for $n > l$ and we get

$$\Psi_{ij}(k) = \sum_{l=0}^{k} \sum_{n=0}^{l} q_{ij}^{(n)}(l) = \sum_{l=0}^{k} \psi(l).$$

Remark 3.7.

1. Note that $\Psi_{ii}(k) = \mathbb{E}_i[1 + \sum_{n=1}^{k} \mathbf{1}_{\{J_n=i;S_n \leq k\}}]$, so $\Psi_{ii}(\cdot)$ is the renewal function of the RC $(S_n^i - S_0^i)_{n \in \mathbb{N}}$ (Equation (2.12) and Remark 2.2). The same remark holds true for $\Psi_{ij}(\cdot), i \neq j$, and the delayed RC $(S_n^j)_{n \in \mathbb{N}}$, with $\mathbb{P}(J_0 = i) = 1$.
2. One can check that a state i is recurrent iff $\Psi_{ii}(\infty) = \infty$ and transient iff $\Psi_{ii}(\infty) < \infty$.

In the sequel, we will see that ψ and $\boldsymbol{\Psi}$ are solutions of Markov renewal equations.

First, the definition of $\psi(k)$ given in (3.14) tells us that $(\boldsymbol{\delta}\mathbf{I} - \mathbf{q}) * \psi(k) = \boldsymbol{\delta}\mathbf{I}(k)$, so $\psi = (\psi(k); k \in \mathbb{N})$ is the solution of the Markov renewal equation

$$\psi(k) = \boldsymbol{\delta}\mathbf{I}(k) + \mathbf{q} * \psi(k), \ k \in \mathbb{N}.$$

Second, writing the previous equation for $k \in \mathbb{N}, 0 \leq k \leq \nu, \ \nu \in \mathbb{N}$ fixed, and taking the sum, we obtain $\sum_{k=0}^{\nu} \psi(k) = \sum_{k=0}^{\nu} \boldsymbol{\delta}\mathbf{I}(k) + \sum_{k=0}^{\nu} \mathbf{q} * \psi(k)$, i.e., the matrix renewal function $\mathbf{\Psi} = (\mathbf{\Psi}(k); k \in \mathbb{N})$ is the solution of the Markov renewal equation

$$\mathbf{\Psi}(\nu) = \mathbf{I} + \mathbf{q} * \mathbf{\Psi}(\nu), \ \nu \in \mathbb{N}.$$

Up to this point we have seen three quantities of interest for a semi-Markov system that verify Markov renewal equations, namely, the transition function $\mathbf{P}$ of the semi-Markov chain Z, the matrix-valued function ψ, and the Markov renewal function $\mathbf{\Psi}$. This phenomenon motivates us to look for a general result on the solution of the Markov renewal equation in discrete time (Equation (3.10)), which is given in the next theorem.

Theorem 3.1. *Using the above notation, the DTMRE (3.10) has a unique solution* $\mathbf{L} = (L_{ij}(k); i, j \in E, k \in \mathbb{N}) \in \mathcal{M}_E(\mathbb{N})$ *given by*

$$\mathbf{L}(k) = \psi * \mathbf{G}(k). \tag{3.31}$$

Proof. First, from the definition of ψ it is obvious that $\psi * \mathbf{G}(k)$ is a solution of Equation (3.10).

Second, we show that $\psi * \mathbf{G}$ is the unique solution of the renewal equation. Let $\mathbf{L}'$ be another solution of Equation (3.10). We obtain

$$(\mathbf{L} - \mathbf{L}')(k) = \mathbf{q}^{(n)} * (\mathbf{L} - \mathbf{L}')(k), \ k \in \mathbb{N}, \tag{3.32}$$

with n an arbitrary positive integer. Taking $n > k$ and using that $\mathbf{q}^{(n)}(k) = 0$ for $n > k$ (Lemma 3.2), we get $L(k) = L'(k), k \in \mathbb{N}$. $\qquad\square$

Using Propositions 3.2 and 3.4 and Theorem 3.1, we obtain that the unique solution of the Markov renewal equation $\mathbf{P} = \mathbf{I} - \mathbf{H} + \mathbf{q} * \mathbf{P}$ is

$$\mathbf{P}(k) = \psi * (\mathbf{I} - \mathbf{H})(k) = (\boldsymbol{\delta}\mathbf{I} - \mathbf{q})^{(-1)} * (\mathbf{I} - \mathbf{H})(k), \ k \in \mathbb{N}. \tag{3.33}$$

Example 3.4. We consider the case of an alternating renewal process given in Example 3.3 and compute the transition function using Relation (3.33). Let us recall that an alternating renewal process is a two-state MRC with the semi-Markov kernel

$$\mathbf{q}(k) = \begin{pmatrix} 0 & f(k) \\ g(k) & 0 \end{pmatrix}, \ k \in \mathbb{N}.$$

The sojourn time cumulative distribution functions in states 1 and 2 are given by

$$H_1(k) = \sum_{l=1}^{k} h_1(l) = \sum_{l=1}^{k} \sum_{i=1}^{2} q_{1i}(l) = \sum_{l=1}^{k} f(l),$$

$$H_2(k) = \sum_{l=1}^{k} h_2(l) = \sum_{l=1}^{k} \sum_{i=1}^{2} q_{2i}(l) = \sum_{l=1}^{k} g(l).$$

Using Relation (3.33), the transition function can be expressed as follows:

$$\mathbf{P}(k) = (\boldsymbol{\delta}\mathbf{I} - \mathbf{q})^{(-1)} * (\mathbf{I} - \mathbf{H})(k)$$

$$= \begin{pmatrix} \mathbf{1}_{\{0\}}(\cdot) & -f(\cdot) \\ -g(\cdot) & \mathbf{1}_{\{0\}}(\cdot) \end{pmatrix}^{(-1)} * \begin{pmatrix} 1 - \sum_{l=1}^{\cdot} f(l) & 0 \\ 0 & 1 - \sum_{l=1}^{\cdot} g(l) \end{pmatrix}(k)$$

$$= \begin{pmatrix} \mathbf{1}_{\{0\}}(\cdot) & -f * g(\cdot) \end{pmatrix}^{(-1)} * \begin{pmatrix} \mathbf{1}_{\{0\}}(\cdot) & f(\cdot) \\ g(\cdot) & \mathbf{1}_{\{0\}}(\cdot) \end{pmatrix}$$

$$* \begin{pmatrix} 1 - \sum_{l=1}^{\cdot} f(l) & 0 \\ 0 & 1 - \sum_{l=1}^{\cdot} g(l) \end{pmatrix}(k).$$

The expression $(\mathbf{1}_{\{0\}}(l) - f * g(l))^{(-1)}$, $0 \leq l \leq k$, represents a renewal function; noting $L(l) = f * g(l)$, we have

$$(\mathbf{1}_{\{0\}} - L)^{(-1)}(l) = \sum_{n=0}^{k} L^{(n)}(l),$$

and we finally obtain the following expression of the transition function:

$$\mathbf{P}(k) = \left(\sum_{n=0}^{\cdot} L^{(n)}(\cdot) \right) * \begin{pmatrix} 1 - \sum_{l=1}^{\cdot} f(l) & f(\cdot) * (1 - \sum_{l=1}^{\cdot} g(l)) \\ g(\cdot) * (1 - \sum_{l=1}^{\cdot} f(l)) & 1 - \sum_{l=1}^{\cdot} g(l) \end{pmatrix}(k).$$

3.3 Limit Theorems for Markov Renewal Chains

In this section we are interested in the asymptotic behavior of Markov renewal chains. First, we investigate the asymptotic behavior of the jump process $(S_n)_{n \in \mathbb{N}}$, of the counting process of the number of jumps $(N(n))_{n \in \mathbb{N}}$ (Lemma 3.3), of the number of visits to a certain state, and of the number of transitions between two states (Propositions 3.7 and 3.8 and Convergences (3.37) and (3.38)). Second, we derive the renewal theorem (Theorem 3.2) and the key renewal theorem for Markov renewal chains (Theorem 3.3). As a direct application of these results, we obtain the limit distribution of a semi-Markov chain (Proposition 3.9). Afterward, we give the discrete-time versions of the strong law of large numbers (SLLN) and of the central limit theorem (CLT) for Markov renewal processes (Theorems 3.4 and 3.5).

The following assumptions concerning the Markov renewal chain will be needed in the rest of the book.

A1 The SMC is irreducible.

A2 The mean sojourn times in all states are finite, i.e.,

$$m_j := \mathbb{E}(S_1 \mid J_0 = j) = \sum_{n \geq 0} (1 - H_j(n)) < \infty, \, j \in E.$$

Note that these assumptions mean that the SMC is positive recurrent.

Lemma 3.3. *For a MRC that satisfies Assumptions A1 and A2, we have:*

1. $\lim_{M \to \infty} S_M = \infty$ *a.s.;*
2. $\lim_{M \to \infty} N(M) = \infty$ *a.s.*

Proof. The proof is identical to that of Lemma 2.1. □

Let us define the number of visits to a certain state and the number of direct transitions between two states.

Definition 3.13. *For all states $i, j \in E$, define:*

1. $N_i(M) := \sum_{n=0}^{N(M)-1} \mathbf{1}_{\{J_n=i\}} = \sum_{n=0}^{M} \mathbf{1}_{\{J_n=i, S_{n+1} \leq M\}}$ *is the number of visits to state i of the EMC, up to time M;*
2. $N_{ij}(M) := \sum_{n=1}^{N(M)} \mathbf{1}_{\{J_{n-1}=i, J_n=j\}} = \sum_{n=1}^{M} \mathbf{1}_{\{J_{n-1}=i, J_n=j, S_n \leq M\}}$ *is the number of transitions of the EMC from i to j, up to time M.*

Remark 3.8. Note that in this definition of the number of visits to a certain state, we have not taken into account the last visited state, $J_{N(M)}$, whereas in (3.29) we did count the last visited state, i.e.,

$$\widetilde{N}_i(M) = N_i(M) + \mathbf{1}_{\{J_{N(M)}=i\}}, \, i \in E.$$

The reason for defining $N_i(M)$ like this comes from a maximum-likelihood computation (Proposition 4.1), where we need to have $\sum_{j \in E} N_{ij}(M) = N_i(M)$, so $N_i(M)$ does not count the last state visited by the EMC $(J_n)_{n \in \mathbb{N}}$. On the other hand, in the definition of the Markov renewal function (Equation (3.28)), we wanted to take into account the last visited state, so we used $\widetilde{N}_i(M)$ instead of $N_i(M)$.

Let us now investigate the asymptotic behavior of $N_i(M), N_{ij}(M)$ and $N(M)$ when M tends to infinity. The first result follows directly from classical Markov chain asymptotic properties (Proposition D.5).

Proposition 3.7. *For an aperiodic MRC that satisfies Assumption A1, we have*

$$N_i(M)/N(M) \xrightarrow[M \to \infty]{a.s.} \nu(i), \tag{3.34}$$

$$N_{ij}(M)/N(M) \xrightarrow[M \to \infty]{a.s.} \nu(i)p_{ij}, \tag{3.35}$$

where $\boldsymbol{\nu}$ is the stationary distribution of the embedded Markov chain $(J_n)_{n \in \mathbb{N}}$.

Proof. Note that $N(M) \to \infty$ as $M \to \infty$. Applying Proposition D.5 for the Markov chain $(J_n)_{n \in \mathbb{N}}$ and using Theorem E.5, we get the desired result. $\quad \square$

The following result will be frequently used in the sequel.

Proposition 3.8. *For an aperiodic MRC that satisfies Assumptions A1 and A2, we have:*

$$N_i(M)/M \xrightarrow[M \to \infty]{a.s.} 1/\mu_{ii}, \tag{3.36}$$

where μ_{ii} is the mean recurrence time of state i for the semi-Markov process $(Z_n)_{n \in \mathbb{N}}$.

Proof. Let $i \in E$ be a fixed arbitrary state. As seen before, the successive passage times in state i form a renewal chain (possibly delayed), denoted by $(S_n^i)_{n \in \mathbb{N}}$ and $(S_n^i - S_0^i)_{n \in \mathbb{N}}$ is a RC (i.e., nondelayed). If $J_0 = i$ a.s., then $S_0^i = 0$ a.s., so we can always work on the renewal chain $(S_n^i - S_0^i)_{n \in \mathbb{N}}$. Note that $\left(\widetilde{N}_i(M) - 1 \right)$ is the counting process of this RC, defined in Equation (2.10) as the number of renewals in the time interval $[1, M]$. Using the relation $\widetilde{N}_i(M) = N_i(M) + \mathbf{1}_{\{J_{N(M)}=i\}}$ (Remark 3.8) and applying the SLLN (Theorem 2.3) to the RC $(S_n^i - S_0^i)_{n \in \mathbb{N}}$, we get

$$\lim_{M \to \infty} \frac{N_i(M)}{M} = \frac{1}{\mathbb{E}_i(S_1^i)} = \frac{1}{\mu_{ii}} \quad \mathbb{P}_j - \text{ a.s., for all } j \in E,$$

which is the desired result. $\quad \square$

It is worth discussing here the structural difference between the convergence relationships obtained in Proposition 3.7 and that obtained in Proposition 3.8. On the one hand, note that Relationships (3.34) and (3.35) from Proposition 3.7 concern only the Markov chain structure of the transitions between states, governed by the chain $(J_n)_{n \in \mathbb{N}}$. On the other hand, Convergence (3.36) from Proposition 3.8 is obtained by taking into account the general Markov renewal structure of $(J_n, S_n)_{n \in \mathbb{N}}$. Note also that similar convergence results can be immediately inferred:

$$N_{ij}(M)/M \xrightarrow[M \to \infty]{a.s.} p_{ij}/\mu_{ii}, \tag{3.37}$$

$$N(M)/M \xrightarrow[M \to \infty]{a.s.} 1/\nu(l)\mu_{ll}, \tag{3.38}$$

for any fixed arbitrary state $l \in E$.

The result presented in Equation (3.38) could seem surprising at first view, because the limit seems to depend on an arbitrary state $l \in E$. When looking carefully at that limit we notice that

$$1/\nu(l)\mu_{ll} = 1/\sum_{i \in E} \nu(i)m_i = 1/\overline{m} \quad (\text{cf. Proposition 3.6}), \tag{3.39}$$

which is independent of state l.

Note also that in the case where the embedded Markov chain $(J_n)_{n\in\mathbb{N}}$ is ergodic (irreducible and aperiodic), for any state $l \in E$ we have the following relationship between the mean recurrence time of l in the Markov chain $(J_n)_{n\in\mathbb{N}}$, denoted by μ_{ll}^*, and the stationary distribution $\boldsymbol{\nu}$ of the chain (see, e.g., Feller, 1993, chapter 15):

$$\mu_{ll}^* = \frac{1}{\nu(l)}.$$

We rewrite Equation (3.39) as

$$\frac{\mu_{ll}}{\mu_{ll}^*} = \sum_{i\in E} \nu(i)m_i = \overline{m}, \tag{3.40}$$

which means that μ_{ll}/μ_{ll}^* is independent of state $l \in E$. We wanted to clearly state this fact because it will be used in next chapter (in the expressions of asymptotic variances, in theorems like Theorem 4.5 or Theorem 4.7).

We investigate now the asymptotic behavior of the matrix-valued function $\boldsymbol{\psi} := (\boldsymbol{\psi}(k); \ k \in \mathbb{N}) \in \mathcal{M}_E(\mathbb{N})$. For continuous-time versions of these results, see, e.g., Çinlar (1975) or Limnios and Oprişan (2001).

Theorem 3.2 (Markov renewal theorem).

1. *If the MRC is aperiodic and satisfies Assumptions A1 and A2, then for fixed arbitrary states $i, j \in E$, we have*

$$\lim_{k\to\infty} \psi_{ij}(k) = \frac{1}{\mu_{jj}}.$$

2. *If the MRC is d-periodic and satisfies Assumptions A1 and A2, then for fixed arbitrary states $i, j \in E$, we have*

$$\lim_{k\to\infty} \psi_{ij}(kd) = \frac{d}{\mu_{jj}}.$$

Proof. In the aperiodic case, let us consider first the case where $i = j$. As we saw above (Equations (3.22) and (3.24)), $\psi_{jj}(k)$ is the probability that a renewal will occur at time k in the RC $(S_n^j - S_0^j)_{n\in\mathbb{N}}$. Then, the result is a direct application of the renewal theorem for an aperiodic recurrent renewal chain (Part 1 of Theorem 2.6).

If $i \neq j$, we have seen in Equations (3.23) and (3.25) that $\psi_{ij}(k)$ is the probability that a renewal will occur at time k in the delayed RC $(S_n^j)_{n\in\mathbb{N}}$, with $\mathbb{P}(J_0 = i) = 1$. Applying the renewal theorem for delayed aperiodic renewal chains (Part 1 of Theorem 2.8), one gets

$$\lim_{k \to \infty} \psi_{ij}(k) = \frac{1}{\mu_{jj}} \sum_{n=0}^{\infty} g_{ij}(n) = \frac{1}{\mu_{jj}},$$

which is the desired result. The aperiodic cases are treated in the same way, applying the corresponding results for periodic renewal chains.   □

Theorem 3.3 (key Markov renewal theorem).

Let us consider an aperiodic MRC, which satisfies Assumptions A1 and A2. Let us associate to each state $j \in E$, a real valued function $v_j(n)$, defined on $\mathbb{N}$, with $\sum_{n \geq 0} |v_j(n)| < \infty$. For any state $i, j \in E$ we have:

$$\psi_{ij} * v_j(k) \xrightarrow[k \to \infty]{} \frac{1}{\mu_{jj}} \sum_{n \geq 0} v_j(n). \tag{3.41}$$

Proof. First, we prove (3.41) for $i = j$. We have

$$\psi_{jj} * v_j(k) = \sum_{n=0}^{k} \psi_{jj}(k - n) v_j(n).$$

As $|\psi_{jj}(k-n)v_j(n)| \leq |v_j(n)|$ $(\psi_{jj}(l) \leq 1, \ l \in \mathbb{N})$, from Theorem 3.2 we infer that $\lim_{k \to \infty} \psi_{jj}(k - n) = 1/\mu_{jj}$. Consequently, applying Proposition E.1 we obtain that

$$\lim_{k \to \infty} \psi_{jj} * v_j(k) = \frac{1}{\mu_{jj}} \sum_{n \geq 0} v_j(n).$$

Let us now prove (3.41) for $i \neq j$. Using Equation (3.23), we obtain

$$\psi_{ij} * v_j(k) = g_{ij} * \psi_{jj} * v_j(k) = \sum_{n=0}^{k} \psi_{jj}(k - n)[g_{ij} * v_j](n).$$

First, we have that $|\psi_{jj}(k - n)(g_{ij} * v_j)(n)| \leq |g_{ij} * v_j(n)|$. Second,

$$\begin{aligned}
\sum_{n \geq 0} g_{ij} * v_j(n) &= \sum_{n \geq 0} \sum_{k=0}^{n} g_{ij}(k) v_j(n - k) \\
&= \sum_{k \geq 0} \sum_{n \geq k} g_{ij}(k) v_j(n - k) \\
&= \sum_{k \geq 0} g_{ij}(k) \sum_{n \geq k} v_j(n - k) \\
&= \sum_{n \geq 0} v_j(n) < \infty.
\end{aligned}$$

Applying Theorem 3.2 and Proposition E.1 we have

$$\lim_{k \to \infty} \psi_{ij} * v_j(k) = \frac{1}{\mu_{jj}} \sum_{n \geq 0} g_{ij} * v_j(n) = \frac{1}{\mu_{jj}} \sum_{n \geq 0} v_j(n),$$

which proves the theorem.   □

Remark 3.9. Note that the preceding theorem is very useful, because it provides a limit, as n tends to infinity, for any solution $\mathbf{L}$ of a Markov renewal equation (3.10), provided that $\sum_{n \geq 0} | G_{ij}(n) | < \infty$, for any states i and j.

Remark 3.10. One can easily see that this theorem can be immediately obtained using the key renewal theorem for renewal chains. Indeed, for $i = j$ we have that $(S_n^j - S_0^j)_{n \in \mathbb{N}}$ is a renewal chain and we can apply the key renewal theorem for recurrent aperiodic renewal chains (Part 1 of Theorem 2.7) to obtain the result. When considering the case $i \neq j$, one has first to derive the key renewal theorem for recurrent delayed renewal chains and then apply it to the delayed recurrent renewal chain $(S_n^j)_{n \in \mathbb{N}}$, with $\mathbb{P}(J_0 = i) = 1$.

One can also easily obtain the key Markov renewal theorem in the periodic case by means of corresponding results for (delayed) renewal chains.

The key Markov renewal theorem can be applied to obtain the limit distribution of a semi-Markov chain, which is defined as follows.

Definition 3.14 (limit distribution of a SMC). *For a semi-Markov chain* $(Z_k)_{k \in \mathbb{N}}$, *the* limit distribution $\boldsymbol{\pi} = (\pi_1, \ldots, \pi_s)^\top$ *is defined, when it exists, by* $\pi_j = \lim_{k \to \infty} P_{ij}(k)$, *for every* $i, j \in E$.

Remark 3.11. We note here that $\boldsymbol{\pi}$ is also the stationary distribution of the SMC $(Z_k)_{k \in \mathbb{N}}$ in the sense that it is the marginal distribution of the stationary distribution $\widetilde{\boldsymbol{\pi}}$ of the Markov chain $(Z_n, U_n)_{n \in \mathbb{N}}$, that is, $\pi_j = \widetilde{\pi}(\{j\}, \mathbb{N}), j \in E$ (see Proposition 6.2, Chapter 6, and Chryssaphinou et al. (2008)).

Proposition 3.9. *For an aperiodic MRC and under Assumptions A1 and A2, the limit distribution is given by*

$$\pi_j = \frac{1}{\mu_{jj}} m_j = \frac{\nu(j) m_j}{\sum_{i \in E} \nu(i) m_i}, \quad j \in E, \tag{3.42}$$

or, in matrix notation,

$$\boldsymbol{\pi} = \frac{1}{m} diag(\boldsymbol{\nu}) \mathbf{m},$$

where $\mathbf{m} = (m_1, \ldots, m_s)^\top$ *is the column vector of mean sojourn times in states* $i \in E$, *the row vector* $\boldsymbol{\nu} = (\nu(1), \ldots, \nu(s))$ *is the stationary distribution of the EMC* $(J_n)_{n \in \mathbb{N}}$, *and* $diag(\boldsymbol{\nu})$ *is the diagonal matrix having* $\boldsymbol{\nu}$ *as main diagonal.*

Proof. The result is a direct application of Theorem 3.3. Indeed, from Equation (3.33) we can write the transition function of the SMC as

$$P_{ij}(k) = \psi_{ij} * (1 - H_j)(k).$$

Applying Theorem 3.3 to the function $v_j := 1 - H_j$ we obtain

$$\lim_{k \to \infty} P_{ij}(k) = \frac{1}{\mu_{jj}} \sum_{n \geq 0} (1 - H_j(n)) = \frac{1}{\mu_{jj}} m_j.$$

From Proposition 3.6 we know that

$$\mu_{jj} = \frac{\sum_{i \in E} \nu(i) m_i}{\nu(j)}, j \in E,$$

and we get the desired result. □

An alternative proof of the previous result, based on generating functions, can be found in Howard (1971).

We present further the SLLN and the CLT for additive functionals of Markov renewal chains. The continuous-time version of these results was first given in Pyke and Schaufele (1964). Notations used here are similar to those of Moore and Pyke (1968).

Let f be a real function defined on $E \times E \times \mathbb{N}$. Define, for each $M \in \mathbb{N}$, the functional $W_f(M)$ as

$$W_f(M) := \sum_{n=1}^{N(M)} f(J_{n-1}, J_n, X_n), \tag{3.43}$$

or, equivalently,

$$W_f(M) := \sum_{i,j=1}^{s} \sum_{n=1}^{N_{ij}(M)} f(i, j, X_{ijn}), \tag{3.44}$$

where X_{ijn} is the nth sojourn time of the chain in state i, before going to state j. Set

$$A_{ij} := \sum_{x=1}^{\infty} f(i, j, x) q_{ij}(x), \quad A_i := \sum_{j=1}^{s} A_{ij},$$

$$B_{ij} := \sum_{x=1}^{\infty} f^2(i, j, x) q_{ij}(x), \quad B_i := \sum_{j=1}^{s} B_{ij},$$

if the sums exist. Define

$$r_i := \sum_{j=1}^{s} A_j \frac{\mu_{ii}^*}{\mu_{jj}^*}, \qquad m_f := \frac{r_i}{\mu_{ii}},$$

$$\sigma_i^2 := -r_i^2 + \sum_{j=1}^{s} B_j \frac{\mu_{ii}^*}{\mu_{jj}^*} + 2 \sum_{r=1}^{s} \sum_{l \neq i} \sum_{k \neq i} A_{rl} A_k \mu_{ii}^* \frac{\mu_{li}^* + \mu_{ik}^* - \mu_{lk}^*}{\mu_{rr}^* \mu_{kk}^*}, \qquad B_f := \frac{\sigma_i^2}{\mu_{ii}}.$$

Theorem 3.4 (strong law of large numbers).

For an aperiodic Markov renewal chain that satisfies Assumptions A1 and A2 we have

$$\frac{W_f(M)}{M} \xrightarrow[M \to \infty]{a.s.} m_f.$$

Theorem 3.5 (central limit theorem).

For an aperiodic Markov renewal chain that satisfies Assumptions A1 and A2 we have

$$\sqrt{M}\left[\frac{W_f(M)}{M} - m_f\right] \xrightarrow[M\to\infty]{\mathcal{D}} \mathcal{N}(0, B_f).$$

We point out that this result will be used for proving the asymptotic normality of different estimators in Chapters 4 and 5.

3.4 Periodic SMC

We shall briefly discuss here the asymptotic behavior of a periodic semi-Markov chain in the sense of Anselone, which uses the standard periodicity notion for Markov chains.

Let $E^* = \{(i, k) \mid i \in E, \overline{H}_i(k) > 0\}$ and consider the Markov chain $(Z_n, U_n), n \geq 0$ with state space E^* (see Equation (6.5) for the definition of $(U_n)_{n\in\mathbb{N}}$).

Definition 3.15. *Let $d > 1$ be a positive integer. The SMC Z is said to be d-periodic in the sense of Anselone if the Markov chain $(Z_n, U_n), n \geq 0$ is d-periodic (in ordinary sense).*

If the SMC Z is d-periodic, let $K_1, K_2, ..., K_d$ be the cyclic classes, i.e., the chain does transitions only from K_m to K_{m+1} (with $K_{d+1} := K_1$). Let us define $\rho(r) := \mathbb{P}((Z_0, U_0) \in K_r)$ and for any $(i, j) \in E^*$ let us define the integers $r_{ij}, 1 \leq r_{ij} \leq d$ and $r_{ij} = r$ if $(i, j) \in K_r$. Then $r_{kl} - r_{ij} = 1 \pmod{d}$ if $p_{(i,j)(k,l)} > 0$. Following Proposition 6.1, for any $i \in E$ there exists integers $k_i, 0 \leq k_i \leq d$ such that $h_i(k) = 0$ if $k \neq k_i$ and $r_{ij} = r_{i0} + k_i - j \pmod{d}$, $j \geq 1$. Then the following limit result holds true (Anselone, 1960).

Theorem 3.6. *Under Assumptions A1 and A2, if the SMC Z is d-periodic in the sense of Anselone, for each integer r, $1 \leq r \leq d$, we have*

$$\lim_{n\to\infty} \mathbb{P}(Z_{nd+r} = i) = \frac{\nu_i}{\sum_{l\in E} m_l\nu(l)}\left[m_i - k_i + d\sum_{l=1}^{k_i} \rho(r_{i0} - r + l)\right].$$

3.5 Monte Carlo Methods

Let us give here a Monte Carlo algorithm in order to simulate a trajectory of a given SMC in the time interval $[0, M]$. The output of the algorithm consists in the successive visited states $(J_0, \ldots, J_n)$ and the jump times $(S_0, \ldots, S_n)$ up to the time $M \in \mathbb{N}$, with $S_n \leq M < S_{n+1}, i.e.$, a sample path of the process up to any arbitrary time M.

For a discrete-time semi-Markov system, we have denoted by $\mathbf{p} = (p_{ij})_{i,j \in E}$ the transition probability matrix of the embedded Markov chain $(J_n)_{n \in \mathbb{N}}$ and by $f_{ij}(\cdot)$ the conditional distribution of the sojourn time in state i, given that the next state is j. The algorithm is based on the EMC.

Algorithm

1. Set $k = 0, S_0 = 0$ and sample J_0 from the initial distribution α;
2. Sample the random variable $J \sim p(J_k, \cdot)$ and set $J_{k+1} = J(\omega)$;
3. Sample the random variable $X \sim F_{J_k J_{k+1}}(\cdot)$
4. Set $S_{k+1} = S_k + X$;
5. If $S_{k+1} \geq M$, then end;
6. Else, set $k = k + 1$ and continue to step 2.

A second Monte Carlo algorithm, based on the unconditional sojourn times, for obtaining a trajectory of a given SMC is proposed in Exercise 3.6.

3.6 Example: a Textile Factory

We present now a practical example where a semi-Markov modeling is appropriate. We obtain explicit expressions for all the quantities of interest of the proposed semi-Markov system. Afterward, we will continue this example in Section 4.3 of Chapter 4, where we show how to estimate these quantities, and in Section 5.3 of Chapter 5, where we compute and estimate the reliability indicators of this system.

Let $\mathcal{F}$ be a textile factory. To avoid river pollution, the factory waste is treated in the unit $\mathcal{U}$ before being thrown in the river $\mathcal{R}$. In order for the factory not to be stopped if a failure occurs in the treatment unit $\mathcal{U}$, the waste is stocked in a tank $\mathcal{T}$. If $\mathcal{U}$ is repaired before $\mathcal{T}$ is full, then the factory starts to work again normally and we suppose that the tank is emptied instantaneously (before another failure of $\mathcal{U}$). Otherwise, the whole factory has to stop and a certain time is necessary to restart it. Figure 3.1 describes the waste treatment system of the textile factory.

As the output of the factory is variable, the filling time of the tank is a random variable. Moreover, if we suppose that the filling time is deterministic, we can still make the assumption that we have a random time that has a Dirac distribution.

We propose a semi-Markov modelization for the evolution of the described system. Let $E = \{1, 2, 3\}$ be the set of possible states of the system, where we set:

State 1 – everything is working, that is, $\mathcal{F}$ and $\mathcal{U}$ are in good state and $\mathcal{T}$ is empty;
State 2 – $\mathcal{U}$ has failed, but $\mathcal{T}$ is not yet full, so the factory is still working;
State 3 – $\mathcal{U}$ has failed and $\mathcal{T}$ is full, so the factory is not working.

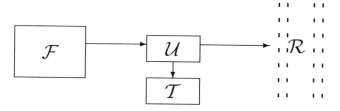

Fig. 3.1. Waste treatment for a textile factory

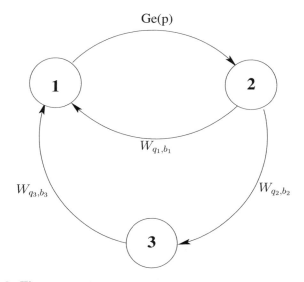

Fig. 3.2. Three-state discrete-time semi-Markov system modelization

The possible transitions between states are given in Figure 3.2. The system is defined by:

- The initial distribution $\boldsymbol{\alpha} = (\alpha_1 \quad \alpha_2 \quad \alpha_3)$;
- The transition matrix $\mathbf{p}$ of the embedded Markov chain $(J_n)_{n \in \mathbb{N}}$

$$\mathbf{p} = \begin{pmatrix} 0 & 1 & 0 \\ a & 0 & b \\ 1 & 0 & 0 \end{pmatrix},$$

  with $0 < a, b < 1$, $a + b = 1$;
- The conditional sojourn time distributions

$$\mathbf{f}(k) = \begin{pmatrix} 0 & f_{12}(k) & 0 \\ f_{21}(k) & 0 & f_{23}(k) \\ f_{31}(k) & 0 & 0 \end{pmatrix}, \ k \in \mathbb{N},$$

where
- $f_{12}(\cdot)$ is the distribution of the failure time of the treatment unit $\mathcal{U}$,
- $f_{21}(\cdot)$ is the distribution of the repairing time of $\mathcal{U}$,
- $f_{23}(\cdot)$ is the distribution of the filling time of the tank $\mathcal{T}$,
- $f_{31}(\cdot)$ is the distribution of time needed to restart the entire factory $\mathcal{F}$,
  after it has been shut down.

This is a irreducible discrete-time Markov renewal system. We chose the following sojourn time distributions:

- f_{12} is the geometric distribution on $\mathbb{N}^*$ of parameter p, $0 < p < 1$, i.e.,
  $f_{12}(k) := p(1-p)^{k-1}$, $k \in \mathbb{N}^*$.
- $f_{21} := W_{q_1,b_1}$, $f_{23} := W_{q_2,b_2}$, $f_{31} := W_{q_3,b_3}$ are discrete-time first-type Weibull distributions (see Nakagawa and Osaki, 1975; Bracquemond and Gaudoin, 2003), defined by

$$W_{q,b}(0) := 0, \ W_{q,b}(k) := q^{(k-1)^b} - q^{k^b}, k \in \mathbb{N}^*.$$

The choice of discrete-time Weibull distribution is motivated by its use in reliability theory. In fact, our main purpose is to compute and estimate the main reliability indicators of this system. As stated above, this will be done in Section 5.3 of Chapter 5.

Equivalently, using Equation (3.5) we can define the system by:

- The initial distribution $\boldsymbol{\alpha}$,
- The semi-Markov kernel $\mathbf{q}$ with

$$\mathbf{q}(k) = \Big(q_{ij}(k)\Big)_{1 \le i,j \le 3} = \begin{pmatrix} 0 & f_{12}(k) & 0 \\ af_{21}(k) & 0 & bf_{23}(k) \\ f_{31}(k) & 0 & 0 \end{pmatrix}, \ k \in \mathbb{N}.$$

In order to obtain the transition matrix of the semi-Markov chain $(Z_n)_{n \in \mathbb{N}}$, first we need to compute all the necessary quantities, like $\mathbf{Q}, \mathbf{H}, \mathbf{I} - \mathbf{H}, \boldsymbol{\psi}, \boldsymbol{\nu}$ and so on. The cumulative kernel $\mathbf{Q}$ is given by

$$\mathbf{Q}(k) = \begin{pmatrix} 0 & \sum_{l=1}^{k} f_{12}(l) & 0 \\ a\sum_{l=1}^{k} f_{21}(l) & 0 & b\sum_{l=1}^{k} f_{23}(l) \\ \sum_{l=1}^{k} f_{31}(l) & 0 & 0 \end{pmatrix}, \quad k \in \mathbb{N}.$$

For $k \in \mathbb{N}$, the sojourn time cumulative distribution function in states $1, 2$, and 3 are computed as follows:

$$H_1(k) = \sum_{l=1}^{k} h_1(l) = \sum_{l=1}^{k} \sum_{i=1}^{3} q_{1i}(l) = \sum_{l=1}^{k} f_{12}(l),$$

$$H_2(k) = \sum_{l=1}^{k} h_2(l) = \sum_{l=1}^{k} \sum_{i=1}^{3} q_{2i}(l) = \sum_{l=1}^{k} (af_{21}(l) + bf_{23}(l)),$$

$$H_3(k) = \sum_{l=1}^{k} h_3(l) = \sum_{l=1}^{k} \sum_{i=1}^{3} q_{3i}(l) = \sum_{l=1}^{k} f_{31}(l).$$

For $k \in \mathbb{N}$ we obtain

$$(\mathbf{I} - \mathbf{H})(k) \tag{3.45}$$

$$= \begin{pmatrix} 1 - \sum_{l=1}^{k} f_{12}(l) & 0 & 0 \\ 0 & 1 - \sum_{l=1}^{k}(af_{21}(l) + bf_{23}(l)) & 0 \\ 0 & 0 & 1 - \sum_{l=1}^{k} f_{31}(l) \end{pmatrix}$$

and

$$\psi(k) = (\boldsymbol{\delta}\mathbf{I} - \mathbf{q})^{(-1)}(k) = \begin{pmatrix} \mathbf{1}_{\{0\}} & -f_{12} & 0 \\ -af_{21} & \mathbf{1}_{\{0\}} & -bf_{23} \\ -f_{31} & 0 & \mathbf{1}_{\{0\}} \end{pmatrix}^{(-1)} (k).$$

The inverse in the convolution sense $(\boldsymbol{\delta}\mathbf{I} - \mathbf{q})^{(-1)}$ can be computed by three methods:

1. By Algorithm (3.16) given in Proposition 3.4;

2. Directly, i.e.,

$$(\boldsymbol{\delta}\mathbf{I} - \mathbf{q})^{(-1)}(k)$$

$$= \begin{pmatrix} \mathbf{1}_{\{0\}} & -f_{12} & 0 \\ -af_{21} & \mathbf{1}_{\{0\}} & -bf_{23} \\ -f_{31} & 0 & \mathbf{1}_{\{0\}} \end{pmatrix}^{(-1)} (k)$$

$$= \left(\mathbf{1}_{\{0\}} - af_{12} * f_{21} - bf_{12} * f_{23} * f_{31} \right)^{(-1)}$$

$$* \begin{pmatrix} \mathbf{1}_{\{0\}} & f_{12} & bf_{12} * f_{23} \\ af_{21} + bf_{23} * f_{31} & \mathbf{1}_{\{0\}} & bf_{23} \\ f_{31} & f_{12} * f_{31} & \mathbf{1}_{\{0\}} - af_{12} * f_{21} \end{pmatrix} (k).$$

The expression

$$\left(\mathbf{1}_{\{0\}}(l) - af_{12} * f_{21}(l) - bf_{12} * f_{23} * f_{31}(l) \right)^{(-1)}, 0 \le l \le k,$$

is a renewal function. By setting

$$T(l) := a f_{12} * f_{21}(l) + b f_{12} * f_{23} * f_{31}(l),$$

which is a discrete distribution on $\mathbb{N}$, we have

$$(1 - T)^{(-1)}(l) = \sum_{n=0}^{l} T^{(n)}(l)$$

and we obtain

$$(\boldsymbol{\delta} \mathbf{I} - \mathbf{q})^{(-1)}(k)$$

$$= \left[\left(\sum_{n=0}^{\cdot} T^{(n)}(\cdot) \right) \right. \tag{3.46}$$

$$\left. * \begin{pmatrix} \mathbf{1}_{\{0\}} & f_{12} & b f_{12} * f_{23} \\ a f_{21} + b f_{23} * f_{31} & \mathbf{1}_{\{0\}} & b f_{23} \\ f_{31} & f_{12} * f_{31} & \mathbf{1}_{\{0\}} - a f_{12} * f_{21} \end{pmatrix} (\cdot) \right] (k).$$

3. Writing $(\boldsymbol{\delta} \mathbf{I} - \mathbf{q})^{(-1)}(k), k \in \mathbb{N}$, as a sum (cf. Equation (3.15))

$$(\boldsymbol{\delta} \mathbf{I} - \mathbf{q})^{(-1)}(k) = \sum_{n=0}^{k} \mathbf{q}^{(n)}(k).$$

The invariant probability $\boldsymbol{\nu}$ for the embedded Markov chain $(J_n)_{n \in \mathbb{N}}$ is obtained by solving the system $\boldsymbol{\nu} \mathbf{p} = \boldsymbol{\nu}$, $\nu(1) + \nu(2) + \nu(3) = 1$. The solution of this system is $\boldsymbol{\nu} = (1/(b+2) \quad 1/(b+2) \quad b/(b+2))$.

The mean sojourn times are obtained as follows:

$$m_1 = \sum_{k \geq 1} (1 - H_1(k)) = \sum_{k \geq 1} \left(1 - \sum_{l=1}^{k} f_{12}(l) \right),$$

$$m_2 = \sum_{k \geq 1} (1 - H_2(k)) = \sum_{k \geq 1} \left(1 - \sum_{l=1}^{k} (a f_{21}(l) + b f_{23}(l)) \right),$$

$$m_3 = \sum_{k \geq 1} (1 - H_3(k)) = \sum_{k \geq 1} \left(1 - \sum_{l=1}^{k} f_{31}(l) \right).$$

The semi-Markov transition matrix $\mathbf{P}$ is obtained from Equation (3.33), i.e.,

$$\mathbf{P}(k) = \boldsymbol{\psi} * (\mathbf{I} - \mathbf{H})(k) = (\boldsymbol{\delta} \mathbf{I} - \mathbf{q})^{(-1)} * (\mathbf{I} - \mathbf{H})(k), \ k \in \mathbb{N},$$

where the convolution inverse $(\boldsymbol{\delta} \mathbf{I} - \mathbf{q})^{(-1)}$ can be computed by one of the three previous methods.

Exercises

Exercise 3.1. Consider $(J_n, S_n)_{n \in \mathbb{N}}$ a Markov renewal chain. Prove the assertions stated in Example 3.1, i.e., $(J_n, S_{n+1})_{n \in \mathbb{N}}$ and $(J_{n+1}, S_n)_{n \in \mathbb{N}}$ are Markov renewal chains with semi-Markov kernels $\widetilde{q}_{ij}(k) = p_{ij} h_j(k)$, respectively $\breve{q}_{ij}(k) = p_{ij} h_i(k)$.

Exercise 3.2. Use Relation (3.33) to obtain the transition function of a two-state Markov chain, regarded as a semi-Markov chain (Example 3.2).

Exercise 3.3. Consider a three-state Markov chain with transition matrix

$$P = \begin{pmatrix} 0.8 & 0.2 & 0 \\ 0.4 & 0.2 & 0.4 \\ 0.5 & 0.4 & 0.1 \end{pmatrix}.$$

1. Recall that, according to Example 3.2, a Markov chain is a particular case of a semi-Markov chain. Give the semi-Markov kernel of the above chain.
2. Compute the mean number of visits to state 1, starting from state 1, up to time 100.
3. Give the semi-Markov transition function of the chain.

Exercise 3.4. Let (J, S) be a MRC with state space E and semi-Markov kernel $\mathbf{q}$. Let F be a proper nonempty subset of E and denote by $(N_n^{(F)})_{n \in \mathbb{N}}$ the successive entrance times in the set F,

$$N_n^{(F)} := \inf\{k > N_{n-1}^{(F)} \mid J_k \in F\}, \; n \geq 1, \quad N_0^{(F)} := 0.$$

Let us define the random sequence $(\tilde{J}_n, \tilde{S}_n)_{n \in \mathbb{N}}$ by

$$\tilde{J}_n := J_{N_n^{(F)}}, \quad \tilde{S}_n := S_{N_n^{(F)}}, \; n \geq 0.$$

Show that this sequence is a MRC with state space $F \cup \{\Delta\}$, where $\Delta \notin E$, and give its semi-Markov kernel $\tilde{\mathbf{q}}$ as a function of $\mathbf{q}$.

Exercise 3.5. Prove Proposition 3.5 by considering the MRC $(\tilde{J}, \tilde{S})$ as in Exercise 3.4, with state space $\{i, j, \Delta\}$.

Exercise 3.6 (algorithm for obtaining a trajectory of a given SMC). In Section 3.5 we have presented a Monte Carlo algorithm that simulates a trajectory of a given SMC. Give another algorithm for generating a trajectory in the time interval $[0, M]$ based on the unconditional sojourn times in successive states. Follow these steps:

1. As in the previous algorithm, set the initial state and the initial jump time (you may sample them if needed).
2. Sample the sojourn time in the current state from $h_i(\cdot)$, the unconditional sojourn time distribution (Definition 3.4).

3. Sample the future visited state, as done in the previous algorithm.
4. Increment the jump time S_n and stop if $S_n \geq M$, otherwise continue from the beginning.

Note that this algorithm simulates a trajectory of a SMC for which the sojourn times in a state depend only on the present visited state, i.e., the semi-Markov kernel is of the type $q_{ij}(k) = p_{ij}h_i(k)$ (see Remark 3.3).

Exercise 3.7. Give a proof of (3.42) by generating function technique.

4

Nonparametric Estimation for Semi-Markov Chains

In this chapter we construct nonparametric estimators for the main characteristics of a discrete-time semi-Markov system, such as the semi-Markov kernel, the transition matrix of the embedded Markov chain, the conditional distributions of the sojourn times, or the semi-Markov transition function. We investigate the asymptotic properties of the estimators, namely, the strong consistency and the asymptotic normality.

There is a growing body of literature on inference problems for continuous-time semi-Markov processes. We can cite, for instance, the works of Moore and Pyke (1968), Lagakos, Sommer and Zelen (1978), Gill (1980), Akritas and Roussas (1980), Ouhbi and Limnios (1999, 2003a), Heutte and Huber-Carol (2002), Alvarez (2005), and Limnios, Ouhbi and Sadek (2005).

To the best of our knowledge, the estimation of discrete-time semi-Markov systems is almost absent in the literature.

In this chapter, we obtain approached maximum-likelihood estimators (MLEs) of the semi-Markov kernel $\mathbf{q}$, of the transition matrix of the embedded Markov chain $\mathbf{p}$, and of the conditional distribution of the sojourn times $\mathbf{f}$. These estimators provide estimators of the matrix-valued function ψ and of the semi-Markov transition function $\mathbf{P}$. Afterward, we prove that the obtained estimators are strongly consistent and asymptotically normal.

Generally speaking, two types of observation procedures are proposed in the statistical literature of stochastic processes, in order to obtain estimators of the quantities of interest: either we observe a single sample path of the process in a time interval $[0, M]$ or we observe K sample paths, with $K > 1$, that are independent and identically distributed. In the first case, the asymptotic properties of the obtained estimators are studied by letting the length M of the path tend to infinity (see, e.g., Basawa and Prakasa Rao, 1980), whereas in the second case we let the number K of sample paths tend to infinity. For example, the papers of Sadek and Limnios (2002, 2005) present both types of observation procedures for Markov chains, respectively, for continuous-time Markov processes.

V.S. Barbu, N. Limnios, *Semi-Markov Chains and Hidden Semi-Markov Models toward Applications*, DOI: 10.1007/978-0-387-73173-5_4,
© Springer Science+Business Media, LLC 2008

The estimation procedure taken into account here consists in considering a sample path of the semi-Markov chain in a time interval $[0, M]$, with M an arbitrarily chosen positive integer. Consequently, the corresponding asymptotic properties of the estimators are obtained when M tends to infinity.

The results presented in this chapter are based on Barbu and Limnios (2006a). Nevertheless, we also present here another method for proving the asymptotic normality of the estimators (Theorems 4.2, 4.5, and 4.7), based on CLT for martingales (Theorem E.4).

4.1 Construction of the Estimators

Let us consider a sample path of an ergodic Markov renewal chain $(J_n, S_n)_{n \in \mathbb{N}}$, censored at fixed arbitrary time $M \in \mathbb{N}^*$,

$$\mathcal{H}(M) := (J_0, X_1, \ldots, J_{N(M)-1}, X_{N(M)}, J_{N(M)}, u_M),$$

where $N(M)$ is the discrete-time counting process of the number of jumps in $[1, M]$, given by (3.7), and $u_M := M - S_{N(M)}$ is the censored sojourn time in the last visited state $J_{N(M)}$.

Starting from the sample path $\mathcal{H}(M)$, we will propose empirical estimators for the quantities of interest of the semi-Markov chain. For any states $i, j \in E$ and positive integer $k \in \mathbb{N}, k \leq M$, we define the empirical estimators of the transition matrix of the embedded Markov chain p_{ij}, of the conditional distribution of the sojourn times $f_{ij}(k)$, and of the discrete-time semi-Markov kernel $q_{ij}(k)$ by

$$\widehat{p}_{ij}(M) := N_{ij}(M)/N_i(M), \text{ if } N_i(M) \neq 0, \tag{4.1}$$

$$\widehat{f}_{ij}(k, M) := N_{ij}(k, M)/N_{ij}(M), \text{ if } N_{ij}(M) \neq 0, \tag{4.2}$$

$$\widehat{q}_{ij}(k, M) := N_{ij}(k, M)/N_i(M), \text{ if } N_i(M) \neq 0, \tag{4.3}$$

where $N_{ij}(k, M)$ is the number of transitions of the EMC from i to j, up to time M, with sojourn time in state i equal to $k, 1 \leq k \leq M$,

$$N_{ij}(k, M) := \sum_{n=1}^{N(M)} \mathbf{1}_{\{J_{n-1}=i, J_n=j, X_n=k\}} = \sum_{n=1}^{M} \mathbf{1}_{\{J_{n-1}=i, J_n=j, X_n=k, S_n \leq M\}}.$$

If $N_i(M) = 0$ we set $\widehat{p}_{ij}(M) := 0$ and $\widehat{q}_{ij}(k, M) := 0$ for any $k \in \mathbb{N}$, and if $N_{ij}(M) = 0$ we set $\widehat{f}_{ij}(k, M) := 0$ for any $k \in \mathbb{N}$.

In the following section it will be show that the empirical estimators proposed in Equations (4.1)–(4.3) have good asymptotic properties. Moreover, they are in fact approached maximum-likelihood estimators (Proposition 4.1). To see this, consider the likelihood function corresponding to the history $\mathcal{H}(M)$

$$L(M) = \alpha_{J_0} \prod_{k=1}^{N(M)} p_{J_{k-1}J_k} f_{J_{k-1}J_k}(X_k) \overline{H}_{J_{N(M)}}(u_M),$$

where $\overline{H}_i(\cdot)$ is the survival function in state i defined by

$$\overline{H}_i(n) := \mathbb{P}(X_1 > n \mid J_0 = i) = 1 - \sum_{j \in E} \sum_{k=1}^{n} q_{ij}(k), \ n \in \mathbb{N}^*, \qquad (4.4)$$

and $u_M = M - S_{N(M)}$. We have the following result concerning the asymptotic behavior of u_M.

Lemma 4.1. *For a semi-Markov chain* $(Z_n)_{n \in \mathbb{N}}$ *we have*

$$u_M / M \xrightarrow[M \to \infty]{a.s.} 0. \qquad (4.5)$$

Proof. First, note that $(J_{n-1}, X_n)_{n \in \mathbb{N}^*}$ is a Markov chain (see, e.g., Limnios and Oprişan, 2001) and its stationary distribution $\widetilde{\nu}$ is given by

$$\widetilde{\nu}(i, k) = \nu(i) h_i(k) = \nu(i) \mathbb{P}(X_1 = k \mid J_0 = i), \quad i \in E, k \in \mathbb{N}^*, \qquad (4.6)$$

where ν is the stationary distribution of the EMC $(J_n)_{n \in \mathbb{N}}$.

Writing $X_n = pr_2(J_{n-1}, X_n)$, with $pr_2 : E \times \mathbb{N}^* \to \mathbb{N}^*$ the projection on the second argument, and applying the SLLN for the Markov chains (Proposition D.10) we obtain

$$\frac{1}{n} \sum_{k=1}^{n} X_k \xrightarrow[n \to \infty]{a.s.} \sum_{k \geq 1} \sum_{i \in E} pr_2(i, k) \widetilde{\nu}(i, k) = \sum_{i \in E} \nu(i) m_i = \overline{m}. \qquad (4.7)$$

As $N(M) \xrightarrow[M \to \infty]{a.s.} \infty$, using Theorem E.5 we obtain

$$\frac{1}{N(M)} \sum_{k=1}^{N(M)} X_k \xrightarrow[M \to \infty]{a.s.} \overline{m}. \qquad (4.8)$$

Second, the Convergence (3.38) states that

$$N(M)/M \xrightarrow[M \to \infty]{a.s.} 1 / \sum_{i \in E} \nu(i) m_i.$$

Consequently,

$$\frac{1}{M} \sum_{k=1}^{N(M)} X_k = \frac{N(M)}{M} \frac{1}{N(M)} \sum_{k=1}^{N(M)} X_k \xrightarrow[M \to \infty]{a.s.} 1.$$

Finally, we obtain

$$\frac{u_M}{M} = \frac{M - S_{N(M)}}{M} = 1 - \frac{1}{M} \sum_{k=1}^{N(M)} X_k \xrightarrow[M \to \infty]{a.s.} 0,$$

which finishes the proof.   □

The result we just proved tells us that, for large M, u_M does not add significant information to the likelihood function. For these reason, we will neglect the term $\overline{H}_{J_{N(M)}}(u_M)$ in the expression of the likelihood function $L(M)$. On the other side, the sample path $\mathcal{H}(M)$ of the MRC $(J_n, S_n)_{n \in \mathbb{N}}$ contains only one observation of the initial distribution α of $(J_n)_{n \in \mathbb{N}}$, so the information on α_{J_0} does not increase with M. As we are interested in large-sample estimation of semi-Markov chains, the term α_{J_0} will be equally neglected in the expression of the likelihood function (see Billingsley, 1961a, page 4, for a similar discussion about Markov chain estimation). Consequently, we will be concerned with the maximization of the *approached likelihood function* defined by

$$L_1(M) = \prod_{k=1}^{N(M)} p_{J_{k-1} J_k} f_{J_{k-1} J_k}(X_k). \tag{4.9}$$

The following result shows that $\widehat{p}_{ij}(M)$, $\widehat{f}_{ij}(k, M)$, and $\widehat{q}_{ij}(k, M)$ defined by Expressions (4.1)–(4.3) are obtained in fact by maximizing $L_1(M)$.

Proposition 4.1. *For a sample path of a semi-Markov chain $(Z_n)_{n \in \mathbb{N}}$, of arbitrarily fixed length $M \in \mathbb{N}$, the empirical estimators of the transition matrix of the embedded Markov chain $(J_n)_{n \in \mathbb{N}}$, of the conditional sojourn time distributions and of the discrete-time semi-Markov kernel, proposed in Equations (4.1)–(4.3), are approached nonparametric maximum-likelihood estimators, i.e., they maximize the approached likelihood function L_1 given by Equation (4.9).*

Proof. Using the equalities

$$\sum_{j=1}^{s} p_{ij} = 1, \ i \in E, \tag{4.10}$$

the approached log-likelihood function can be written in the form

$$\log(L_1(M)) \tag{4.11}$$

$$= \sum_{k=1}^{M} \sum_{i,j=1}^{s} [N_{ij}(M) \log(p_{ij}) + N_{ij}(k, M) \log(f_{ij}(k)) + \lambda_i (1 - \sum_{j=1}^{s} p_{ij})],$$

where the Lagrange multipliers λ_i are arbitrarily chosen constants. In order to obtain the approached MLE of p_{ij}, we maximize (4.11) with respect to p_{ij} and get $p_{ij} = N_{ij}(M)/\lambda_i$. Equation (4.10) becomes

$$1 = \sum_{j=1}^{s} p_{ij} = \sum_{j=1}^{s} \frac{N_{ij}(M)}{\lambda_i} = \frac{N_i(M)}{\lambda_i}.$$

Finally, we infer that the λ_i that maximize (4.11) with respect to p_{ij} are given by $\lambda_i(M) = N_i(M)$, and we obtain

$$\widehat{p}_{ij}(M) = \frac{N_{ij}(M)}{N_i(M)}.$$

The expression of $\widehat{f}_{ij}(k, M)$ can be obtained by the same method. Indeed, using the equality

$$\sum_{k=1}^{\infty} f_{ij}(k) = 1, \qquad (4.12)$$

we write the approached log-likelihood function in the form

$$\log(L_1(M)) = \sum_{k=1}^{M} \sum_{i,j=1}^{s} [N_{ij}(M) \log(p_{ij}) + N_{ij}(k, M) \log(f_{ij}(k))$$

$$+\lambda_{ij}(1 - \sum_{k=1}^{\infty} f_{ij}(k))], \qquad (4.13)$$

where λ_{ij} are arbitrarily chosen constants. Maximizing (4.13) with respect to $f_{ij}(k)$ we get $\widehat{f}_{ij}(k, M) = N_{ij}(k, M)/\lambda_{ij}$, and from Equation (4.12) we obtain $\lambda_{ij} = N_{ij}(M)$. Thus $\widehat{f}_{ij}(k, M) = N_{ij}(k, M)/N_{ij}(M)$.

In an analogous way we can prove that the expression of the approached MLE of the kernel $q_{ij}(k)$ is given by Equation (4.3).   □

4.2 Asymptotic Properties of the Estimators

In this section, we first study the asymptotic properties of the proposed estimators $\widehat{p}_{ij}(M)$, $\widehat{f}_{ij}(k, M)$, and $\widehat{q}_{ij}(k, M)$. Afterward, we will see how estimators for the matrix-valued function ψ and for the semi-Markov transition function $\mathbf{P}$ are immediately obtained. Their asymptotic properties (consistency and asymptotic normality) will be equally studied.

For B a quantity to be estimated, we will denote by ΔB the difference between the estimator of B and the true value of B. For instance, we set $\Delta q_{ij}(\cdot) := \widehat{q}_{ij}(\cdot, M) - q_{ij}(\cdot)$.

Note that the consistency of $\widehat{p}_{ij}(M)$ is a direct consequence of Proposition 3.7.

Corollary 4.1. *For any $i, j \in E$, under A1, we have*

$$\widehat{p}_{ij}(M) = N_{ij}(M)/N_i(M) \xrightarrow[M \to \infty]{a.s.} p_{ij}.$$

For $i, j \in E$ two fixed states, let us also define the empirical estimator of the conditional cumulative distribution of $(X_n)_{n \in \mathbb{N}^*}$ (Definition 3.3):

$$\widehat{F}_{ij}(k, M) := \sum_{l=0}^{k} \widehat{f}_{ij}(l, M) = \sum_{l=0}^{k} N_{ij}(l, M) / N_{ij}(M). \qquad (4.14)$$

The following result concerns the convergence of $\widehat{f}_{ij}(k, M)$ and $\widehat{F}_{ij}(k, M)$.

Proposition 4.2. *For any fixed arbitrary states $i, j \in E$, the empirical estimators $\widehat{f}_{ij}(k, M)$ and $\widehat{F}_{ij}(k, M)$, proposed in Equations (4.2) and (4.14), are uniformly strongly consistent, i.e.*

1. $\displaystyle \max_{i,j \in E} \max_{0 \leq k \leq M} \left| \widehat{F}_{ij}(k, M) - F_{ij}(k) \right| \xrightarrow[M \to \infty]{a.s.} 0,$

2. $\displaystyle \max_{i,j \in E} \max_{0 \leq k \leq M} \left| \widehat{f}_{ij}(k, M) - f_{ij}(k) \right| \xrightarrow[M \to \infty]{a.s.} 0.$

Proof. We first prove the strong consistency of the estimators using the SLLN. Second, we show the uniform consistency, i.e., that the convergence does not depend on the chosen $k, 0 \leq k \leq M$. This second part is done by means of the Glivenko–Cantelli theorem.

Obviously, the strong consistency can be directly obtained using Glivenko–Cantelli theorem. Anyway, we prefer to derive separately the consistency result because it is easy and constructive.

Let us denote by $\{n_1, n_2, \ldots, n_{N_{ij}(M)}\}$ the transition times from state i to state j, up to time M. Note that we have

$$\widehat{F}_{ij}(k, M) = \frac{1}{N_{ij}(M)} \sum_{l=1}^{N_{ij}(M)} \mathbf{1}_{\{X_{n_l} \leq k\}}$$

and

$$\widehat{f}_{ij}(k, M) = \frac{1}{N_{ij}(M)} \sum_{l=1}^{N_{ij}(M)} \mathbf{1}_{\{X_{n_l} = k\}}.$$

For any $l \in \{1, 2, \ldots, N_{ij}(M)\}$ we have

$$\mathbb{E}[\mathbf{1}_{\{X_{n_l} \leq k\}}] = \mathbb{P}(X_{n_l} \leq k) = F_{ij}(k)$$

and

$$\mathbb{E}[\mathbf{1}_{\{X_{n_l} = k\}}] = \mathbb{P}(X_{n_l} = k) = f_{ij}(k).$$

Since $N_{ij}(M) \xrightarrow[M \to \infty]{a.s.} \infty$, applying the SLLN to the sequences of i.i.d. random variables $\{\mathbf{1}_{\{X_{n_l} \leq k\}}\}_{l \in \{1,2,\ldots,N_{ij}(M)\}}$ and $\{\mathbf{1}_{\{X_{n_l} = k\}}\}_{l \in \{1,\ldots,N_{ij}(M)\}}$, and using Theorem E.5, we get

$$\widehat{F}_{ij}(k, M) = \frac{1}{N_{ij}(M)} \sum_{l=1}^{N_{ij}(M)} \mathbf{1}_{\{X_{n_l} \leq k\}} \xrightarrow[M \to \infty]{a.s.} \mathbb{E}[\mathbf{1}_{\{X_{n_l} \leq k\}}] = F_{ij}(k)$$

and

$$\widehat{f}_{ij}(k, M) = \frac{1}{N_{ij}(M)} \sum_{l=1}^{N_{ij}(M)} \mathbf{1}_{\{X_{n_l}=k\}} \xrightarrow[M \to \infty]{a.s.} \mathbb{E}[\mathbf{1}_{\{X_{n_l}=k\}}] = f_{ij}(k).$$

In order to obtain uniform consistency, from the Glivenko–Cantelli theorem (see, e.g., Billingsley, 1995) we have

$$\max_{0 \le k \le m} \left| \frac{1}{m} \sum_{l=1}^{m} \mathbf{1}_{\{X_{n_l} \le k\}} - F_{ij}(k) \right| \xrightarrow[m \to \infty]{a.s.} 0.$$

Let us define $\xi_m := \max_{0 \le k \le m} \left| \frac{1}{m} \sum_{l=1}^{m} \mathbf{1}_{\{X_{n_l} \le k\}} - F_{ij}(k) \right|$. The previous convergence tells us that $\xi_m \xrightarrow[m \to \infty]{a.s.} 0$. As $N(M) \xrightarrow[m \to \infty]{a.s.} \infty$ (Lemma 3.3), applying Theorem E.5 we obtain $\xi_{N(M)} \xrightarrow[M \to \infty]{a.s.} 0$, which reads

$$\max_{0 \le k \le M} \left| \widehat{F}_{ij}(k, M) - F_{ij}(k) \right| \xrightarrow[M \to \infty]{a.s.} 0.$$

As the state space E is finite, we take the maximum with respect to $i, j \in E$ and the desired result for $\widehat{F}_{ij}(k, M)$ follows.

Concerning the uniform consistency of $\widehat{f}_{ij}(k, M)$, note that we have

$$\max_{i,j \in E} \max_{0 \le k \le M} \left| \widehat{f}_{ij}(k, M) - f_{ij}(k) \right|$$
$$= \max_{i,j \in E} \max_{0 \le k \le M} \left| \widehat{F}_{ij}(k, M) - \widehat{F}_{ij}(k-1, M) - F_{ij}(k) + F_{ij}(k-1) \right|$$
$$\le \max_{i,j \in E} \max_{0 \le k \le M} \left| \widehat{F}_{ij}(k, M) - F_{ij}(k) \right|$$
$$+ \max_{i,j \in E} \max_{0 \le k \le M} \left| \widehat{F}_{ij}(k-1, M) - F_{ij}(k-1) \right|,$$

and the result follows from the uniform strong consistency of $\widehat{F}_{ij}(k, M)$.   □

Using the estimator of the semi-Markov kernel (Equation (4.3)) we immediately obtain estimators of the sojourn time distribution in state i, $h_i(\cdot)$, and of the sojourn time cumulative distribution function in state i, $H_i(\cdot)$ (Definition 3.4). The uniform strong consistency of these estimators can be easily proved following the same method as used for proving Proposition 4.2.

The following result is on the consistency of the semi-Markov kernel estimator.

Theorem 4.1. *The empirical estimator of the semi-Markov kernel proposed in Equation (4.3) is uniformly strongly consistent, i.e.,*

$$\max_{i,j \in E} \max_{0 \le k \le M} \left| \widehat{q}_{ij}(k, M) - q_{ij}(k) \right| \xrightarrow[M \to \infty]{a.s.} 0.$$

Proof. We have

$$\max_{i,j \in E} \max_{0 \le k \le M} |\widehat{q}_{ij}(k, M) - q_{ij}(k)|$$

$$= \max_{i,j \in E} \max_{0 \le k \le M} |\widehat{p}_{ij}(M)\widehat{f}_{ij}(k, M) - \widehat{p}_{ij}(M)f_{ij}(k)$$

$$+ \widehat{p}_{ij}(M)f_{ij}(k) - p_{ij}f_{ij}(k)|$$

$$\le \max_{i,j \in E} \widehat{p}_{ij}(M) \max_{i,j \in E} \max_{0 \le k \le M} |\widehat{f}_{ij}(k, M) - f_{ij}(k)|$$

$$+ \max_{i,j \in E} \max_{0 \le k \le M} f_{ij}(k) \max_{i,j \in E} |\widehat{p}_{ij}(M) - p_{ij}|$$

$$\le \max_{i,j \in E} |\widehat{p}_{ij}(M) - p_{ij}| + \max_{i,j \in E} \max_{0 \le k \le M} |\widehat{f}_{ij}(k, M) - f_{ij}(k)|.$$

The conclusion follows from the consistency of $\widehat{p}_{ij}(M)$ and $\widehat{f}_{ij}(k, M)$ (Corollary 4.1 and Proposition 4.2). □

In the next theorem we prove the asymptotic normality of the semi-Markov kernel estimator. The same type of result is true for the estimators of the matrix-valued function ψ and of the semi-Markov transition function $\mathbf{P}$ (Theorems 4.5 and 4.7). We present two different proofs of the theorem. The first one is based on the CLT for Markov renewal chains (Theorem 3.5). The second one relies on the Lindeberg–Lévy CLT for martingales (Theorem E.4). Note that this second method is used in Limnios (2004) for deriving the asymptotic normality of a continuous-time semi-Markov kernel estimator.

Theorem 4.2. *For any fixed arbitrary states $i, j \in E$ and any fixed arbitrary positive integer $k \in \mathbb{N}, k \le M$, we have*

$$\sqrt{M}[\widehat{q}_{ij}(k, M) - q_{ij}(k)] \xrightarrow[M \to \infty]{\mathcal{D}} \mathcal{N}(0, \sigma_q^2(i, j, k)), \tag{4.15}$$

with the asymptotic variance

$$\sigma_q^2(i, j, k) = \mu_{ii}q_{ij}(k)[1 - q_{ij}(k)]. \tag{4.16}$$

Proof (1).

$$\sqrt{M}[\widehat{q}_{ij}(k, M) - q_{ij}(k)]$$

$$= \frac{M}{N_i(M)} \left(\frac{1}{\sqrt{M}} \sum_{n=1}^{N(M)} [\mathbf{1}_{\{J_n=j, X_n=k\}} - q_{ij}(k)]\mathbf{1}_{\{J_{n-1}=i\}} \right).$$

Let us consider the function

$$f(m, l, u) := \mathbf{1}_{\{m=i, l=j, u=k\}} - q_{ij}(k)\mathbf{1}_{\{m=i\}}.$$

Using the notation from the Pyke and Schaufele's CLT (Theorem 3.5), we have

$$W_f(M) := \sum_{n=1}^{N(M)} f(J_{n-1}, J_n, X_n) = \sum_{n=1}^{N(M)} [\mathbf{1}_{\{J_n=j, X_n=k\}} - q_{ij}(k)]\mathbf{1}_{\{J_{n-1}=i\}}.$$

In order to apply the CLT, we need to compute A_{ml}, A_m, B_{ml}, B_m, m_f, and B_f for all $m, l \in E$, using Lemmas A.1 and A.2.

$$A_{ml} := \sum_{u=1}^{\infty} f(m, l, u)q_{ml}(u)$$

$$= \sum_{u=1}^{\infty} \mathbf{1}_{\{m=i, l=j, u=k\}} q_{ml}(u) - \sum_{u=1}^{\infty} \mathbf{1}_{\{m=i\}} q_{ij}(k) q_{ml}(u)$$

$$= \delta_{mi}\delta_{lj} \sum_{u=1}^{\infty} \mathbf{1}_{\{u=k\}} q_{ij}(u) - \delta_{mi} q_{ij}(k) \sum_{u=1}^{\infty} q_{il}(u) = q_{ij}(k)\delta_{mi}(\delta_{lj} - p_{il}).$$

$$A_m := \sum_{l=1}^{s} A_{ml} = q_{ij}(k)\delta_{mi}[\sum_{l=1}^{s} \delta_{lj} - \sum_{l=1}^{s} p_{il}] = 0.$$

$$B_{ml} := \sum_{u=1}^{\infty} f^2(m, l, u) q_{ml}(u)$$

$$= \sum_{u=1}^{\infty} \mathbf{1}_{\{m=i, l=j, u=k\}} q_{ml}(u) + \sum_{u=1}^{\infty} \mathbf{1}_{\{m=i\}} q_{ij}^2(k) q_{ml}(u)$$

$$-2 \sum_{u=1}^{\infty} \mathbf{1}_{\{m=i, l=j, u=k\}} q_{ij}(k) q_{ml}(u)$$

$$= q_{ij}(k)\delta_{mi}\delta_{lj} + q_{ij}^2(k)\delta_{mi}p_{il} - 2q_{ij}^2(k)\delta_{mi}\delta_{lj}.$$

$$B_m := \sum_{l=1}^{s} B_{ml} = \delta_{mi}q_{ij}(k)[1 - q_{ij}(k)].$$

Finally, we obtain

$$r_i := \sum_{m=1}^{s} A_m \frac{\mu_{ii}^*}{\mu_{mm}^*} = 0, \qquad m_f := \frac{r_i}{\mu_{ii}} = 0,$$

$$\sigma_i^2 = \sum_{m=1}^{s} B_m \frac{\mu_{ii}^*}{\mu_{mm}^*} = q_{ij}(k)[1 - q_{ij}(k)], \quad B_f = \frac{\sigma_i^2}{\mu_{ii}} = \frac{q_{ij}(k)[1 - q_{ij}(k)]}{\mu_{ii}}.$$

Since $N_i(M)/M \xrightarrow[M \to \infty]{a.s.} 1/\mu_{ii}$ (Proposition 3.8), we get the desired result. $\square$

Proof (2). For $i, j \in E$ arbitrarily fixed states and $k \in \mathbb{N}$ arbitrarily fixed positive integer, we write the random variable $\sqrt{M}[\widehat{q}_{ij}(k, M) - q_{ij}(k)]$ as

$$\sqrt{M}[\widehat{q}_{ij}(k, M) - q_{ij}(k)] = \frac{M}{N_i(M)} \frac{1}{\sqrt{M}} \sum_{n=1}^{N(M)} [\mathbf{1}_{\{J_n=j, X_n=k\}} - q_{ij}(k)]\mathbf{1}_{\{J_{n-1}=i\}}.$$

Let $\mathcal{F}_n$ be the σ-algebra defined by $\mathcal{F}_n := \sigma(J_l, X_l; l \leq n), n \geq 0$, and let Y_n be the random variable

$$Y_n = \mathbf{1}_{\{J_{n-1}=i, J_n=j, X_n=k\}} - q_{ij}(k)\mathbf{1}_{\{J_{n-1}=i\}}.$$

Obviously, Y_n is $\mathcal{F}_n$-measurable and $\mathcal{F}_n \subseteq \mathcal{F}_{n+1}$, for all $n \in \mathbb{N}$. Moreover, we have

$$
\begin{aligned}
\mathbb{E}(Y_n \mid \mathcal{F}_{n-1}) &= \mathbb{P}(J_{n-1}=i, J_n=j, X_n=k \mid \mathcal{F}_{n-1}) \\
&\quad - q_{ij}(k)\mathbb{P}(J_{n-1}=i \mid \mathcal{F}_{n-1}) \\
&= \mathbf{1}_{\{J_{n-1}=i\}}\mathbb{P}(J_n=j, X_n=k \mid J_{n-1}=i) - q_{ij}(k)\mathbf{1}_{\{J_{n-1}=i\}} \\
&= 0.
\end{aligned}
$$

Therefore, $(Y_n)_{n\in\mathbb{N}}$ is an $\mathcal{F}_n$-martingale difference and $(\sum_{l=1}^n Y_l)_{n\in\mathbb{N}}$ is an $\mathcal{F}_n$- martingale. Note also that, as Y_l is bounded for all $l \in \mathbb{N}$, we have

$$\frac{1}{n}\sum_{l=1}^n \mathbb{E}(Y_l^2 \mathbf{1}_{\{|Y_l|>\epsilon\sqrt{n}\}}) \xrightarrow[n\to\infty]{} 0$$

for any $\epsilon > 0$. Using the CLT for martingales (Theorem E.4) we obtain

$$\frac{1}{\sqrt{n}}\sum_{l=1}^n Y_l \xrightarrow[n\to\infty]{\mathcal{D}} \mathcal{N}(0,\sigma^2), \tag{4.17}$$

where $\sigma^2 > 0$ is given by

$$\sigma^2 = \lim_{n\to\infty} \frac{1}{n}\sum_{l=1}^n \mathbb{E}(Y_l^2 \mid \mathcal{F}_{l-1}) > 0. \tag{4.18}$$

As $N(M)/M \xrightarrow[M\to\infty]{a.s.} 1/\nu(l)\mu_{ll}$ (Convergence (3.38)), applying Anscombe's theorem (Theorem E.6) we obtain

$$\frac{1}{\sqrt{N(M)}}\sum_{l=1}^{N(M)} Y_l \xrightarrow[M\to\infty]{\mathcal{D}} \mathcal{N}(0,\sigma^2). \tag{4.19}$$

To obtain σ^2, we need to compute Y_l^2 and $\mathbb{E}(Y_l^2 \mid \mathcal{F}_{l-1})$. First,

$$Y_l^2 = \mathbf{1}_{\{J_{l-1}=i, J_l=j, X_l=k\}} + (q_{ij}(k))^2 \mathbf{1}_{\{J_{l-1}=i\}} - 2q_{ij}(k)\mathbf{1}_{\{J_{l-1}=i, J_l=j, X_l=k\}}.$$

Second,

$$
\begin{aligned}
\mathbb{E}(Y_l^2 \mid \mathcal{F}_{l-1}) &= \mathbf{1}_{\{J_{l-1}=i\}}\mathbb{P}(J_l=j, X_l=k \mid J_{l-1}=i) + (q_{ij}(k))^2 \mathbf{1}_{\{J_{l-1}=i\}} \\
&\quad - 2\mathbf{1}_{\{J_{l-1}=i\}}q_{ij}(k)\mathbb{P}(J_l=j, X_l=k \mid J_{l-1}=i) \\
&= \mathbf{1}_{\{J_{l-1}=i\}}q_{ij}(k) + (q_{ij}(k))^2 \mathbf{1}_{\{J_{l-1}=i\}} - 2(q_{ij}(k))^2 \mathbf{1}_{\{J_{l-1}=i\}} \\
&= \mathbf{1}_{\{J_{l-1}=i\}}q_{ij}(k)[1 - q_{ij}(k)].
\end{aligned}
$$

Thus, σ^2 is given by

$$\sigma^2 = \lim_{n\to\infty} \left(\frac{1}{n}\sum_{l=1}^{n} \mathbf{1}_{\{J_{l-1}=i\}}\right) q_{ij}(k)[1-q_{ij}(k)] = \nu(i)q_{ij}(k)[1-q_{ij}(k)],$$

where $\boldsymbol{\nu}$ is the stationary distribution of the embedded Markov chain $(J_n)_{n\in\mathbb{N}}$. The random variable of interest $\sqrt{M}[\widehat{q}_{ij}(k,M) - q_{ij}(k)]$ can be written as

$$\sqrt{M}[\widehat{q}_{ij}(k,M) - q_{ij}(k)] = \frac{M}{N_i(M)}\frac{1}{\sqrt{M}}\sqrt{N(M)}\frac{1}{\sqrt{N(M)}}\sum_{l=1}^{N(M)} Y_l$$

$$= \frac{M}{N_i(M)}\sqrt{\frac{N(M)}{M}}\frac{1}{\sqrt{N(M)}}\sum_{l=1}^{N(M)} Y_l.$$

Note that we have

$$N_i(M)/M \xrightarrow[M\to\infty]{a.s.} 1/\mu_{ii} \quad \text{(cf. Proposition 3.8)},$$

$$N(M)/M \xrightarrow[M\to\infty]{a.s.} 1/\nu(i)\mu_{ii} \quad \text{(Convergence (3.38))}.$$

Using these results and Convergence (4.19), we obtain that $\sqrt{M}[\widehat{q}_{ij}(k,M) - q_{ij}(k)]$ converges in distribution, as M tends to infinity, to a zero-mean normal random variable, of variance

$$\sigma_q^2(i,j,k) = \left(\mu_{ii}\sqrt{1/\mu_{ii}\nu(i)}\right)^2 \nu(i)q_{ij}(k)[1-q_{ij}(k)]$$

$$= \mu_{ii}q_{ij}(k)[1-q_{ij}(k)],$$

which is the desired result. $\qquad\square$

Let $\widehat{\mathbf{Q}}(k,M)$ and $\widehat{\boldsymbol{\psi}}(k,M)$ be the estimators of $\mathbf{Q}(k)$ and of $\boldsymbol{\psi}(k)$ defined by

$$\widehat{\mathbf{Q}}(k,M) := \sum_{l=1}^{k} \widehat{\mathbf{q}}(l,M), \qquad (4.20)$$

$$\widehat{\boldsymbol{\psi}}(k,M) := \sum_{n=0}^{k} \widehat{\mathbf{q}}^{(n)}(k,M), \qquad (4.21)$$

where $\widehat{\mathbf{q}}^{(n)}(k,M)$ is the n-fold convolution of $\widehat{\mathbf{q}}(k,M)$ (Definition 3.6). We can easily prove the consistency of $\widehat{\mathbf{Q}}(k,M)$ as M tends to infinity.

Theorem 4.3. $\widehat{\mathbf{Q}}$ *is uniformly strongly consistent, i.e.,*

$$\max_{i,j\in E}\max_{0\leq k\leq M} |\widehat{Q}_{ij}(k,M) - Q_{ij}(k)| \xrightarrow[M\to\infty]{a.s.} 0.$$

Proof. From the definition of $\mathbf{Q}$ (Equation (3.2)) and that of $\widehat{\mathbf{Q}}$, note that

$$Q_{ij}(k) = \mathbb{P}(J_{n+1} = j, X_{n+1} \leq k \mid J_n = i) = p_{ij}F_{ij}(k),$$
$$\widehat{Q}_{ij}(k, M) = \widehat{p}_{ij}(M)\widehat{F}_{ij}(k, M).$$

Now, the proof is similar to that of Theorem 4.1. Indeed, we have

$$\max_{i,j \in E} \max_{0 \leq k \leq M} |\widehat{Q}_{ij}(k, M) - Q_{ij}(k)|$$
$$= \max_{i,j \in E} \max_{0 \leq k \leq M} |\widehat{p}_{ij}(M)\widehat{F}_{ij}(k, M) - \widehat{p}_{ij}(M)F_{ij}(k)$$
$$+ \widehat{p}_{ij}(M)F_{ij}(k) - p_{ij}F_{ij}(k)|$$
$$\leq \max_{i,j \in E} |\widehat{p}_{ij}(M) - p_{ij}| + \max_{i,j \in E} \max_{0 \leq k \leq M} |\widehat{F}_{ij}(k, M) - F_{ij}(k)|,$$

and from the consistency of $\widehat{p}_{ij}(M)$ and $\widehat{F}_{ij}(k, M)$ (Corollary 4.1 and Proposition 4.2) the result follows. $\square$

Using Relation (3.30), the estimator $\widehat{\boldsymbol{\Psi}}(k, M)$ of $\boldsymbol{\Psi}(k)$ is defined by

$$\widehat{\boldsymbol{\Psi}}(k, M) := \sum_{l=0}^{k} \widehat{\boldsymbol{\psi}}(l, M) = \sum_{l=0}^{k} \sum_{n=0}^{l} \widehat{\mathbf{q}}^{(n)}(l, M).$$

The following proposition is an intermediate result, namely, the consistency of the estimator of the n-fold convolution of the semi-Markov kernel. This property will be used to obtain the consistency of the estimators of $\boldsymbol{\psi}$ and $\boldsymbol{\Psi}$.

Proposition 4.3. *For any fixed $n \in \mathbb{N}$ we have*

$$\max_{i,j \in E} \max_{0 \leq k \leq M} |\widehat{q}_{ij}^{(n)}(k, M) - q_{ij}^{(n)}(k)| \xrightarrow[M \to \infty]{a.s.} 0.$$

Proof. We will prove the result by induction. For $n = 1$, the convergence was obtained in Theorem 4.1. Suppose the result holds true for $n - 1$. Then,

$$\max_{i,j \in E} \max_{0 \leq k \leq M} |\widehat{q}_{ij}^{(n)}(k, M) - q_{ij}^{(n)}(k)|$$
$$= \max_{i,j \in E} \max_{0 \leq k \leq M} |\sum_{r=1}^{s} (\widehat{q}_{ir} * \widehat{q}_{rj}^{(n-1)})(k, M) - \sum_{r=1}^{s} (q_{ir} * q_{rj}^{(n-1)})(k)|$$
$$\leq \max_{i,j \in E} \max_{0 \leq k \leq M} \sum_{r=1}^{s} |(\widehat{q}_{ir} * \widehat{q}_{rj}^{(n-1)})(k, M) - (q_{ir} * q_{rj}^{(n-1)})(k)$$
$$- (q_{ir} * \widehat{q}_{rj}^{(n-1)})(k, M) + (q_{ir} * \widehat{q}_{rj}^{(n-1)})(k, M)|$$
$$\leq \max_{i,j \in E} \max_{0 \leq k \leq M} \sum_{r=1}^{s} |q_{ir} * (\widehat{q}_{rj}^{(n-1)} - q_{rj}^{(n-1)})(k, M)|$$
$$+ \max_{i,j \in E} \max_{0 \leq k \leq M} \sum_{r=1}^{s} |(\widehat{q}_{ir} - q_{ir}) * \widehat{q}_{rj}^{(n-1)}(k, M)|$$

$$\leq \max_{r,j \in E} \max_{0 \leq k \leq M} |(\widehat{q}_{rj}^{(n-1)} - q_{rj}^{(n-1)})(k,M)| \max_{i,r \in E} \max_{0 \leq k \leq M} \sum_{r=1}^{s} q_{ir}(k)$$

$$+ \max_{i,r \in E} \max_{0 \leq k \leq M} |(\widehat{q}_{ir} - q_{ir})(k,M)| \max_{r,j \in E} \max_{0 \leq k \leq M} \sum_{r=1}^{s} \widehat{q}_{rj}^{(n-1)}(k,M)$$

$$\leq \max_{r,j \in E} \max_{0 \leq k \leq M} |(\widehat{q}_{rj}^{(n-1)} - q_{rj}^{(n-1)})(k,M)|$$

$$+ s \max_{i,r \in E} \max_{0 \leq k \leq M} |(\widehat{q}_{ir} - q_{ir})(k,M)|.$$

Using the induction hypothesis and Theorem 4.1, the proof is achieved. □

Theorem 4.4. *The estimators of the matrix function $\boldsymbol{\psi} = (\boldsymbol{\psi}(k); \ k \in \mathbb{N}) \in \mathcal{M}_E(\mathbb{N})$ and of the Markov renewal matrix $\boldsymbol{\Psi} = (\boldsymbol{\Psi}(k); \ k \in \mathbb{N}) \in \mathcal{M}_E(\mathbb{N})$ are strongly consistent, i.e.,*

$$\max_{i,j \in E} |\widehat{\psi}_{ij}(k,M) - \psi_{ij}(k)| \xrightarrow[M \to \infty]{a.s.} 0,$$

$$\max_{i,j \in E} |\widehat{\Psi}_{ij}(k,M) - \Psi_{ij}(k)| \xrightarrow[M \to \infty]{a.s.} 0.$$

Proof. Since $\psi_{ij}(k)$ (resp. $\widehat{\psi}_{ij}(k,M)$) is a finite sum of $q_{ij}^{(n)}(k,M)$ (resp. of $\widehat{q}_{ij}^{(n)}(k,M)$), the first convergence is a direct application of Proposition 4.3. The second convergence is obtained following the same method. □

Remark 4.1. Let us denote by $\| \cdot \|$ the matrix norm defined by $\|\mathbf{A}(k)\| := \max_{i,j \in E} | A_{ij}(k) |$, where $\mathbf{A}$ is a matrix-valued function, $\mathbf{A} \in \mathcal{M}_E(\mathbb{N})$. The previous result reads

$$\|\widehat{\boldsymbol{\psi}}(k,M) - \boldsymbol{\psi}(k)\| \xrightarrow[M \to \infty]{a.s.} 0,$$

$$\|\widehat{\boldsymbol{\Psi}}(k,M) - \boldsymbol{\Psi}(k)\| \xrightarrow[M \to \infty]{a.s.} 0.$$

The same remark holds true for the almost sure convergence of the semi-Markov transition matrix estimator (Theorem 4.6).

The following theorem describes the asymptotic normality of the estimator $\widehat{\boldsymbol{\psi}}$. The result can be proved by the two methods used for the asymptotic normality of the kernel estimator (Theorem 4.2), namely the CLT for Markov renewal chains (Theorem 3.5) and the Lindeberg–Lévy CLT for martingales (Theorem E.4). Our proof here is based on the second technique.

Theorem 4.5. *For any fixed $k \in \mathbb{R}$ and $i,j \in E$, we have*

$$\sqrt{M}[\widehat{\psi}_{ij}(k,M) - \psi_{ij}(k)] \xrightarrow[M \to \infty]{\mathcal{D}} \mathcal{N}(0, \sigma_{\psi}^2(i,j,k)),$$

where

$$\sigma_\psi^2(i,j,k) = \sum_{m=1}^s \mu_{mm} \left\{ \sum_{r=1}^s \left[(\psi_{im} * \psi_{rj})^2 * q_{mr} \right](k) - \left[\sum_{r=1}^s (\psi_{im} * q_{mr} * \psi_{rj})(k) \right]^2 \right\}.$$

$$(4.22)$$

Proof. From Chapter 3 we know that the matrix function ψ is the solution of the Markov renewal equation $\psi = \delta \mathbf{I} + \mathbf{q} * \psi$ and can be written in the form $\psi = (\delta \mathbf{I} - \mathbf{q})^{(-1)}$, as stated in Proposition 3.4. Using these results we have

$$
\begin{aligned}
\sqrt{M}[\widehat{\psi}_{ij}(k,M) - \psi_{ij}(k)] &= \sqrt{M}[\widehat{\psi}_{ij} - (\widehat{\psi} * \psi)_{ij} + (\widehat{\psi} * \psi)_{ij} - \psi_{ij}](k) \\
&= \sqrt{M}\left[[\widehat{\psi} * (\delta \mathbf{I} - \psi)]_{ij} + [(\widehat{\psi} - \delta \mathbf{I}) * \psi]_{ij} \right](k) \\
&= \sqrt{M}\left[-[\widehat{\psi} * \mathbf{q} * \psi]_{ij} + [\widehat{\psi} * \widehat{\mathbf{q}} * \psi]_{ij} \right](k) \\
&= \sqrt{M}[\widehat{\psi} * (\widehat{\mathbf{q}} - \mathbf{q}) * \psi]_{ij}(k) \\
&= \sqrt{M}[(\widehat{\psi} - \psi) * (\widehat{\mathbf{q}} - \mathbf{q}) * \psi]_{ij}(k) \\
&\quad + \sqrt{M}[\psi * (\widehat{\mathbf{q}} - \mathbf{q}) * \psi]_{ij}(k).
\end{aligned}
$$

For every $t \in \mathbb{N}, t \leq M$, and for every $r, l, u, v \in E$, $\sqrt{M}(\widehat{q}_{rl}(t,M) - q_{rl}(t))$ converges in distribution to a normal random variable (Theorem 4.2), as M tends to infinity, and $\widehat{\psi}_{uv}(t,M) - \psi_{uv}(t)$ converges in probability to zero (Theorem 4.4) as M tends to infinity. Thus, using Slutsky's theorem (Theorem E.10) we obtain that $\sqrt{M}[(\widehat{\psi} - \psi) * (\widehat{\mathbf{q}} - \mathbf{q}) * \psi]_{ij}(k)$ converges in probability to zero when M tends to infinity. Consequently, applying again Slutsky's theorem we get that $\sqrt{M}[\widehat{\psi}_{ij}(k,M) - \psi_{ij}(k)]$ has the same limit in distribution as $\sqrt{M}(\psi * (\widehat{\mathbf{q}} - \mathbf{q}) * \psi)_{ij}(k)$. The last term can be written as follows:

$$
\begin{aligned}
&\sqrt{M}(\psi * (\widehat{\mathbf{q}} - \mathbf{q}) * \psi)_{ij}(k) \\
&= \sqrt{M} \sum_{l=1}^s \sum_{r=1}^s (\psi_{il} * (\widehat{\mathbf{q}}(\cdot, M) - \mathbf{q})_{lr} * \psi_{rj})(k) \\
&= \sqrt{M} \sum_{l=1}^s \sum_{r=1}^s (\psi_{il} * \widehat{q}(\cdot, M)_{lr} * \psi_{rj})(k) - \sqrt{M} \sum_{l=1}^s \sum_{r=1}^s (\psi_{il} * q_{lr} * \psi_{rj})(k) \\
&= \frac{1}{\sqrt{M}} \sum_{n=1}^{N(M)} \sum_{l=1}^s \frac{M}{N_l(M)} \sum_{r=1}^s \Big[(\psi_{il} * \mathbf{1}_{\{J_{n-1}=l, J_n=r, X_n=\cdot\}} * \psi_{rj})(k) \\
&\quad - (\psi_{il} * q_{lr} \mathbf{1}_{\{J_{n-1}=l\}} * \psi_{rj})(k) \Big].
\end{aligned}
$$

Since $N_l(M)/M \xrightarrow[M\to\infty]{a.s.} 1/\mu_{ll}$ (cf. Proposition 3.8), using Slutsky's Theorem we obtain that $\sqrt{M}[\widehat{\psi}_{ij}(k,M) - \psi_{ij}(k)]$ has the same limit in distribution as

$$\frac{1}{\sqrt{M}} \sum_{n=1}^{N(M)} \sum_{l=1}^{s} \mu_{ll} \sum_{r=1}^{s} \Big[(\psi_{il} * \mathbf{1}_{\{J_{n-1}=l, J_n=r, X_n=\cdot\}} * \psi_{rj})(k)$$

$$- (\psi_{il} * q_{lr} \mathbf{1}_{\{J_{n-1}=l\}} * \psi_{rj})(k) \Big]$$

$$= \sqrt{\frac{N(M)}{M}} \frac{1}{\sqrt{N(M)}} \sum_{n=1}^{N(M)} Y_n,$$

where the random variables Y_n are defined by

$$Y_n := \sum_{l=1}^{s} \mu_{ll} \sum_{r=1}^{s} \Big[(\psi_{il} * \mathbf{1}_{\{J_{n-1}=l, J_n=r, X_n=\cdot\}} * \psi_{rj})(k)$$

$$- (\psi_{il} * q_{lr} \mathbf{1}_{\{J_{n-1}=l\}} * \psi_{rj})(k) \Big].$$

Let $\mathcal{F}_n$ be the σ-algebra defined by $\mathcal{F}_n := \sigma(J_l, X_l; l \leq n)$. Note that Y_n is $\mathcal{F}_n$-measurable, for all $n \in \mathbb{N}$, and it can be easily checked that $\mathbb{E}(Y_n \mid \mathcal{F}_{n-1}) = 0$. Therefore, $(Y_n)_{n \in \mathbb{N}}$ is an $\mathcal{F}_n$-martingale difference and $(\sum_{l=1}^{n} Y_l)_{n \in \mathbb{N}}$ is an $\mathcal{F}_n$-martingale. As Y_m is bounded for $m \in \mathbb{N}$ fixed, we have for any $\epsilon > 0$ that

$$\frac{1}{n} \sum_{m=1}^{n} \mathbb{E}(Y_m^2 \mathbf{1}_{\{|Y_m| > \epsilon \sqrt{n}\}}) \xrightarrow[n \to \infty]{} 0.$$

Using the CLT for martingales (Theorem E.4) we obtain

$$\frac{1}{\sqrt{n}} \sum_{m=1}^{n} Y_m \xrightarrow[n \to \infty]{\mathcal{D}} \mathcal{N}(0, \sigma^2), \qquad (4.23)$$

where $\sigma^2 > 0$ is given by

$$\sigma^2 = \lim_{n \to \infty} \frac{1}{n} \sum_{m=1}^{n} \mathbb{E}(Y_m^2 \mid \mathcal{F}_{m-1}). \qquad (4.24)$$

As $N(M)/M \xrightarrow[M \to \infty]{a.s.} 1/\nu(l)\mu_{ll}$ (Convergence (3.38)), from (4.23) and Anscombe's theorem (Theorem E.6) we can infer

$$\frac{1}{\sqrt{N(M)}} \sum_{n=1}^{N(M)} Y_n \xrightarrow[M \to \infty]{\mathcal{D}} \mathcal{N}(0, \sigma^2). \qquad (4.25)$$

Since $N(M)/M \xrightarrow[M \to \infty]{a.s.} 1/\nu(l)\mu_{ll}$ for any state $l \in E$ (Convergence (3.38)), we obtain that

$$\sqrt{M}[\widehat{\psi}_{ij}(k, M) - \psi_{ij}(k)] \xrightarrow[n \to \infty]{\mathcal{D}} \mathcal{N}(0, \sigma^2/\nu(l)\mu_{ll}). \qquad (4.26)$$

To obtain the asymptotic variance $\sigma_\psi^2(i,j,k)$, we need to compute Y_l^2, $\mathbb{E}(Y_l^2 \mid \mathcal{F}_{l-1})$, and σ^2.

$$Y_n^2 = \sum_{l=1}^{s} \mu_{ll}^2 \left[\sum_{r=1}^{s} \sum_{k_1=0}^{k} (\psi_{il} * \psi_{rj})^2 (k - k_1) \mathbf{1}_{\{J_{n-1}=l, J_n=r, X_n=k_1\}} \right.$$

$$+ \mathbf{1}_{\{J_{n-1}=l\}} \sum_{r_1,r_2=1}^{s} \sum_{k_1,k_2=0}^{k} (\psi_{il} * \psi_{r_1 j})(k - k_1)(\psi_{il} * \psi_{r_2 j})(k - k_2)$$

$$\times q_{lr_1}(k_1) q_{lr_2}(k_2)$$

$$+ \sum_{r_1,r_2=1}^{s} \sum_{k_1,k_2=0}^{k} (\psi_{il} * \psi_{r_1 j})(k - k_1)(\psi_{il} * \psi_{r_2 j})(k - k_2)$$

$$\left. \times \mathbf{1}_{\{J_{n-1}=l, J_n=r_1, X_n=k_1\}} q_{lr_2}(k_2) \right].$$

Consequently, we get

$$\mathbb{E}(Y_n^2 \mid \mathcal{F}_{n-1})$$

$$= \sum_{l=1}^{s} \mu_{ll}^2 \mathbf{1}_{\{J_{n-1}=l\}} \left[\sum_{r=1}^{s} (\psi_{il} * \psi_{rj})^2 * q_{lr}(k) - \left(\sum_{r=1}^{s} \psi_{il} * \psi_{rj} * q_{lr}(k) \right)^2 \right]$$

and, using Equation (4.24) and Proposition D.5, we obtain

$$\sigma^2 = \sum_{l=1}^{s} \mu_{ll}^2 \left(\lim_{n\to\infty} \frac{1}{n} \sum_{m=1}^{n} \mathbf{1}_{\{J_{m-1}=l\}} \right)$$

$$\times \left[\sum_{r=1}^{s} (\psi_{il} * \psi_{rj})^2 * q_{lr}(k) - \left(\sum_{r=1}^{s} \psi_{il} * \psi_{rj} * q_{lr}(k) \right)^2 \right]$$

$$= \sum_{l=1}^{s} \mu_{ll}^2 \nu(l) \left[\sum_{r=1}^{s} (\psi_{il} * \psi_{rj})^2 * q_{lr}(k) - \left(\sum_{r=1}^{s} \psi_{il} * \psi_{rj} * q_{lr}(k) \right)^2 \right].$$

Finally, this form of σ^2, together with Convergence (4.26), yields the convergence of $\sqrt{M}[\widehat{\psi}_{ij}(k, M) - \psi_{ij}(k)]$, as M tends to infinity, to a zero-mean normal random variable of variance $\sigma_\psi^2(i, j, k)$ given by Equation (4.22). □

In the previous chapter we obtained that the semi-Markov transition matrix $\mathbf{P}$ is the unique solution of the associated renewal Markov equation (Equation 3.2) and we showed that it can be written as

$$\mathbf{P}(k) = (\boldsymbol{\delta}\mathbf{I} - \mathbf{q})^{(-1)} * (\mathbf{I} - \mathbf{H})(k) = \boldsymbol{\psi} * (\mathbf{I} - diag(\mathbf{Q} \cdot \mathbf{1}))(k), \qquad (4.27)$$

with $\mathbf{1}$ an s-dimensional column vector all of whose elements are equal to 1. Consequently, we have the following estimator of $\mathbf{P}(\cdot)$:

$$\widehat{\mathbf{P}}(k, M) := \left(\delta \mathbf{I} - \widehat{\mathbf{q}}(\cdot, M) \right)^{(-1)} * \left(\mathbf{I} - diag(\widehat{\mathbf{Q}}(\cdot, M) \cdot \mathbf{1}) \right)(k)$$
$$= \widehat{\boldsymbol{\psi}}(\cdot, M) * \left(\mathbf{I} - diag(\widehat{\mathbf{Q}}(\cdot, M) \cdot \mathbf{1}) \right)(k). \tag{4.28}$$

The following results concern the asymptotic properties of this estimator.

Theorem 4.6. *The estimator of the semi-Markov transition matrix is strongly consistent, i.e., for any fixed arbitrary states $i, j \in E$ and any fixed arbitrary positive integer $k \in \mathbb{N}, k \leq M$, we have*

$$\max_{i,j \in E} \left| \widehat{P}_{ij}(k, M) - P_{ij}(k) \right| \xrightarrow[M \to \infty]{a.s.} 0.$$

Proof.

$$\left| \widehat{P}_{ij}(k, M) - P_{ij}(k) \right|$$
$$= \left| \left[\widehat{\boldsymbol{\psi}} * (\mathbf{I} - diag(\widehat{\mathbf{Q}} \cdot \mathbf{1})) \right]_{ij}(k, M) - \left[\boldsymbol{\psi} * (\mathbf{I} - diag(\mathbf{Q} \cdot \mathbf{1})) \right]_{ij}(k) \right|$$
$$= \left| \widehat{\psi}_{ij} * (\mathbf{I} - diag(\widehat{\mathbf{Q}} \cdot \mathbf{1}))_{jj}(k, M) - \psi_{ij} * (\mathbf{I} - diag(\mathbf{Q} \cdot \mathbf{1}))_{jj}(k) \right.$$
$$\left. + \psi_{ij} * diag(\widehat{\mathbf{Q}} \cdot \mathbf{1})_{jj}(k, M) - \psi_{ij} * diag(\widehat{\mathbf{Q}} \cdot \mathbf{1})_{jj}(k, M) \right|$$
$$= \left| (\widehat{\psi}_{ij} - \psi_{ij}) * I_{jj}(k, M) - (\widehat{\psi}_{ij} - \psi_{ij}) * diag(\widehat{\mathbf{Q}} \cdot \mathbf{1})_{jj}(k, M) \right.$$
$$\left. - \psi_{ij} * \left(diag(\widehat{\mathbf{Q}} \cdot \mathbf{1})_{jj} - diag(\mathbf{Q} \cdot \mathbf{1})_{jj} \right)(k, M) \right|$$
$$\leq \left| \widehat{\psi}_{ij}(k, M) - \psi_{ij}(k) \right| + \left| (\widehat{\psi}_{ij} - \psi_{ij}) * diag(\widehat{\mathbf{Q}} \cdot \mathbf{1})_{jj}(k, M) \right|$$
$$+ \left| \psi_{ij} * diag((\widehat{\mathbf{Q}} - \mathbf{Q}) \cdot \mathbf{1})_{jj}(k, M) \right|.$$

Since $diag(\widehat{\mathbf{Q}}(k, M) \cdot \mathbf{1})_{jj} = \sum_{l=1}^{s} \widehat{Q}_{jl}(k, M) = \widehat{H}_j(k, M) \leq 1$ and the estimators of the matrix functions $\mathbf{Q}$ and $\boldsymbol{\psi}$ are strongly consistent, we obtain the strong consistency of the estimator of the semi-Markov matrix. □

Theorem 4.7. *For any fixed $k \in \mathbb{N}, k \leq M$, and $i, j \in E$, we have*

$$\sqrt{M} [\widehat{P}_{ij}(k, M) - P_{ij}(k)] \xrightarrow[M \to \infty]{\mathcal{D}} \mathcal{N}(0, \sigma_P^2(i, j, k)),$$

where

$$\sigma_P^2(i, j, k) = \sum_{m=1}^{s} \mu_{mm} \left\{ \sum_{r=1}^{s} \left[\delta_{mj} \Psi_{ij} - (1 - H_j) * \psi_{im} * \psi_{rj} \right]^2 * q_{mr}(k) \right.$$

$$\left. - \left[\delta_{mj} \psi_{ij} * H_m(k) - \sum_{r=1}^{s} (1 - H_j) * \psi_{im} * \psi_{rj} * q_{mr} \right]^2 (k) \right\}. \tag{4.29}$$

Proof. As the proof is similar to that of Theorems 4.2 and 4.5, we give only the main steps. We have

$$\sqrt{M}[\widehat{P}_{ij}(k, M) - P_{ij}(k)]$$

$$= \sqrt{M}\Big[\widehat{\psi}_{ij} * \Big(\mathbf{I} - diag(\widehat{\mathbf{Q}} \cdot \mathbf{1})\Big)_{jj} - \psi_{ij} * \Big(\mathbf{I} - diag(\mathbf{Q} \cdot \mathbf{1})\Big)_{jj}$$

$$+ \widehat{\psi}_{ij} * \Big(\mathbf{I} - diag(\mathbf{Q} \cdot \mathbf{1})\Big)_{jj} - \widehat{\psi}_{ij} * \Big(\mathbf{I} - diag(\mathbf{Q} \cdot \mathbf{1})\Big)_{jj}$$

$$+ \psi_{ij} * diag(\widehat{\mathbf{Q}} \cdot \mathbf{1})_{jj} - \psi_{ij} * diag(\mathbf{Q} \cdot \mathbf{1})_{jj}$$

$$+ \psi_{ij} * diag(\mathbf{Q} \cdot \mathbf{1})_{jj} - \psi_{ij} * diag(\mathbf{Q} \cdot \mathbf{1})_{jj}\Big](k, M)$$

$$= \sqrt{M}\Big[(\widehat{\psi}_{ij} - \psi_{ij}) * \Big(\mathbf{I} - diag(\mathbf{Q} \cdot \mathbf{1})\Big)_{jj} - \psi_{ij} * \Big(diag(\widehat{\mathbf{Q}} - \mathbf{Q}) \cdot \mathbf{1}\Big)_{jj}\Big]$$

$$- \sqrt{M}\Big[(\widehat{\psi}_{ij} - \psi_{ij}) * \Big(diag(\widehat{\mathbf{Q}} - \mathbf{Q}) \cdot \mathbf{1}\Big)_{jj}\Big](k, M).$$

The term $(\widehat{\psi}_{ij} - \psi_{ij}) * \Big(diag(\widehat{\mathbf{Q}} - \mathbf{Q}) \cdot \mathbf{1}\Big)_{jj}(k, M)$ can be written

$$(\widehat{\psi}_{ij} - \psi_{ij}) * \Big(diag(\widehat{\mathbf{Q}} - \mathbf{Q}) \cdot \mathbf{1}\Big)_{jj}(k, M)$$

$$= \sum_{r=1}^{s}\sum_{l=0}^{k}(\widehat{\psi}_{ij} - \psi_{ij})(l, M)\Delta Q_{jr}(k - l, M).$$

For every $l \in \mathbb{N}, l \leq M, \sqrt{M}(\widehat{\psi}_{ij} - \psi_{ij})(l, M)$ converges in distribution to a normal random variable (Theorem 4.5) and $\Delta Q_{jr}(k - l, M)$ converges in probability to zero (Theorem 4.3). Thus, using Slutsky's theorem (Theorem E.10) we obtain that $\sqrt{M}\Big[(\widehat{\psi}_{ij} - \psi_{ij}) * \Big(diag(\widehat{\mathbf{Q}} - \mathbf{Q}) \cdot \mathbf{1}\Big)_{jj}\Big](k, M)$ converges in probability to zero when M tends to infinity. Thus, $\sqrt{M}[\widehat{P}_{ij}(k, M) - P_{ij}(k)]$ has the same limit in distribution as

$$\sqrt{M}\Big[(\widehat{\psi}_{ij} - \psi_{ij}) * \Big(\mathbf{I} - diag(\mathbf{Q} \cdot \mathbf{1})\Big)_{jj} - \psi_{ij} * \Big(diag(\widehat{\mathbf{Q}} - \mathbf{Q}) \cdot \mathbf{1}\Big)_{jj}\Big](k, M).$$

From the proof of Theorem 4.5 we know that $\sqrt{M}[\widehat{\psi}_{ij}(k, M) - \psi_{ij}(k)]$ has the same limit in distribution as $\sqrt{M}[\psi * \Delta \mathbf{q} * \psi]_{ij}(k, M)$ when M tends to infinity. Writing

$$\Delta q_{ij}(\cdot) = \frac{1}{N_i(M)} \sum_{l=1}^{N(M)} [\mathbf{1}_{\{J_{l-1}=i, J_l=j, X_l=\cdot\}} - q_{ij}(\cdot)\mathbf{1}_{\{J_{l-1}=i\}}],$$

$$\Delta Q_{ij}(\cdot) = \frac{1}{N_i(M)} \sum_{l=1}^{N(M)} [\mathbf{1}_{\{J_{l-1}=i, J_l=j, X_l\leq\cdot\}} - Q_{ij}(\cdot)\mathbf{1}_{\{J_{l-1}=i\}}],$$

and

$$\sum_{u=1}^{s} \psi_{ij} * \Delta Q_{ju} = \sum_{u=1}^{s}\sum_{n=1}^{s} \delta_{nj}\psi_{ij} * \Delta Q_{nu},$$

we obtain that $\sqrt{M}[\widehat{P}_{ij}(k, M) - P_{ij}(k)]$ has the same limit in distribution as

$$\sqrt{M}\Big\{\sum_{l=1}^{s}\sum_{u=1}^{s}\Big[\Big(1 - H_j\Big) * \psi_{il} * \psi_{uj} * \Delta q_{lu}\Big](k) - \sum_{u=1}^{s}\psi_{ij} * \Delta Q_{ju}(k)\Big\}$$

$$= \frac{1}{\sqrt{M}}\sum_{n=1}^{N(M)}\sum_{l=1}^{s}\sum_{u=1}^{s}\frac{M}{N_l(M)}$$

$$\times\Big[-\delta_{lj}\psi_{ij} * \Big(\mathbf{1}_{\{J_{n-1}=l, J_n=u, X_n\le\cdot\}} - Q_{lu}(\cdot)\mathbf{1}_{\{J_{n-1}=l\}}\Big)(k)$$

$$+(1 - H_j) * \psi_{il} * \psi_{uj} * \Big(\mathbf{1}_{\{J_{n-1}=l, J_n=u, X_n=\cdot\}} - q_{lu}(\cdot)\mathbf{1}_{\{J_{n-1}=l\}}\Big)(k)\Big].$$

Since $N_l(M)/M \xrightarrow[M\to\infty]{a.s.} 1/\mu_{ll}$ (cf. Proposition 3.8), using Slutsky's theorem we obtain that $\sqrt{M}[\widehat{P}_{ij}(k, M) - P_{ij}(k)]$ has the same limit in distribution as

$$\frac{1}{\sqrt{M}}\sum_{n=1}^{N(M)}\sum_{l=1}^{s}\sum_{u=1}^{s}\mu_{ll}\Big[-\delta_{lj}\psi_{ij} * \Big(\mathbf{1}_{\{J_{n-1}=l, J_n=u, X_n\le\cdot\}} - Q_{lu}(\cdot)\mathbf{1}_{\{J_{n-1}=l\}}\Big)$$

$$+(1 - H_j) * \psi_{il} * \psi_{uj} * \Big(\mathbf{1}_{\{J_{n-1}=l, J_n=u, X_n=\cdot\}} - q_{lu}(\cdot)\mathbf{1}_{\{J_{n-1}=l\}}\Big)\Big](k)$$

$$= \sqrt{\frac{N(M)}{M}}\frac{1}{\sqrt{N(M)}}\sum_{n=1}^{N(M)}Y_n,$$

where we have defined the random variables Y_n by

$$Y_n := \sum_{l=1}^{s}\sum_{u=1}^{s}\mu_{ll}\Big[-\delta_{lj}\psi_{ij} * \Big(\mathbf{1}_{\{J_{n-1}=l, J_n=u, X_n\le\cdot\}} - Q_{lu}(\cdot)\mathbf{1}_{\{J_{n-1}=l\}}\Big)(k)$$

$$+(1 - H_j) * \psi_{il} * \psi_{uj} * \Big(\mathbf{1}_{\{J_{n-1}=l, J_n=u, X_n=\cdot\}} - q_{lu}(\cdot)\mathbf{1}_{\{J_{n-1}=l\}}\Big)(k)\Big].$$

As we did before, let $\mathcal{F}_n$ be the σ-algebra $\mathcal{F}_n := \sigma(J_l, X_l; l \le n)$, with Y_n $\mathcal{F}_n$-measurable for all $n \in \mathbb{N}$. Using the fact that for all $l, u \in E$ and $k \in \mathbb{N}$ we have

$$\mathbb{E}(\mathbf{1}_{\{J_{n-1}=l, J_n=u, X_n=k\}} - q_{lu}(k)\mathbf{1}_{\{J_{n-1}=l\}} \mid \mathcal{F}_{n-1}) = 0,$$

$$\mathbb{E}(\mathbf{1}_{\{J_{n-1}=l, J_n=u, X_n\le k\}} - Q_{lu}(k)\mathbf{1}_{\{J_{n-1}=l\}} \mid \mathcal{F}_{n-1}) = 0,$$

it can be easily seen that $\mathbb{E}(Y_n \mid \mathcal{F}_{n-1}) = 0$. Therefore, $(Y_n)_{n\in\mathbb{N}}$ is an $\mathcal{F}_n$-martingale difference and $(\sum_{l=1}^{n}Y_l)_{n\in\mathbb{N}}$ is an $\mathcal{F}_n$-martingale.

As $N(M)/M \xrightarrow[M\to\infty]{a.s.} 1/\nu(l)\mu_{ll}$ (Convergence (3.38)) using the CLT for martingales (Theorem E.4) and Anscombe's theorem (Theorem E.6) we obtain

$$\frac{1}{\sqrt{N(M)}}\sum_{n=1}^{N(M)}Y_n \xrightarrow[M\to\infty]{\mathcal{D}} \mathcal{N}(0, \sigma^2), \qquad (4.30)$$

where $\sigma^2 > 0$ is given by

$$\sigma^2 = \lim_{n \to \infty} \frac{1}{n} \sum_{m=1}^{n} \mathbb{E}(Y_m^2 \mid \mathcal{F}_{m-1}). \tag{4.31}$$

Using again Convergence (3.38) we get

$$\sqrt{M}[\widehat{P}_{ij}(k, M) - P_{ij}(k)] \xrightarrow[M \to \infty]{\mathcal{D}} \mathcal{N}(0, \sigma^2/\nu(l)\mu_{ll}). \tag{4.32}$$

In order to obtain the asymptotic variance $\sigma_P^2(i, j, k)$, we need to compute Y_l^2, $\mathbb{E}(Y_l^2 \mid \mathcal{F}_{l-1})$ and σ^2. Using Proposition D.5 we get

$$\lim_{n \to \infty} \frac{1}{n} \sum_{m=1}^{n} \mathbf{1}_{\{J_{m-1}=l\}}) = \nu(l),$$

and a routine but long computation leads to

$$\sigma^2 = \sum_{l=1}^{s} \mu_{ll}^2 \nu(l) \left\{ \sum_{u=1}^{s} \left[\delta_{lj} \Psi_{ij} - (1 - H_j) * \psi_{il} * \psi_{uj} \right]^2 * q_{lu}(k) \right.$$
$$\left. - \left[\delta_{lj} \psi_{ij} * H_l(k) - \sum_{u=1}^{s} (1 - H_j) * \psi_{il} * \psi_{uj} * q_{lu}(k) \right]^2 \right\}.$$

This result, together with Convergence (4.32), yields the convergence of $\sqrt{M}[\widehat{P}_{ij}(k, M) - P_{ij}(k)]$, as M tends to infinity, to a zero-mean normal random variable of variance $\sigma_P^2(i, j, k)$ given by Equation (4.29). $\qquad \square$

Asymptotic Confidence Intervals

The asymptotic results obtained above allow one to construct asymptotic confidence intervals for the semi-Markov kernel $\mathbf{q}$, the matrix-valued function $\boldsymbol{\psi}$, and the semi-Markov transition matrix $\mathbf{P}$. Here we present asymptotic confidence intervals for the semi-Markov transition matrix $\mathbf{P}$.

First, we need to construct a consistent estimator of the asymptotic variance of $\mathbf{P}$. For $i, j \in E$ and $k \leq M$, replacing $\mathbf{q}(k), \mathbf{Q}(k), \boldsymbol{\psi}(k), \boldsymbol{\Psi}(k)$, respectively by $\widehat{\mathbf{q}}(k, M), \widehat{\mathbf{Q}}(k, M), \widehat{\boldsymbol{\psi}}(k, M), \widehat{\boldsymbol{\Psi}}(k, M)$ in (4.29), we obtain an estimator $\widehat{\sigma}_P^2(i, j, k, M)$ of the variance $\sigma_P^2(i, j, k)$. Second, from the strong consistency of $\widehat{\mathbf{q}}(k, M), \widehat{\mathbf{Q}}(k, M), \widehat{\boldsymbol{\psi}}(k, M)$, and $\widehat{\boldsymbol{\Psi}}(k, M)$ (Theorems 4.2, 4.3, and 4.4), we obtain that $\widehat{\sigma}_P^2(i, j, k, M)$ converges almost surely to $\sigma_P^2(i, j, k)$ as M tends to infinity. Finally, the asymptotic confidence interval of $P_{ij}(k)$ at level $100(1 - \gamma)\%$, $\gamma \in (0, 1)$, is

$$\widehat{P}_{ij}(k, M) - u_{1-\gamma/2} \frac{\widehat{\sigma}_P(i, j, k, M)}{\sqrt{M}} \leq P_{ij}(k) \leq \widehat{P}_{ij}(k, M) + u_{1-\gamma/2} \frac{\widehat{\sigma}_P(i, j, k, M)}{\sqrt{M}},$$

where u_γ is the γ-quantile of the $\mathcal{N}(0, 1)$ distribution. In the same way, we obtain the confidence intervals for the other quantities.

4.3 Example: a Textile Factory (Continuation)

Let us consider the example of the textile factory presented in Chapter 3, Section 3.6, for which we will show how to estimate all the quantities related to the semi-Markov system. We will also illustrate in some figures the asymptotic properties of the semi-Markov transition matrix estimator.

We take the initial distribution $\alpha = (1 \quad 0 \quad 0)$, the transition matrix of the embedded Markov chain $(J_n)_{n \in \mathbb{N}}$,

$$
\mathbf{p} = \left(
\begin{array}{ccc}
0 & 1 & 0 \\
0.95 & 0 & 0.05 \\
1 & 0 & 0
\end{array}
\right),
$$

and the following conditional sojourn time distributions:

- f_{12} is the geometric distribution on $\mathbb{N}^*$ with parameter $p = 0.8$;
- $f_{21} = W_{q_1,b_1}, f_{23} = W_{q_2,b_2}, f_{31} = W_{q_3,b_3}$ are discrete-time first-type Weibull distributions with parameters

$$q_1 = 0.3, b_1 = 0.5, q_2 = 0.5, b_2 = 0.7, q_3 = 0.6, b_3 = 0.9.$$

For our particular semi-Markov system a simulation was made (using the Matlab software) for obtaining a sample path of length M of the semi-Markov chain $(Z_n)_{n \in \mathbb{N}}$ for different values of M. Starting from this sample path, we can count the different transitions and obtain the $N_i(M), N_{ij}(M)$, and $N_{ij}(k, M)$ (cf. Definition 3.13). These quantities allow us to obtain the empirical estimators of the transition matrix of the embedded Markov chain p_{ij}, of the conditional distributions of the sojourn times $f_{ij}(k)$, and of the semi-Markov kernel $q_{ij}(k)$, defined in Equations (4.1)–(4.3). Afterward, estimators for all the characteristics of the semi-Markov system $(\mathbf{Q}, \mathbf{H}, \psi, \mathbf{P},$ etc.) are obtained.

The true values of the elements $P_{13}(k), P_{23}(k), P_{33}(k)$ of the semi-Markov transition matrix $\mathbf{P}(k)$, their estimators, and the confidence intervals at level 95% are given in Figure 4.1, for a sample size $M = 8000$. Note that the confidence intervals cover the true values of the estimators, which is in accordance with Theorem 4.7. Note also that all the $P_{i3}(k), i = 1, 2, 3$, tend to the same limit, as k increases. In fact, this common limit represents the semi-Markov limit distribution π_3, as stated in Proposition 3.9.

Figure 4.2 gives a comparison between the semi-Markov transition matrix estimators obtained for different sample sizes ($M = 2000, M = 6000,$ and $M = 8000$). We see that the estimators approach the true value as M increases, which is in accordance with the strong consistency of the estimator $\widehat{\mathbf{P}}$ (Theorem 4.6).

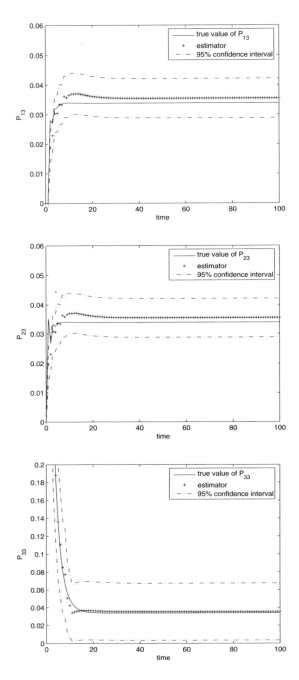

Fig. 4.1. Confidence interval of **P**

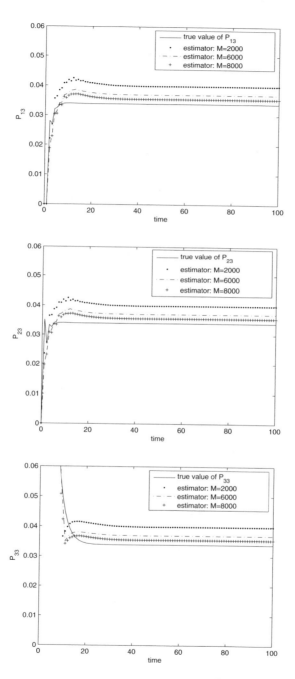

Fig. 4.2. Consistency of $\widehat{\mathbf{P}}$

Exercises

Exercise 4.1. Prove that the limit distribution π of an ergodic semi-Markov chain (Equation (3.42)) is the marginal stationary distribution of the Markov chain $(Z_n, n - S_{N(n)})_{n \in \mathbb{N}}$ (see Remark 3.11).

Exercise 4.2. Prove Proposition D.5.

Exercise 4.3. Prove Proposition 4.1 without using Lagrange multipliers.

Hint: For obtaining the estimators of p_{ij} that maximize the approached log-likelihood function $\log(L_1(M))$ given by (4.9), one can use the s equations $\sum_{j=1}^{s} p_{ij} = 1$, $i \in E$, in order to express s parameters p_{ij} as functions of the others. Then, $\log(L_1(M))$ can be written only in terms of independent parameters and the maximization is straightforward. The same technique can be used for obtaining the estimators of $f_{ij}(k)$ and $q_{ij}(k)$, $i, j \in E, k \in \mathbb{N}$.

Exercise 4.4. Let us consider $(J_n, S_n)_{n \in \mathbb{N}}$ a MRC. Show that $(J_{n-1}, X_n)_{n \in \mathbb{N}^*}$ is a Markov chain, find its transition matrix, and show that its stationary distribution $\widetilde{\nu}$, if it exists, is given by

$$\widetilde{\nu}(i, k) = \nu(i) h_i(k) \quad i \in E, k \in \mathbb{N}^*, \tag{4.33}$$

where ν is the stationary distribution of the EMC $(J_n)_{n \in \mathbb{N}}$.

Exercise 4.5.

1. Prove that $(Y_n)_{n \in \mathbb{N}}$ defined in the proof of Theorem 4.5 is a martingale difference.
2. The same question for $(Y_n)_{n \in \mathbb{N}}$ defined in the proof of Theorem 4.7.

Exercise 4.6. Let us consider an ergodic SMC Z with state space $E = \{1, ..., s\}$, limit distribution $\pi = (\pi_1, ..., \pi_s)$, and mean sojourn time $\overline{m}$ (see Equation 3.27). Suppose that we have an observation $(Z_0, \ldots, Z_M)$, $M \geq 1$, of this SMC.

In case $\overline{m}$ is known, we propose the following estimator for the limit distribution:

$$\widehat{\pi}_i(M) = \frac{1}{\overline{m} N(M)} \sum_{k=1}^{N_i(M)} X_{ik}, \quad i \in E, \tag{4.34}$$

where X_{ik} is the kth sojourn time of the SMC in state i.

Prove that this estimator is consistent and asymptotically normal, as M tends to infinity, and give its asymptotic variance.

Exercise 4.7. In the setting of Exercise 4.6, we consider the following estimator for the mean time $\overline{m}$:

$$\widehat{\overline{m}}(M) = \frac{1}{N(M)} \sum_{j \in E} \sum_{k=1}^{N_j(M)} X_{jk}, \quad i \in E. \tag{4.35}$$

Show that this estimator is consistent and asymptotically normal as M tends to infinity.

Exercise 4.8. In the setting of Exercise 4.6, if we replace in (4.34) $\overline{m}$ by its estimator given in (4.35), then we obtain a new estimator for the limit distribution of the SMC Z. Show that this new estimator is consistent and asymptotically normal as M tends to infinity.

5

Reliability Theory for Discrete-Time Semi-Markov Systems

In this chapter, we study the reliability of a reparable discrete-time semi-Markov system. First, we obtain explicit expressions for different measures of the reliability of a system: reliability, availability, maintainability, failure rates, and mean hitting times. Second, using the estimator of the semi-Markov kernel obtained in the previous chapter, we propose estimators for reliability measures and we study their asymptotic properties.

In the last 50 years, much work has been carried out in the field of probabilistic and statistical methods in reliability. We do not intend to provide here an overview of the field, but only to point out some bibliographical references that are close to the work presented in this chapter. More precisely, we are interested in discrete-time models for reliability (with an i.i.d. or Markovian approach) and in models based on semi-Markov processes. Elements of reliability theory for a Markov system are provided in Appendix D.

It is worth noting here that most of the mathematical models for reliability consider the time to be continuous. But there are real situations when systems have natural discrete lifetimes. We can cite those systems that work on demand, those working on cycles, or those monitored only at certain discrete times (once a month, say). In such situations, the lifetimes are expressed in terms of the number of working periods, the number of working cycles, or the number of months before failure. In other words, all these lifetimes are intrinsically discrete. A good overview of discrete probability distributions used in reliability theory can be found in Bracquemond and Gaudoin (2003) (see also Johnson et al., 2005, for a general reference for discrete probability distributions).

Several authors have studied discrete-time models for reliability in a general i.i.d. context (see Roy and Gupta, 1992; Bracquemond, 2001; Xie et al., 2002). Discrete-time reliability modeling via homogeneous and nonhomogeneous Markov chains can be found in Balakrishnan et al. (2001) and Platis et al. (1998). Statistical estimation and asymptotic properties for reliability metrics, using discrete-time homogeneous Markov chains, are presented

V.S. Barbu, N. Limnios, *Semi-Markov Chains and Hidden Semi-Markov Models toward Applications*, DOI: 10.1007/978-0-387-73173-5_5,
© Springer Science+Business Media, LLC 2008

in Sadek and Limnios (2002). The continuous-time semi-Markov model in reliability can be found in Limnios and Oprişan (2001), Ouhbi and Limnios (2003a), Lisnianski and Levitin (2003), and Limnios and Ouhbi (2003). Among the few works which consider also the discrete-time semi-Markov framework for reliability we can cite Csenki (2002) and Barbu et al. (2004), while the discrete-time semi-Markov reliability estimation can be found in Barbu and Limnios (2006a,b).

The present chapter is structured as follows. In the first part, we consider a reparable discrete-time semi-Markov system and we obtain explicit forms for reliability measurements: reliability, availability, maintainability, failure rates, and mean hitting times (mean time to repair, mean time to failure, mean up time, mean down time). Two different definitions of failure rate function are taken into account: the first one is the usual failure rate, defined in Barlow et al. (1963), which is generally used in continuous-time reliability analysis. The second one is more recent, proposed in Roy and Gupta (1992) as being adapted to reliability studies carried out in discrete time (see also Bracquemond, 2001; Xie et al., 2002; Lai and Xie, 2006). A detailed discussion justifies the use of this discrete-time adapted failure rate. In the second part of the chapter, we propose estimators for measures of reliability using the estimation results of Chapter 4 and we investigate the asymptotic properties of these estimators. We illustrate these results using the textile factory example presented in Sections 3.6 and 4.3.

Most of the results presented in this chapter can be found in Barbu et al. (2004) and Barbu and Limnios (2006a,b). Nevertheless, the proofs of the asymptotic normality of the estimators presented here use also a different approach than in the cited articles, based on the CLT for martingales (Theorem E.4).

5.1 Reliability Function and Associated Measures

Consider a system (or a component) S whose possible states during its evolution in time are $E = \{1, \ldots, s\}$. Denote by $U = \{1, \ldots, s_1\}$ the subset of operational states of the system (the up states) and by $D = \{s_1 + 1, \ldots, s\}$ the subset of failure states (the down states), with $0 < s_1 < s$ (obviously, $E = U \cup D$ and $U \cap D = \emptyset$, $U \neq \emptyset$, $D \neq \emptyset$). One can think of the states of U as different operating modes or performance levels of the system, whereas the states of D can be seen as failures of the systems with different modes.

We are interested in investigating the reliability theory of a discrete-time semi-Markov system S. Consequently, we suppose that the evolution in time of the system is governed by an E-state space semi-Markov chain $(Z_k)_{k \in \mathbb{N}}$. The system starts to work at instant 0 and the state of the system is given at each instant $k \in \mathbb{N}$ by Z_k: the event $\{Z_k = i\}$, for a certain $i \in U$, means that the system S is in operating mode i at time k, whereas $\{Z_k = j\}$, for a certain $j \in D$, means that the system is not operational at time k due to the mode of failure j or that the system is under the repairing mode j.

According to the partition of the state space in up states and down states, we will partition the vectors, matrices, or matrix functions we are working with.

First, for $\boldsymbol{\alpha}$, $\boldsymbol{\nu}$, $\boldsymbol{\pi}$, $\mathbf{p}$, $\mathbf{q}(k)$, $\mathbf{f}(k)$, $\mathbf{Q}(k)$, $\mathbf{F}(k)$, $\mathbf{H}(k)$, and $\mathbf{m}$ we consider the natural matrix partition corresponding to the state space partition U and D. For example, we have

$$
\mathbf{p} = \begin{bmatrix} \mathbf{p}_{11} & \mathbf{p}_{12} \\ \mathbf{p}_{21} & \mathbf{p}_{22} \end{bmatrix} \begin{matrix} U \\ D \end{matrix}, \quad \mathbf{q}(k) = \begin{bmatrix} \mathbf{q}_{11}(k) & \mathbf{q}_{12}(k) \\ \mathbf{q}_{21}(k) & \mathbf{q}_{22}(k) \end{bmatrix} \begin{matrix} U \\ D \end{matrix},
$$

$$
\mathbf{H}(k) = \begin{bmatrix} \mathbf{H}_1(k) & \mathbf{0} \\ \mathbf{0} & \mathbf{H}_2(k) \end{bmatrix} \begin{matrix} U \\ D \end{matrix},
$$

where $\mathbf{H}_1(k) = diag(H_i(k))_{i \in U}$ and $\mathbf{H}_2(k) = diag(H_i(k))_{i \in D}$.

Second, for $\boldsymbol{\psi}(k)$ and $\mathbf{P}(k)$ we consider the restrictions to $U \times U$ and $D \times D$ induced by the corresponding restrictions of the semi-Markov kernel $\mathbf{q}(k)$. To be more specific, recall that the matrix function $\boldsymbol{\psi}$ was expressed in terms of the semi-Markov kernel as follows (cf. Equations (3.14) and (3.15)):

$$
\boldsymbol{\psi}(k) = (\boldsymbol{\delta}\mathbf{I} - \mathbf{q})^{(-1)}(k) = \sum_{n=0}^{k} \mathbf{q}^{(n)}(k)
$$

and that the semi-Markov transition matrix $\mathbf{P}$ was given in Equation (3.33) as

$$
\mathbf{P}(k) = \boldsymbol{\psi} * (\mathbf{I} - \mathbf{H})(k) = (\boldsymbol{\delta}\mathbf{I} - \mathbf{q})^{(-1)} * (\mathbf{I} - diag(\mathbf{Q} \cdot \mathbf{1}))(k), \ k \in \mathbb{N}.
$$

Using the partition given above for the kernel $\mathbf{q}$, we note:

- $\boldsymbol{\psi}_{11}(k) := (\boldsymbol{\delta}\mathbf{I} - \mathbf{q}_{11})^{(-1)}(k) = \sum_{n=0}^{k} \mathbf{q}_{11}^{(n)}(k)$,
- $\boldsymbol{\psi}_{22}(k) := (\boldsymbol{\delta}\mathbf{I} - \mathbf{q}_{22})^{(-1)}(k) = \sum_{n=0}^{k} \mathbf{q}_{22}^{(n)}(k)$,
- $\mathbf{P}_{11}(k) := \boldsymbol{\psi}_{11} * (\mathbf{I} - \mathbf{H}_1)(k) = (\boldsymbol{\delta}\mathbf{I} - \mathbf{q}_{11})^{(-1)} * (\mathbf{I} - diag(\mathbf{Q} \cdot \mathbf{1})_{11})(k)$,
- $\mathbf{P}_{22}(k) := \boldsymbol{\psi}_{22} * (\mathbf{I} - \mathbf{H}_2)(k) = (\boldsymbol{\delta}\mathbf{I} - \mathbf{q}_{22})^{(-1)} * (\mathbf{I} - diag(\mathbf{Q} \cdot \mathbf{1})_{22})(k)$.

Remark 5.1. First, note that $\boldsymbol{\psi}_{11}(k)$ (resp. $\boldsymbol{\psi}_{22}(k)$), just defined above, does not represent the natural matrix partition of $\boldsymbol{\psi}(k)$, corresponding to $U \times U$ (resp. to $D \times D$). This can be immediately seen by checking that the restriction of the matrix $\mathbf{q}^{(n)}(k)$ to $U \times U$ is not equal to $\mathbf{q}_{11}^{(n)}(k)$. The same remark holds true for $\mathbf{P}_{11}(k)$ (resp. $\mathbf{P}_{22}(k)$).

Second, one can ask why we defined $\boldsymbol{\psi}_{11}(k), \boldsymbol{\psi}_{22}(k), \mathbf{P}_{11}(k)$, and $\mathbf{P}_{22}(k)$ as we did above and not as a natural matrix restriction. The reasons will become clear when expressing the reliability of a semi-Markov system in terms of the basic quantities of the system (Proposition 5.1, second proof). Anyway, we can already state that $\mathbf{P}_{11}(k)$ (as defined above) is given by

$$(\mathbf{P}_{11})_{ij}(k) = \mathbb{P}(Z_k = j, Z_l \in U, l = 1, \ldots, k-1 \mid Z_0 = i), i, j \in U. \qquad (5.1)$$

See also Exercise 5.1 for more details.

For $m, n \in \mathbb{N}^*$ such that $m > n$, let $\mathbf{1}_{m,n}$ denote the m-dimensional column vector whose n first elements are 1 and last $m-n$ elements are 0; for $m \in \mathbb{N}^*$, let $\mathbf{1}_m$ denote the m-column vector whose elements are all 1, that is, $\mathbf{1}_m = \mathbf{1}_{m,m}$.

5.1.1 Reliability

Definition 5.1. *Consider a system $\mathcal{S}$ starting to function at time $k = 0$. The reliability of $\mathcal{S}$ at time $k \in \mathbb{N}$ is the probability that the system has functioned without failure in the period $[0, k]$.*

Let T_D denote the first passage time in subset D, called the *lifetime of the system*, i.e.,

$$T_D := \inf\{n \in \mathbb{N}; \quad Z_n \in D\} \text{ and } \inf \emptyset := \infty.$$

The reliability at time $k \in \mathbb{N}$ of a discrete-time semi-Markov system is

$$R(k) := \mathbb{P}(T_D > k) = \mathbb{P}(Z_n \in U, n = 0, \ldots, k).$$

For any state $i \in U$ and $k \in \mathbb{N}$, we introduce:

- $R_i(k) := \mathbb{P}(T_D > k \mid Z_0 = i)$ is the system conditional reliability at time k, given that it starts in state $i \in U$;
- $\mathbf{R}(k) := (R_1(k), \ldots, R_{s_1}(k))^\top$ is the column vector of all conditional reliabilities.

Note that the reliability $R(k)$, the conditional reliabilities $R_i(k)$, and the reliability vector $\mathbf{R}$ are related by

$$R(k) = \sum_{i \in U} \mathbb{P}(Z_0 = i)\mathbb{P}(T_D > k \mid Z_0 = i) = \sum_{i \in U} \alpha_i R_i(k) = \boldsymbol{\alpha}_1 \mathbf{R}(k). \ (5.2)$$

The following result gives the reliability of the system in terms of the basic quantities of the semi-Markov chain. We present two different ways of deriving a closed form for the reliability.

The first one, based on the fundamental idea that can be generally used for finding explicit forms of quantities associated to semi-Markov systems, is the following: first, find the Markov renewal equation (i.e., of the type (3.10)) verified by the quantity of interest; second, solve the corresponding equation (using Theorem 3.1) and get the quantity of interest in terms of the basic quantities of the semi-Markov system ($\mathbf{q}$, $\boldsymbol{\psi}$, $\mathbf{P}$, etc.).

The second one is directly related to the meaning of the reliability function. We want to present this second method as well, because it is intuitive and helps in understanding the system's evolution.

Proposition 5.1. *The reliability of a discrete-time semi-Markov system at time $k \in \mathbb{N}$ is given by*

$$R(k) = \boldsymbol{\alpha}_1 \, \mathbf{P}_{11}(k) \, \mathbf{1}_{s_1} = \boldsymbol{\alpha}_1 \boldsymbol{\psi}_{11} * (\mathbf{I} - \mathbf{H}_1)(k) \mathbf{1}_{s_1}. \tag{5.3}$$

Proof (1). We want to find a Markov renewal equation that has $\mathbf{R}$ as its unique solution. We proceed as in the proof of Proposition 3.2. For any up state $i \in U$ and any positive integer $k \in \mathbb{N}$ we have

$$R_i(k) = \mathbb{P}_i(T_D > k) = \mathbb{P}_i(T_D > k, S_1 > k) + \mathbb{P}_i(T_D > k, S_1 \leq k). \tag{5.4}$$

First, note that

$$\mathbb{P}_i(T_D > k, S_1 > k) = 1 - H_i(k). \tag{5.5}$$

Second,

$$
\begin{aligned}
\mathbb{P}_i(T_D > k, S_1 \leq k) &= \mathbb{E}_i[\mathbb{P}_i(T_D > k, S_1 \leq k \mid S_1)] \\
&= \mathbb{E}_i[\mathbf{1}_{\{S_1 \leq k\}} \mathbb{P}_{Z_{S_1}}(T_D > k - S_1)] \\
&= \sum_{j \in U} \sum_{m \geq 1} \mathbf{1}_{\{m \leq k\}} \mathbb{P}_j(T_D > k - m) \mathbb{P}_i(J_1 = j, S_1 = m) \\
&= \sum_{j \in U} \sum_{m=1}^{k} q_{ij}(m) R_j(k - m).
\end{aligned}
\tag{5.6}
$$

For any up state $i \in U$ and any positive integer $k \in \mathbb{N}$, Equations (5.4)–(5.6) yield the Markov renewal equation associated to R_i

$$R_i(k) = 1 - H_i(k) + \sum_{j \in U} \sum_{m=1}^{k} q_{ij}(m) R_j(k - m). \tag{5.7}$$

Equation (5.7) can be written in matrix form as follows:

$$\mathbf{R}(k) = (\mathbf{I} - \mathbf{H}_1)(k) \mathbf{1}_{s_1} + \mathbf{q}_{11} * \mathbf{R}(k). \tag{5.8}$$

Solving this vector Markov renewal equation, we get

$$\mathbf{R}(k) = (\boldsymbol{\delta} \mathbf{I} - \mathbf{q}_{11})^{(-1)} * (\mathbf{I} - \mathbf{H}_1)(k) \mathbf{1}_{s_1}. \tag{5.9}$$

Thus, from Equation (5.2) we obtain the desired result. $\qquad\square$

Proof (2). Consider a new chain $Y = (Y_k)_{k \in \mathbb{N}}$ with state space $E_Y = U \cup \{\Delta\}$, where Δ is an absorbing state, defined for all $k \in \mathbb{N}$ as follows:

$$Y_k = \begin{cases} Z_k, & \text{if } k < T_D, \\ \Delta, & \text{if } k \geq T_D. \end{cases}$$

One can check that $Y = (Y_k)_{k \in \mathbb{N}}$ is a semi-Markov chain with the semi-Markov kernel

$$\mathbf{q_Y}(k) = \begin{pmatrix} \mathbf{q}_{11}(k) & \mathbf{q}_{12}(k)\, \mathbf{1}_{s-s_1} \\ \mathbf{0}_{1s_1} & 0 \end{pmatrix}, \; k \in \mathbb{N}.$$

Let us denote by $\mathbf{P_Y}$ the semi-Markov transition matrix of Y. The reliability of the system at time $k \in \mathbb{N}$ can be written as

$$
\begin{aligned}
R(k) &= \mathbb{P}(Z_n \in U, n = 0, \ldots, k) = \mathbb{P}(Y_k \in U) \\
&= \sum_{j \in U} \sum_{i \in U} \mathbb{P}(Y_k = j \mid Y_0 = i)\, \mathbb{P}(Y_0 = i) \\
&= [\boldsymbol{\alpha}_1, 0]\, \mathbf{P_Y}(k)\, \mathbf{1}_{s_1+1, s_1}.
\end{aligned}
$$

Note further that for any up states $i, j \in U$ and positive integer k, we have

$$
\begin{aligned}
(P_Y)_{ij}(k) &= \mathbb{P}(Y_k = j, Y_l \in U, l = 1, \ldots, k-1 \mid Y_0 = i) \\
&= \mathbb{P}(Z_k = j, Z_l \in U, l = 1, \ldots, k-1 \mid Z_0 = i),
\end{aligned}
$$

so we have here the probability that starting from the up state i, the system gets to the up state j and the passage is done only through up states. As described in Exercise 5.1, one can show that

$$\mathbb{P}(Z_k = j, Z_l \in U, l = 1, \ldots, k-1 \mid Z_0 = i) = (\boldsymbol{\delta}\mathbf{I} - \mathbf{q}_{11})^{(-1)} * (\mathbf{I} - \mathbf{H}_1)(k) = \mathbf{P}_{11}(k).$$

So the restriction of the matrix $\mathbf{P_Y}(k)$ to $U \times U$ equals $\mathbf{P}_{11}$ and we obtain the desired expression of reliability. $\square$

Remark 5.2.

1. The function $F(k) = 1 - R(k) = 1 - \boldsymbol{\alpha}_1 \boldsymbol{\psi}_{11} * (\mathbf{I} - \mathbf{H}_1)(k) \mathbf{1}_{s_1}$ is the distribution function of the hitting time T_D. This formula is a generalization of the phase-type distribution in discrete time (see Neuts (1981); see also Limnios and Oprişan (2001) for the continuous-time generalization).
2. It is worth noticing that $R(k) = 0$ for all $k \geq 0$ on the event $\{Z_0 \in D\}$.

5.1.2 Availability

Definition 5.2. *The pointwise (or instantaneous) availability of a system $\mathcal{S}$ at time $k \in \mathbb{N}$ is the probability that the system is operational at time k (independently of the fact that the system has failed or not in $[0, k)$).*

Thus, the pointwise availability of a semi-Markov system at time $k \in \mathbb{N}$ is

$$A(k) = \mathbb{P}(Z_k \in U) = \sum_{i \in E} \alpha_i A_i(k),$$

where we have denoted by $A_i(k)$ the conditional availability of the system at time $k \in \mathbb{N}$, given that it starts in state $i \in E$,

$$A_i(k) = P(Z_k \in U \mid Z_0 = i).$$

Proposition 5.2. *The pointwise availability of a discrete-time semi-Markov system at time $k \in \mathbb{N}$ is given by*

$$A(k) = \boldsymbol{\alpha} \, \mathbf{P}(k) \, \mathbf{1}_{s,s_1} = \boldsymbol{\alpha} \, \boldsymbol{\psi} * (\mathbf{I} - \mathbf{H})(k) \mathbf{1}_{s,s_1}. \tag{5.10}$$

Proof. We immediately have

$$A(k) = \sum_{i \in E} \alpha_i P(Z_k \in U \mid Z_0 = i) = \sum_{i \in E} \alpha_i \sum_{j \in U} P_{ij}(k) = \boldsymbol{\alpha} \, \mathbf{P}(k) \, \mathbf{1}_{s,s_1},$$

which proves the result. □

Another reliability indicator related to the pointwise availability is the *steady-state* (or *limit*) *availability*, defined as the limit of the pointwise availability (when this limit exists), as time tends to infinity,

$$A := \lim_{k \to \infty} A(k). \tag{5.11}$$

Thus, the steady-state availability gives an equilibrium point of the pointwise availability.

Proposition 5.3. *For an irreducible and aperiodic discrete-time semi-Markov system, with finite mean sojourn times $m_i, i \in E$, the steady-state availability is given by*

$$A = \frac{1}{\boldsymbol{\nu} \, \mathbf{m}} \mathbf{m}^\top \, diag(\boldsymbol{\nu}) \, \mathbf{1}_{s,s_1}. \tag{5.12}$$

Proof. Using Proposition 3.9 for the limit distribution $\boldsymbol{\pi}$ of a semi-Markov system and the expression of pointwise availability given in Proposition 5.2, we have

$$A = \lim_{k \to \infty} A(k) = \sum_{i \in E} \alpha_i \sum_{j \in U} \lim_{k \to \infty} P_{ij}(k)$$

$$= \sum_{i \in E} \alpha_i \sum_{j \in U} \pi_j = \frac{\sum_{j \in U} \nu(j) m_j}{\sum_{i \in E} \nu(i) m_i},$$

which gives (5.12) when writing it in matrix form. □

Remark 5.3.

1. In the modeling of reliability and related measures, we can consider two cases: the case of nonergodic (i.e., nonreparable) systems and that of ergodic (i.e., reparable) systems.

   For nonergodic systems, the set of down states D is a final class ("absorbing set") and the system can be observed only up to the absorbing time (failure time). In this case, the availability of the system is identical to the reliability, so it is computed using Equation (5.3).

On the other hand, for ergodic systems, the situation is different and we no longer have the availability equal to the reliability, but they are computed using the formulas given above.

This discussion is important in order to stress that, although the systems considered in this chapter are ergodic (reparable), explicit forms of the reliability indicators can be immediately obtained even in the nonergodic case.

2. Let us consider a system, reparable or not, and let us take $i \in U$ an up state. If we suppose that state i is absorbing, then we obviously have the reliability

$$R(k) = 1, \quad \text{for all } k \in \mathbb{N}, \ \mathbb{P}_i \ a.s.$$

As this case is not interesting, we will always suppose that the up states are not absorbing.

5.1.3 Maintainability

Definition 5.3. *For a reparable system $\mathcal{S}$ for which the failure occurs at time $k = 0$, its* maintainability *at time $k \in \mathbb{N}$ is the probability that the system is repaired up to time k, given that it has failed at time $k = 0$.*

Thus, we take $\boldsymbol{\alpha}_1 = \mathbf{0}$ and we denote by T_U the first hitting time of subset U, called the *duration of repair* or *repair time*, that is,

$$T_U = \inf\{n \in \mathbb{N}; \ Z_n \in U\}.$$

The maintainability at time $k \in \mathbb{N}$ of a discrete-time semi-Markov system is

$$M(k) := \mathbb{P}(T_U \leq k) = 1 - \mathbb{P}(T_U > k) = 1 - \mathbb{P}(Z_n \in D, n = 0, \ldots, k).$$

As we already did for reliability, we can express the maintainability of the system in terms of the basic quantities of the semi-Markov chain.

Proposition 5.4. *The maintainability of a discrete-time semi-Markov system at time $k \in \mathbb{N}$ is given by*

$$M(k) = 1 - \boldsymbol{\alpha}_2 \, \mathbf{P}_{22}(k) \, \mathbf{1}_{s-s_1} = 1 - \boldsymbol{\alpha}_2 \boldsymbol{\psi}_{22} * (\mathbf{I} - \mathbf{H}_2)(k)\mathbf{1}_{s-s_1}. \tag{5.13}$$

Proof. The result is straightforward, using the fact that $\mathbb{P}(T_U > k)$ can be computed exactly like $\mathbb{P}(T_D > k) = R(k)$ (cf. Proposition 5.1) by swapping U and D. See also Exercise 5.4. $\qquad\square$

5.1.4 Failure Rates

As pointed out in the introduction, we consider two different definitions of the failure rate function. The first one is the usual failure rate, introduced by Barlow, Marshall and Prochan (1963). We call it the *BMP-failure rate* and denote it by $\lambda(k), k \in \mathbb{N}$. The second one is a discrete-time adapted failure rate, proposed by Roy and Gupta (1992). We call it the *RG-failure rate* and denote it by $r(k), k \in \mathbb{N}$.

1. BMP-Failure Rate Function $\lambda(k)$

Consider a system $\mathcal{S}$ starting to work at time $k = 0$. The BMP-failure rate at time $k \in \mathbb{N}$ is the conditional probability that the failure of the system occurs at time k, given that the system has worked until time $k - 1$.

For a discrete-time semi-Markov system, the failure rate at time $k \geq 1$ has the expression

$$\lambda(k) := \mathbb{P}(T_D = k \mid T_D \geq k)$$

$$= \begin{cases} 1 - \frac{R(k)}{R(k-1)}, & \text{if } R(k-1) \neq 0, \\ 0, & \text{otherwise}, \end{cases}$$

$$= \begin{cases} 1 - \frac{\alpha_1 \, \mathbf{P}_{11}(k) \, \mathbf{1}_{s_1}}{\alpha_1 \, \mathbf{P}_{11}(k-1) \, \mathbf{1}_{s_1}}, & \text{if } R(k-1) \neq 0, \\ 0, & \text{otherwise}. \end{cases} \tag{5.14}$$

The failure rate at time $k = 0$ is defined by $\lambda(0) := 1 - R(0)$.

It is worth noticing that the failure rate $\lambda(k)$ in the discrete-time case is a probability function and not a general positive function as in the continuous-time case.

2. RG-Failure Rate Function $r(k)$

In the sequel, we give the main reasons for introducing a discrete-time adapted failure rate, following the works of Roy and Gupta (1992), Bracquemond (2001), Xie et al. (2002), and Lai and Xie (2006). Basically, the idea is that the usual failure rate $\lambda(k)$ (the one we denoted by BMP-failure rate) is not adapted for discrete-time systems. Let us address briefly the main problems raised by the use of the classical failure rate in discrete time.

- In discrete time, the classical failure rate is bounded by 1, $\lambda(k) \leq 1$, which is not the case in continuous time. For this reason, we have shape restrictions for the failure rate $\lambda(k)$ in discrete time. For instance, it cannot be linearly or exponentially increasing, or, more generally, it cannot be a convex increasing function (which is usually the case in reliability studies).
- Consider a series system of n independent components, denote by $\lambda_i(k)$ and $R_i(k)$ respectively the failure rate function and the reliability function of component i at time k, with $i = 1, \ldots, n$. Denote also by $\lambda(k)$ and $R(k)$ respectively the failure rate function and the reliability function of the system at time k.

For a continuous-time system, it is easy to check the well-known property of failure rate additivity:

$$\lambda(t) = \sum_{i=1}^{n} \lambda_i(t), \ t \in \mathbb{R}_+.$$

On the other hand, in discrete time, the failure rate $\lambda(k)$ is no longer additive for series systems. Indeed, we first note that $R(k) = \prod_{i=1}^{n} R_i(k)$, and second we can prove that

$$\lambda(k) = 1 - \prod_{i=1}^{n} [1 - \lambda_i(k)] \neq \sum_{i=1}^{n} \lambda_i(k), \ k \in \mathbb{N}.$$

- In *continuous-time* reliability studies, one type of system aging is based on failure rate average defined, whenever the failure rate $\lambda(t)$ exists, as

$$\Lambda(t) := \frac{1}{t} \int_0^t \lambda(s) ds = -\frac{1}{t} \ln[R(t)], \ t \in \mathbb{R}_+. \tag{5.15}$$

The continuous lifetime distributions can then be classified in an Increasing Failure Rate Average (IFRA) class and a Decreasing Failure Rate Average (DFRA) class. The properties of systems whose components have lifetimes belonging to the IFRA or DFRA class have been widely studied (see, e.g., Barlow and Prochan, 1975; Roy and Gupta, 1992). For example, for a continuous-time parallel system, with independent components whose lifetimes belong to the IFRA class, it can be shown that the lifetime distribution of the system also belongs to the IFRA class. This example (a particular case of the *IFRA Closure Theorem*) shows the importance of having a classification of lifetime distribution based on the failure rate average.

When working in *discrete-time* reliability, one can immediately check that the two expressions that give the failure rate average in continuous time (cf. Equation (5.15)) are no longer equal. Indeed, set

$$\Lambda(k) := \frac{1}{k+1} \sum_{i=0}^{k} \lambda(i),$$

$$\widetilde{\Lambda}(k) := -\frac{1}{k+1} \ln[R(k)] = -\frac{1}{k+1} \sum_{i=0}^{k} \ln[1 - \lambda(i)],$$

and note that $\Lambda(k) < \widetilde{\Lambda}(k)$ for all $k \in \mathbb{N}^*$. Consequently, in discrete time there are two different ways of defining the IFRA and DFRA lifetime distribution classes, depending on the definition taken into account for failure rate average ($\Lambda(k)$ or $\widetilde{\Lambda}(k)$).

All these reasons led Roy and Gupta (1992) to propose a new discrete-time failure rate function $r(k)$ such that

$$\sum_{i=0}^{k} r(i) = -\ln[R(k)], \ k \in \mathbb{N}.$$

Expressing $r(k)$ in terms of the reliability R we obtain that the RG-failure rate function for a discrete-time system is given by

$$r(k) := \begin{cases} -\ln \frac{R(k)}{R(k-1)}, & \text{if } k \geq 1, \\ -\ln R(0), & \text{if } k = 0, \end{cases} \tag{5.16}$$

for $R(k) \neq 0$. If $R(k) = 0$, we set $r(k) := 0$.

Note that the two failure rate functions are related by

$$r(k) = -\ln(1 - \lambda(k)), \ k \in \mathbb{N}. \tag{5.17}$$

Note also that for the RG-failure rate $r(k)$, the problems raised by the use of $\lambda(k)$ in discrete time are solved: $r(k)$ is not bounded and it can be a convex function; $r(k)$ is additive for series systems; defining $\Lambda(k) := 1/(k + 1) \sum_{i=0}^{k} r(i)$, we have $\Lambda(k) = \widetilde{\Lambda}(k)$ and there is only one way of defining IFRA (DFRA) classes of distributions with respect to the RG-failure rate.

5.1.5 Mean Hitting Times

There are various mean times that are interesting for the reliability analysis of a system. We will be concerned here only with the mean time to failure, mean time to repair, mean up time, mean down time, and mean time between failures.

The *mean time to failure* (MTTF) is defined as the mean lifetime, i.e., the expectation of the hitting time to down set D,

$$MTTF = \mathbb{E}[T_D].$$

For any state $i \in U$, we introduce:

- $MTTF_i := \mathbb{E}_i[T_D]$ is the MTTF of the system, given that it starts in state $i \in U$;
- $\mathbf{MTTF} := (MTTF_1, \ldots, MTTF_{s_1})^\top$ is the column vector of the conditional MTTFs.

As we showed for reliability, the following relation holds true for the previous quantities:

$$MTTF = \sum_{i \in U} \alpha_i MTTF_i = \boldsymbol{\alpha}_1 \mathbf{MTTF}. \tag{5.18}$$

Symmetrically, the *mean time to repair* (MTTR) is defined as the mean of the repair duration, i.e., the expectation of the hitting time to up set U,

$$MTTR = \mathbb{E}[T_U].$$

Remark 5.4. Note that, when defining the MTTF, we can consider two cases: either the initial distribution is concentrated on the up states, $\boldsymbol{\alpha}_2 = \mathbf{0}$, which means that the system is in perfect state when starting to work (or, equivalently, the repair was perfect) or $\boldsymbol{\alpha}_2 \neq \mathbf{0}$. In the first case, the reliability at the initial moment equals 1, $R(0) = 1$, whereas in the second case we have $R(0) < 1$. Similarly, when defining the MTTR we also have two cases: either $\boldsymbol{\alpha}_1 = \mathbf{0}$ or $\boldsymbol{\alpha}_1 \neq \mathbf{0}$.

Proposition 5.5. *If the matrices* $\mathbf{I} - \mathbf{p}_{11}$ *and* $\mathbf{I} - \mathbf{p}_{22}$ *are nonsingular, then*

$$MTTF = \boldsymbol{\alpha}_1(\mathbf{I} - \mathbf{p}_{11})^{-1}\mathbf{m}_1, \tag{5.19}$$

$$MTTR = \boldsymbol{\alpha}_2(\mathbf{I} - \mathbf{p}_{22})^{-1}\mathbf{m}_2, \tag{5.20}$$

where $\mathbf{m} = (\mathbf{m}_1 \quad \mathbf{m}_2)^\top$ *is the partition of the mean sojourn times vector corresponding to the partition of state space* E *in up states* U *and down states* D. *If the matrices* $\mathbf{I} - \mathbf{p}_{11}$ *or* $\mathbf{I} - \mathbf{p}_{22}$ *are singular, we set* $MTTF = \infty$ *or* $MTTR = \infty$.

Necessary conditions of nonsingularity of matrices $\mathbf{I} - \mathbf{p}_{11}$ and $\mathbf{I} - \mathbf{p}_{22}$ are provided in Exercise 5.7.

Proof. The idea of the proof is based on the Markov renewal argument (like the proof of Proposition 5.1): first we find the Markov renewal equation associated to MTTF; second we solve this equation (cf. Theorem 3.1) and we obtain the explicit form of the MTTF. A different proof is proposed in Exercise 5.6. For any state $i \in U$ we have

$$MTTF_i := \mathbb{E}_i[T_D] = \sum_{k \geq 0} \mathbb{P}_i[T_D > k] = \sum_{k \geq 0} R_i(k).$$

For any up state $i \in U$ and positive integer k, consider the Markov renewal equation associated to $R_i(k)$ (Equation (5.7)) and sum over all $k \in \mathbb{N}$. We get

$$\sum_{k \geq 0} R_i(k) = \sum_{k \geq 0}(1 - H_i(k)) + \sum_{k \geq 0}\sum_{j \in U}\sum_{m=1}^{k} q_{ij}(m)R_j(k - m), \tag{5.21}$$

and by changing the order of summation we get

$$MTTF_i = m_i + \sum_{j \in U}\sum_{m \geq 0} q_{ij}(m)\sum_{k \geq m} R_j(k - m),$$

$$MTTF_i = m_i + \sum_{j \in U} p_{ij} MTTF_j. \tag{5.22}$$

Writing Equation (5.22) for all $i \in U$ we obtain the vector equation of **MTTF**

$$\mathbf{MTTF} = \mathbf{m}_1 + \mathbf{p}_{11}\mathbf{MTTF}. \qquad (5.23)$$

For $\mathbf{I} - \mathbf{p}_{11}$ nonsingular we solve this equation and we obtain

$$\mathbf{MTTF} = (\mathbf{I} - \mathbf{p}_{11})^{-1}\mathbf{m}_1,$$

which, using Equation (5.18), yields the desired result for MTTF. A symmetric argument is used for obtaining the MTTR. □

Note that Equation (5.22) has a very intuitive meaning and could have been obtained directly: $MTTF_i$ equals the mean time in state i, m_i, plus the mean time to failure counting from the first jump time to an up state.

Define the row vector $\boldsymbol{\beta} = (\beta(1), \ldots, \beta(s_1))$, with $\beta(j)$ the probability that the system will be in state $j \in U$, given that it has just entered U, when the system is in steady state. We have

$$\beta(j) = \mathbb{P}_\nu(J_n = j | J_{n-1} \in D, J_n \in U) = \frac{\displaystyle\sum_{i \in D} p_{ij}\nu(i)}{\displaystyle\sum_{l \in U}\sum_{i \in D} p_{il}\nu(i)}, \ j \in U,$$

where $\boldsymbol{\nu}$ is the stationary distribution of the EMC $(J_n)_{n \in \mathbb{N}}$. In matrix form, $\boldsymbol{\beta}$ can be written as

$$\boldsymbol{\beta} = \frac{\boldsymbol{\nu}_2\mathbf{p}_{21}}{\boldsymbol{\nu}_2\mathbf{p}_{21}\mathbf{1}_{s_1}}. \qquad (5.24)$$

Similarly, we consider the row vector $\boldsymbol{\gamma} = (\gamma(1), \ldots, \gamma(s - s_1))$, with $\gamma(j)$ the probability that the system is in state $j \in D$, given that it has just entered D, when the system is in steady state. We have

$$\gamma(j) = \mathbb{P}_\nu(J_n = j | J_{n-1} \in U, J_n \in D) = \frac{\displaystyle\sum_{i \in U} p_{ij}\nu(i)}{\displaystyle\sum_{l \in D}\sum_{i \in U} p_{il}\nu(i)}, j \in D.$$

In matrix form, $\boldsymbol{\gamma}$ can be written as

$$\boldsymbol{\gamma} = \frac{\boldsymbol{\nu}_1\mathbf{p}_{12}}{\boldsymbol{\nu}_1\mathbf{p}_{12}\mathbf{1}_{s-s_1}}. \qquad (5.25)$$

The *mean up time* (MUT) and the *mean down time* (MDT) are defined by

$$MUT = \mathbb{E}_{\boldsymbol{\beta}}(T_D)$$

and

$$MDT = \mathbb{E}_{\boldsymbol{\gamma}}(T_U).$$

Proposition 5.6. *For a discrete-time semi-Markov system the MUT and MDT can be expressed as follows:*

$$MUT = \frac{\boldsymbol{\nu}_1 \mathbf{m}_1}{\boldsymbol{\nu}_2 \mathbf{p}_{21} \mathbf{1}_{s_1}} \tag{5.26}$$

and

$$MDT = \frac{\boldsymbol{\nu}_2 \mathbf{m}_2}{\boldsymbol{\nu}_1 \mathbf{p}_{12} \mathbf{1}_{s-s_1}}. \tag{5.27}$$

Proof. From equation

$$\begin{pmatrix} \boldsymbol{\nu}_1 & \boldsymbol{\nu}_2 \end{pmatrix} \begin{pmatrix} \mathbf{p}_{11} & \mathbf{p}_{12} \\ \mathbf{p}_{21} & \mathbf{p}_{22} \end{pmatrix} = \begin{pmatrix} \boldsymbol{\nu}_1 & \boldsymbol{\nu}_2 \end{pmatrix}$$

we get

$$\boldsymbol{\nu}_2 \mathbf{p}_{21} = \boldsymbol{\nu}_1 (\mathbf{I} - \mathbf{p}_{11}), \quad \boldsymbol{\nu}_1 \mathbf{p}_{12} = \boldsymbol{\nu}_2 (\mathbf{I} - \mathbf{p}_{22}).$$

Thus, we can write the probabilities $\boldsymbol{\beta}$ and $\boldsymbol{\gamma}$ as follows:

$$\boldsymbol{\beta} = \frac{\boldsymbol{\nu}_1 (\mathbf{I} - \mathbf{p}_{11})}{\boldsymbol{\nu}_2 \mathbf{p}_{21} \mathbf{1}_{s_1}}, \; \boldsymbol{\gamma} = \frac{\boldsymbol{\nu}_2 (\mathbf{I} - \mathbf{p}_{22})}{\boldsymbol{\nu}_1 \mathbf{p}_{12} \mathbf{1}_{s-s_1}}.$$

Finally, replacing the initial distribution $\boldsymbol{\alpha}_1$ by $\boldsymbol{\beta}$ in the expression of MTTF, we obtain the desired result for the MUT. Replacing the initial distribution $\boldsymbol{\alpha}_2$ by $\boldsymbol{\gamma}$ in the expression of MTTR and using the same argument as above, we get the result for MDT. □

Another mean time of interest in reliability studies is the *mean time between failures* (MTBF) defined by

$$MTBF = MUT + MDT = \mathbb{E}_{\boldsymbol{\beta}}(T_D) + \mathbb{E}_{\boldsymbol{\gamma}}(T_U).$$

The explicit form for the MTBF of a discrete-time semi-Markov system can be immediately obtained from Proposition 5.6.

Remark 5.5. For defining the MUT, we introduced the initial probabilities $\beta(j), j \in E$, which actually represent the probabilities that the system is in state $j \in U$, given that it has just entered U. As we supposed that the system is in steady state, a natural question is why not use the time-stationary probabilities $\pi(j)$ instead of $\beta(j)$. The answer is that the initial probabilities we need for defining the MUT must keep track of the fact that the passage from the down-set D to the up-set U has just happened. Obviously, the probabilities $\boldsymbol{\beta}$ that we introduced answer this need, whereas the time-stationary probabilities are not appropriate.

For symmetrical reasons, the same remark holds true for the probabilities $\gamma(j), j \in E$, introduced for defining the MDT.

Remark 5.6. At the end of this part of reliability modeling, we would like to say a few words about the systems that we took into consideration. As stated at the beginning of the chapter, we considered systems whose state space E is partitioned into the up states U and the down states D. Note that we do not have here just a binary system, and all the theory can be applied to more general situations. For instance, if one is interested in the the time spent by the system before entering some specific degraded states, one just needs to define the down states D as the set of degraded states of interest, and the reliability (computed in this chapter) provides a way of computing this time, i.e., $1 - R(k) = \mathbb{P}(T_D \leq k)$ is just the cumulative distribution function of T_D. In other words, although we have chosen to speak of the system in terms of up states U and down states D, it actually covers a larger class of systems and of practical problems.

5.2 Nonparametric Estimation of Reliability and Asymptotic Properties

In this section we present estimators for reliability indicators and we study their asymptotic properties. As we already saw in Chapter 4, the asymptotic normality of the estimators can be proved by two different techniques, namely, the CLT for Markov renewal chains or the Lindeberg–Lévy CLT for martingales. We choose here to use the second method. The reader interested in deriving the results by means of the first method, can do it using Lemmas A.1–A.4.

As we have seen thus far, all the reliability indicators are expressed as functions of the semi-Markov kernel $\mathbf{q}$ and of its powers in the convolution sense. Consequently, estimators of reliability indicators are immediately obtained using the proposed estimators of Chapter 4.

Using Equation (5.3) we have the following estimator for the system's reliability

$$\widehat{R}(k, M) := \boldsymbol{\alpha}_1 \cdot \widehat{\mathbf{P}}_{11}(k, M) \cdot \mathbf{1}_{s_1}$$
$$= \boldsymbol{\alpha}_1 \left[\widehat{\boldsymbol{\psi}}_{11}(\cdot, M) * \left(\mathbf{I} - diag(\widehat{\mathbf{Q}}(\cdot, M) \cdot \mathbf{1})_{11} \right) \right](k) \mathbf{1}_{s_1}, \quad (5.28)$$

where the estimators $\widehat{\boldsymbol{\psi}}$ and $\widehat{\mathbf{Q}}$ are respectively defined in (4.21) and (4.20).

The following theorem concerns the consistency and the asymptotic normality of the reliability estimator.

Theorem 5.1. *For any fixed arbitrary positive integer $k \in \mathbb{N}$, the estimator of the reliability of a discrete-time semi-Markov system at instant k is strongly consistent and asymptotically normal, in the sense that*

$$\widehat{R}(k, M) \xrightarrow[M \to \infty]{a.s.} R(k)$$

and

$$\sqrt{M}[\widehat{R}(k, M) - R(k)] \xrightarrow[M \to \infty]{\mathcal{D}} \mathcal{N}(0, \sigma_R^2(k)),$$

with the asymptotic variance

$$\sigma_R^2(k) = \sum_{i=1}^{s} \mu_{ii} \Big\{ \sum_{j=1}^{s} \Big[D_{ij}^U - \mathbf{1}_{\{i \in U\}} \sum_{t \in U} \alpha_t \Psi_{ti} \Big]^2 * q_{ij}(k)$$

$$- \Big[\sum_{j=1}^{s} \Big(D_{ij}^U * q_{ij} - \mathbf{1}_{\{i \in U\}} \sum_{t \in U} \alpha_t \psi_{ti} * Q_{ij} \Big) \Big]^2 (k) \Big\}, \qquad (5.29)$$

where

$$D_{ij}^U := \sum_{n \in U} \sum_{r \in U} \alpha_n \psi_{ni} * \psi_{jr} * \Big(\mathbf{I} - diag(\mathbf{Q} \cdot \mathbf{1}) \Big)_{rr}. \qquad (5.30)$$

Proof. The strong consistency of the reliability estimator is obtained from the strong consistency of the semi-Markov transition function estimator $\widehat{P}_{ij}(k, M)$ (Theorem 4.6) and from the following inequality:

$$\Big| \widehat{R}(k, M) - R(k) \Big| = \Big| \sum_{i \in U} \sum_{j \in U} \alpha_i \widehat{P}_{ij}(k, M) - \sum_{i \in E} \sum_{j \in U} \alpha_i P_{ij}(k) \Big|$$

$$\leq \sum_{i \in U} \sum_{j \in U} \alpha_i \Big| \widehat{P}_{ij}(k, M) - P_{ij}(k) \Big|.$$

The proof of the asymptotic normality is similar to the proof of Theorems 4.5 and 4.7, so we will give only the main steps. First, using the proof of Theorem 4.5 we can show that $\sqrt{M}[\widehat{R}(k, M) - R(k)]$ has the same limit in distribution as

$$\sqrt{M} \sum_{r=1}^{s} \sum_{l=1}^{s} \Big[\sum_{i \in U} \sum_{j \in U} \alpha_i \psi_{ir} * \psi_{lj} * \Big(\mathbf{I} - diag(\mathbf{Q} \cdot \mathbf{1}) \Big)_{jj} \Big] * \Delta q_{rl}(k)$$

$$- \sqrt{M} \sum_{i \in U} \sum_{j \in U} \alpha_i \sum_{l=1}^{s} \psi_{ij} * \Delta Q_{jl}(k)$$

$$= \sqrt{M} \sum_{r=1}^{s} \sum_{l=1}^{s} D_{rl}^U * \Delta q_{rl}(k) - \sqrt{M} \sum_{r \in U} \sum_{l=1}^{s} \Big(\sum_{n \in U} \alpha_n \psi_{nr} \Big) * \Delta Q_{rl}(k).$$

Since $N_l(M)/M \xrightarrow[M \to \infty]{a.s.} 1/\mu_{ll}$ (Proposition 3.8), $\sqrt{M}[\widehat{R}(k, M) - R(k)]$ has the same limit in distribution as

$$\frac{1}{\sqrt{M}} \sum_{n=1}^{N(M)} \sum_{r=1}^{s} \sum_{l=1}^{s} \mu_{rr} \Big[D_{rl}^{U} * \Big(\mathbf{1}_{\{J_{n-1}=r, J_n=l, X_n=\cdot\}} - q_{rl}(\cdot)\mathbf{1}_{\{J_{n-1}=r\}} \Big)(k)$$

$$-\mathbf{1}_{\{r\in U\}} \Big(\sum_{t\in U} \alpha_t \psi_{tr} \Big) * \Big(\mathbf{1}_{\{J_{n-1}=r, J_n=l, X_n\le\cdot\}} - Q_{rl}(\cdot)\mathbf{1}_{\{J_{n-1}=r\}} \Big)(k) \Big]$$

$$= \sqrt{\frac{N(M)}{M}} \frac{1}{\sqrt{N(M)}} \sum_{n=1}^{N(M)} Y_n,$$

where we have defined the random variables Y_n by

$$Y_n := \sum_{r=1}^{s} \sum_{l=1}^{s} \mu_{rr} \Big[D_{rl}^{U} * \Big(\mathbf{1}_{\{J_{n-1}=r, J_n=l, X_n=\cdot\}} - q_{rl}(\cdot)\mathbf{1}_{\{J_{n-1}=r\}} \Big)(k)$$

$$-\mathbf{1}_{\{r\in U\}} \Big(\sum_{t\in U} \alpha_t \psi_{tr} \Big) * \Big(\mathbf{1}_{\{J_{n-1}=r, J_n=l, X_n\le\cdot\}} - Q_{rl}(\cdot)\mathbf{1}_{\{J_{n-1}=r\}} \Big)(k) \Big].$$

Set $\mathcal{F}_n$ for the σ-algebra $\mathcal{F}_n := \sigma(J_l, X_l; l \le n)$, $n \in \mathbb{N}$. Using the fact that for all $l, u \in E$ and $k \in \mathbb{N}$ we have

$$\mathbb{E}(\mathbf{1}_{\{J_{n-1}=l, J_n=u, X_n=k\}} - q_{lu}(k)\mathbf{1}_{\{J_{n-1}=l\}} \mid \mathcal{F}_{n-1}) = 0,$$
$$\mathbb{E}(\mathbf{1}_{\{J_{n-1}=l, J_n=u, X_n\le k\}} - Q_{lu}(k)\mathbf{1}_{\{J_{n-1}=l\}} \mid \mathcal{F}_{n-1}) = 0,$$

we immediately obtain that $\mathbb{E}(Y_n \mid \mathcal{F}_{n-1}) = 0$. Thus, $(Y_n)_{n\in\mathbb{N}}$ is an $\mathcal{F}_n$-martingale difference and $(\sum_{l=1}^{n} Y_l)_{n\in\mathbb{N}}$ is an $\mathcal{F}_n$-martingale.

As $N(M)/M \xrightarrow[M\to\infty]{a.s.} 1/\nu(l)\mu_{ll}$ (Convergence (3.38)) and

$$\frac{1}{n} \sum_{l=1}^{n} \mathbb{E}(Y_l^2 \mathbf{1}_{\{|Y_l|>\epsilon\sqrt{n}\}}) \xrightarrow[n\to\infty]{} 0$$

for any $\epsilon > 0$, using the CLT for martingales (Theorem E.4) and Anscombe's theorem (Theorem E.6) we obtain that

$$\sqrt{M}[\widehat{R}(k, M) - R(k)] \xrightarrow[M\to\infty]{\mathcal{D}} \mathcal{N}(0, \sigma^2/(\nu(r)\mu_{rr})), \qquad (5.31)$$

with σ^2 given by

$$\sigma^2 = \lim_{n\to\infty} \frac{1}{n} \sum_{m=1}^{n} \mathbb{E}(Y_m^2 \mid \mathcal{F}_{m-1}). \qquad (5.32)$$

After computing Y_m^2, $\mathbb{E}(Y_m^2 \mid \mathcal{F}_{m-1})$ and taking the limit in (5.32), we obtain that $\sqrt{M}[\widehat{R}(k, M) - R(k)]$ converges in distribution, as M tends to infinity, to a normal random variable with zero mean and variance $\sigma_R^2(k)$ given by (5.29). $\qquad \square$

We propose the following estimator for the system's availability:

$$\widehat{A}(k, M) := \boldsymbol{\alpha} \, \widehat{\mathbf{P}}(k, M) \, \mathbf{1}_{s,s_1}$$
$$= \boldsymbol{\alpha} \left[\widehat{\psi}(\cdot, M) * \left(\mathbf{I} - diag(\widehat{\mathbf{Q}}(\cdot, M) \cdot \mathbf{1}) \right) \right] (k) \, \mathbf{1}_{s,s_1}, \quad (5.33)$$

where the estimators $\widehat{\psi}$ and $\widehat{\mathbf{Q}}$ are respectively defined in (4.21) and (4.20).

The following result concerns the consistency and the asymptotic normality of the availability estimator. The proof of the consistency is almost identical to the proof of the consistency of $\widehat{R}$ (Theorem 5.1), whereas the asymptotic normality is obtained following the same steps as the proof of Theorem 5.1, so we do not prove these results here. The proof of the normality part is given in Appendix C.

Theorem 5.2. *For any fixed arbitrary positive integer $k \in \mathbb{N}$, the estimator of the availability of a discrete-time semi-Markov system at instant k is strongly consistent and asymptotically normal, in the sense that*

$$\widehat{A}(k, M) \xrightarrow[M \to \infty]{a.s.} A(k)$$

and

$$\sqrt{M} \left[\widehat{A}(k, M) - A(k) \right] \xrightarrow[M \to \infty]{\mathcal{D}} \mathcal{N}(0, \sigma_A^2(k)),$$

with the asymptotic variance

$$\sigma_A^2(k) = \sum_{i=1}^{s} \mu_{ii} \Big\{ \sum_{j=1}^{s} \Big[D_{ij} - \mathbf{1}_{\{i \in U\}} \sum_{t=1}^{s} \alpha_t \Psi_{ti} \Big]^2 * q_{ij}(k)$$
$$- \Big[\sum_{j=1}^{s} \Big(D_{ij} * q_{ij} - \mathbf{1}_{\{i \in U\}} \sum_{t=1}^{s} \alpha_t \psi_{ti} * Q_{ij} \Big) \Big]^2 (k) \Big\}, \quad (5.34)$$

where

$$D_{ij} := \sum_{n=1}^{s} \sum_{r \in U} \alpha_n \psi_{ni} * \psi_{jr} * \left(\mathbf{I} - diag(\mathbf{Q} \cdot \mathbf{1}) \right)_{rr}.$$

Using the expressions of the failure rates for a discrete-time semi-Markov system obtained in Equations (5.14) and (5.16), we have the following expressions for the estimators of the failure rates at time $k \in \mathbb{N}$:

$$\widehat{\lambda}(k, M) := \begin{cases} 1 - \dfrac{\widehat{R}(k,M)}{\widehat{R}(k-1,M)}, & \text{if } k \geq 1 \text{ and } \widehat{R}(k-1, M) \neq 0, \\ 1 - \widehat{R}(0, M), & \text{if } k = 0, \\ 0, & \text{otherwise.} \end{cases}$$

and

$$\widehat{r}(k, M) := \begin{cases} -\ln \dfrac{\widehat{R}(k,M)}{\widehat{R}(k-1,M)}, & \text{if } k \geq 1 \text{ and } \widehat{R}(k, M) \neq 0, \\ -\ln \widehat{R}(0, M), & \text{if } k = 0 \text{ and } \widehat{R}(0, M) \neq 0, \\ 0, & \text{otherwise.} \end{cases}$$

The following results concern the strong consistency and the asymptotic normality of failure rate estimators.

Theorem 5.3. *For any fixed arbitrary positive integer $k \in \mathbb{N}$, the estimators of the failure rates of a discrete-time semi-Markov system at instant k are strongly consistent, in the sense that*

$$\widehat{\lambda}(k, M) \xrightarrow[M\to\infty]{a.s.} \lambda(k), \quad \widehat{r}(k, M) \xrightarrow[M\to\infty]{a.s.} r(k).$$

Proof. We have

$$\left| \widehat{\lambda}(k, M) - \lambda(k) \right|$$

$$= \left| \frac{\widehat{R}(k, M) - R(k)}{\widehat{R}(k-1, M)} - \frac{R(k)}{R(k-1)} \frac{\widehat{R}(k-1, M) - R(k-1)}{\widehat{R}(k-1, M)} \right|$$

$$\leq \frac{\left| \widehat{R}(k, M) - R(k) \right|}{\widehat{R}(k-1, M)} + \frac{R(k)}{R(k-1)} \frac{\left| \widehat{R}(k-1, M) - R(k-1) \right|}{\widehat{R}(k-1, M)}.$$

From the strong consistency of the reliability estimator (Theorem 5.1) we obtain the desired result for $\widehat{\lambda}(k, M)$. Using Relation (5.17) between the BMP-failure rate and the RG-failure rate we infer the consistency of $\widehat{r}(k, M)$. $\square$

For a matrix function $\mathbf{A} \in \mathcal{M}_E(\mathbb{N})$, we denote by $\mathbf{A}^+ \in \mathcal{M}_E(\mathbb{N})$ the matrix function defined by $\mathbf{A}^+(k) := \mathbf{A}(k+1)$, $k \in \mathbb{N}$.

Theorem 5.4. *For any fixed arbitrary positive integer $k \in \mathbb{N}$, we have*

$$\sqrt{M}[\widehat{\lambda}(k, M) - \lambda(k)] \xrightarrow[M\to\infty]{\mathcal{D}} \mathcal{N}(0, \sigma_\lambda^2(k)),$$

with the asymptotic variance

$$\sigma_\lambda^2(k) = \frac{1}{R^4(k-1)} \sigma_1^2(k),$$

$$\sigma_1^2(k) = \sum_{i=1}^{s} \mu_{ii} \Bigg\{ R^2(k) \sum_{j=1}^{s} \Big[D_{ij}^U - \mathbf{1}_{\{i \in U\}} \sum_{t \in U} \alpha_t \Psi_{ti} \Big]^2 * q_{ij}(k-1)$$

$$+ R^2(k-1) \sum_{j=1}^{s} \Big[D_{ij}^U - \mathbf{1}_{\{i \in U\}} \sum_{t \in U} \alpha_t \Psi_{ti} \Big]^2 * q_{ij}(k) - T_i^2(k)$$

$$+ 2R(k-1)R(k) \sum_{j=1}^{s} \Big[\mathbf{1}_{\{i \in U\}} D_{ij}^U \sum_{t \in U} \alpha_t \Psi_{ti}^+ + \mathbf{1}_{\{i \in U\}} (D_{ij}^U)^+ \sum_{t \in U} \alpha_t \Psi_{ti}$$

$$- (D_{ij}^U)^+ D_{ij}^U - \mathbf{1}_{\{i \in U\}} \Big(\sum_{t \in U} \alpha_t \Psi_{ti} \Big) \Big(\sum_{t \in U} \alpha_t \Psi_{ti}^+ \Big) \Big] * q_{ij}(k-1) \Bigg\} \quad (5.35)$$

where

$$T_i(k) := \sum_{j=1}^{s} \Big[R(k)D_{ij}^{U} * q_{ij}(k-1) - R(k-1)D_{ij}^{U} * q_{ij}(k)$$

$$- R(k)\mathbf{1}_{\{i \in U\}} \sum_{t \in U} \alpha_t \psi_{ti} * Q_{ij}(k-1)$$

$$+ R(k-1)\mathbf{1}_{\{i \in U\}} \sum_{t \in U} \alpha_t \psi_{ti} * Q_{ij}(k) \Big],$$

and D_{ij}^{U} is given in Equation (5.30).

Proof. We give here only the main steps of the proof, based on the CLT for martingales. A complete proof, using the CLT for Markov renewal chains, can be found in Appendix C.

First, using the consistency of the reliability estimator (Theorem 5.1), we can show that we need only to prove that

$$\sqrt{M}\Big[R(k)\Big(\widehat{R}(k-1, M) - R(k-1)\Big) - \Big(\widehat{R}(k, M) - R(k)\Big)R(k-1)\Big]$$

converges in distribution, as M tends to infinity, to a zero-mean normal random variable of variance $\sigma_1^2(k)$. Second, we obtain that

$$\sqrt{M}\Big[R(k)\Big(\widehat{R}(k-1, M) - R(k-1)\Big) - \Big(\widehat{R}(k, M) - R(k)\Big)R(k-1)\Big]$$

has the same limit in distribution as

$$\sqrt{\frac{N(M)}{M}} \frac{1}{\sqrt{N(M)}} \sum_{n=1}^{N(M)} Y_n,$$

where the random variables Y_n are defined by

$$Y_n := \sum_{r=1}^{s} \sum_{l=1}^{s} \mu_{rr} \Big[R(k)D_{rl}^{U} * \Big(\mathbf{1}_{\{J_{n-1}=r, J_n=l, X_n=\cdot\}} - q_{rl}(\cdot)\mathbf{1}_{\{J_{n-1}=r\}} \Big)(k-1)$$

$$- R(k)\mathbf{1}_{\{r \in U\}} \Big(\sum_{t \in U} \alpha_t \psi_{tr} \Big)$$

$$* \Big(\mathbf{1}_{\{J_{n-1}=r, J_n=l, X_n \le \cdot\}} - Q_{rl}(\cdot)\mathbf{1}_{\{J_{n-1}=r\}} \Big)(k-1)$$

$$- R(k-1)D_{rl}^{U} * \Big(\mathbf{1}_{\{J_{n-1}=r, J_n=l, X_n=\cdot\}} - q_{rl}(\cdot)\mathbf{1}_{\{J_{n-1}=r\}} \Big)(k)$$

$$- R(k-1)\mathbf{1}_{\{r \in U\}} \Big(\sum_{t \in U} \alpha_t \psi_{tr} \Big)$$

$$* \Big(\mathbf{1}_{\{J_{n-1}=r, J_n=l, X_n \le \cdot\}} - Q_{rl}(\cdot)\mathbf{1}_{\{J_{n-1}=r\}} \Big)(k) \Big].$$

We consider the filtration $\mathcal{F}_n := \sigma(J_l, X_l; l \leq n), n \in \mathbb{N}$, and we show that $(Y_n)_{n \in \mathbb{N}}$ is an $\mathcal{F}_n$-martingale difference, i.e., $(\sum_{l=1}^n Y_l)_{n \in \mathbb{N}}$ is an $\mathcal{F}_n$-martingale. After computing Y_l^2 and $\mathbb{E}(Y_l^2 \mid \mathcal{F}_{l-1})$ we apply the CLT for martingales (Theorem E.4) and Anscombe's theorem (Theorem E.6) and we obtain the desired result.   □

The asymptotic normality of the RG-failure rate estimation is a direct consequence of the previous result.

Corollary 5.1. *For any fixed arbitrary positive integer $k \in \mathbb{N}$ we have*

$$\sqrt{M}[\widehat{r}(k, M) - r(k)] \xrightarrow[M \to \infty]{\mathcal{D}} \mathcal{N}(0, \sigma_r^2(k)),$$

with the asymptotic variance

$$\sigma_r^2(k) = \frac{1}{(1 - \lambda(k))^2} \sigma_\lambda^2(k) = \frac{1}{R^2(k-1)R^2(k)} \sigma_1^2(k),$$

where $\sigma_1^2(k)$ is given in Equation (5.35).

Proof. Relation (5.17) between the BMP-failure rate and the RG-failure rate can be written in the form $r(k) = \phi(\lambda(k))$, where ϕ is defined by $\phi(x) := -\ln(1-x)$. Using the delta method (Theorem E.7) and the asymptotic normality of the BMP-failure rate (Theorem 5.4), we obtain that $\sqrt{M}[\widehat{r}(k, M) - r(k)]$ converges in distribution, as M tends to infinity, to a zero-mean normal random variable, with the variance $\sigma_r^2(k)$ given by

$$\sigma_r^2(k) = \left(\phi'(\lambda(k))\right)^2 \sigma_\lambda^2(k) = \frac{1}{(1 - \lambda(k))^2} \sigma_\lambda^2(k),$$

which finishes the proof.   □

From the expressions of MTTF and MTTR obtained in Equations (5.19) and (5.20), we get the corresponding plug-in estimators. Indeed, for $(\mathbf{I} - \widehat{\mathbf{p}}_{11}(M))$ and $(\mathbf{I} - \widehat{\mathbf{p}}_{11}(M))$ nonsingular, the MTTF and MTTR estimators are given by:

- $\widehat{MTTF}(M) := \boldsymbol{\alpha}_1(\mathbf{I} - \widehat{\mathbf{p}}_{11}(M))^{-1}\widehat{\mathbf{m}}_1(M),$
- $\widehat{MTTR}(M) := \boldsymbol{\alpha}_2(\mathbf{I} - \widehat{\mathbf{p}}_{22}(M))^{-1}\widehat{\mathbf{m}}_2(M),$

where, for all state $i \in E$,

$$\widehat{m}_i(M) := \sum_{n \geq 0} \left(1 - \widehat{H}_i(n, M)\right) = \sum_{n \geq 0} \left(1 - \sum_{j \in E} \sum_{k=1}^n \widehat{q}_{ij}(k, M)\right). \quad (5.36)$$

We present in the sequel the consistency and asymptotic normality of the MTTF estimator. Using the symetry between MTTF and MTTR, one can immediately state the analogous result which holds true for the MTTR estimator.

Theorem 5.5. *Consider a discrete-time semi-Markov system and suppose that there exists an integer $M_1 \in \mathbb{N}$ such that $(I - \widehat{\mathbf{p}}_{11}(M))$ is nonsingular for all $M \geq M_1$. Then, the MTTF estimator defined above is strongly consistent and asymptotically normal, in the sense that for all $M \geq M_1$ we have*

$$\widehat{MTTF}(M) \xrightarrow[M \to \infty]{a.s.} MTTF$$

and

$$\sqrt{M}[\widehat{MTTF}(M) - MTTF] \xrightarrow[M \to \infty]{\mathcal{D}} \mathcal{N}(0, \sigma^2_{MTTF})$$

with the asymptotic variance

$$\sigma^2_{MTTF} = \sum_{i \in U, j \in U} \alpha_i^2 m_i^2 \left[\sum_{r \in U} a_{ir}^2 p_{rj}(1 - p_{rj}) + \sum_{r \in U} a_{rj}^2 p_{ir}(1 - p_{ir}) \right.$$

$$\left. + \sum_{l,r \in U} a_{il}^2 a_{rj}^2 p_{lr}(1 - p_{lr}) + a_{ij}^2 \sigma_j^2 \right], \tag{5.37}$$

where, for all $i, j \in E$, σ_j^2 is the variance of the sojourn time in state j and $a_{ij} = (I - \mathbf{p}_{11})_{ij}^{-1}$.

Proof. For the proof of the theorem, see Limnios and Ouhbi (2006).   □

The asymptotic normality of the estimators of reliability, availability, failure rates, and MTTF (Theorems 5.1, 5.2, 5.4, 5.5, and Corollary 5.1) allows us to construct the corresponding asymptotic confidence intervals. The mechanism is exactly like that used for the asymptotic confidence interval of the semi-Markov transition matrix $\mathbf{P}$ in Chapter 4, and the commentary made there on the estimator of the asymptotic variance holds true here too.

5.3 Example: a Textile Factory (Continuation)

Let us continue the example of the textile factory presented in Chapter 3, Section 3.6 and Chapter 4, Section 4.3. We have already proposed a semi-Markov modelization for this system and we have shown how to obtain the quantities of interest in terms of the semi-Markov kernel and how to estimate these quantities. We are concerned now with the reliability analysis of the textile factory.

Consider that the set of working states is $U = \{1, 2\}$ and the set of failure states is $D = \{3\}$. First, we obtain explicit expressions for the reliability indicators. Second, we illustrate via simulations the good qualities of the estimators.

Using the expression of the reliability given in Proposition 5.1, we obtain that the reliability of the system at time $k, k \in \mathbb{N}$ is

$$R(k) = \boldsymbol{\alpha}_1 (\boldsymbol{\delta I} - \mathbf{q}_{11})^{(-1)} * (\mathbf{I} - \mathbf{H}_1)(k) \begin{pmatrix} 1 \\ 1 \end{pmatrix}$$

$$= \begin{pmatrix} \alpha_1 & \alpha_2 \end{pmatrix} \begin{pmatrix} \mathbf{1}_{\{0\}} & -f_{12} \\ -af_{21} & \mathbf{1}_{\{0\}} \end{pmatrix}^{(-1)} (\cdot)$$

$$* \begin{pmatrix} 1 - \displaystyle\sum_{l=1}^{\cdot} f_{12}(l) & 0 \\ 0 & 1 - \displaystyle\sum_{l=1}^{\cdot}(af_{21}(l) + bf_{23}(l)) \end{pmatrix} (k) \begin{pmatrix} 1 \\ 1 \end{pmatrix}.$$

The inverse in the convolution sense $(\boldsymbol{\delta I} - \mathbf{q}_{11})^{(-1)} = \boldsymbol{\psi}_{11}$ can be computed by means of one of the three methods described in Section 3.6. The second of them provides an explicit form of the reliability. Indeed, writing Equation (3.46) for $\mathbf{q}_{11}$ we obtain

$$R(k) = \left(\sum_{n=0}^{\cdot} L^{(n)}(\cdot) \right)$$
$$* \left\{ \alpha_1 \left[1 - \sum_{l=1}^{\cdot} f_{12}(l) + f_{12}(\cdot) * \left(1 - \sum_{l=1}^{\cdot}(af_{12}(l) + bf_{23}(l)) \right) \right] \right.$$
$$\left. + \alpha_2 \left[af_{21}(\cdot) * \left(1 - \sum_{l=1}^{\cdot} f_{12}(l) \right) + 1 - \sum_{l=1}^{\cdot}(af_{21}(l) + bf_{23}(l)) \right] \right\}(k),$$

where we have set $L(k) := af_{12} * f_{21}(k), k \in \mathbb{N}$.

In Equations (5.14) and (5.16) we have expressed the BMP-failure rate and the RG-failure rate in terms of the reliability. Consequently, using the previous explicit form of the reliability, we get explicit forms for the two failure rates.

The availability at time $k \in \mathbb{N}$ has the expression

$$A(k) = \boldsymbol{\alpha} \begin{pmatrix} \mathbf{1}_{\{0\}} & -f_{12} & 0 \\ -af_{21} & \mathbf{1}_{\{0\}} & -bf_{23} \\ -f_{31} & 0 & \mathbf{1}_{\{0\}} \end{pmatrix}^{(-1)} (\cdot) * \big(\mathbf{I} - \mathbf{H}(\cdot)\big)(k) \begin{pmatrix} 1 \\ 1 \\ 0 \end{pmatrix},$$

with $\mathbf{I} - \mathbf{H}(\cdot)$ given in Equation (3.45).

As we did before for the reliability, we compute the inverse in the convolution sense $(\boldsymbol{\delta I} - \mathbf{q})^{(-1)} = \boldsymbol{\psi}$ with the second method and we obtain the following explicit form of the availability:

$$A(k) = \left(\sum_{n=0}^{\cdot} T^{(n)}(\cdot) \right)$$

$$* \left\{ \alpha_1 \left[1 - \sum_{l=1}^{\cdot} f_{12}(l) + f_{12}(\cdot) * \left(1 - \sum_{l=1}^{\cdot} (af_{12}(l) + bf_{23}(l)) \right) \right] \right.$$

$$+ \alpha_2 \left[af_{21}(\cdot) * \left(1 - \sum_{l=1}^{\cdot} f_{12}(l) \right) + 1 - \sum_{l=1}^{\cdot} (af_{21}(l) + bf_{23}(l)) \right]$$

$$+ \alpha_3 \left[f_{31}(\cdot) * \left(1 - \sum_{l=1}^{\cdot} f_{12}(l) \right) \right.$$

$$\left. \left. + f_{12}(\cdot) * f_{31}(\cdot) * \left(1 - \sum_{l=1}^{\cdot} (af_{21}(l) + bf_{23}(l)) \right) \right] \right\} (k),$$

where $T = (T(k))$, with $T(k) := af_{12} * f_{21}(k) + bf_{12} * f_{23} * f_{31}(k), k \in \mathbb{N}$.
In the same way, the maintainability at time $k \in \mathbb{N}$ is

$$M(k) = 1 - \boldsymbol{\alpha}_2(\boldsymbol{\delta I} - \mathbf{q}_{22})^{(-1)} * (\mathbf{I} - \mathbf{H}_2)(k)$$

$$= 1 - \alpha_3 (1 - \sum_{l=1}^{k} f_{31}(l)).$$

Using Equations (5.19) and (5.20), we obtain the mean time to failure and the mean time to repair expressed as

$$MTTF = \begin{pmatrix} \alpha_1 & \alpha_2 \end{pmatrix} (\mathbf{I} - \mathbf{p}_{11})^{-1} \begin{pmatrix} m_1 \\ m_2 \end{pmatrix}$$

$$= \frac{1}{1-a} \begin{pmatrix} \alpha_1 & \alpha_2 \end{pmatrix} \begin{pmatrix} 1 & 1 \\ a & 1 \end{pmatrix} \begin{pmatrix} m_1 \\ m_2 \end{pmatrix}$$

$$= \frac{1}{b} \Big(m_1(\alpha_1 + a\alpha_2) + m_2(\alpha_1 + \alpha_2) \Big),$$

$$MTTR = \alpha_3 (\mathbf{I} - \mathbf{p}_{22})^{-1} m_3 = \alpha_3 m_3.$$

After having obtained these explicit formulas for the reliability measures of this system, we have simulated a sample path of the semi-Markov chain in order to illustrate the theoretical properties of the estimators of reliability, availability, and failure rates obtained in this chapter.

The consistency of the reliability estimator is illustrated in Figure 5.1, where reliability estimators obtained for several values of the sample size M are drawn. In Figure 5.2 we present the confidence interval of the reliability. Note that the confidence interval covers the true value of the reliability. In Figure 5.3 we present the estimators of the asymptotic variance of the reliability $\sigma_R^2(k)$, obtained for different sample sizes. Note that the estimator approaches the true value as the sample size M increases.

The same type of figures are drawn for the availability and RG-failure rate. Thus, in Figures 5.4–5.6 we have illustrated the consistency of the availability estimator, its asymptotic normality, and the consistency of the estimator of the asymptotic variance $\sigma_A^2(k)$. Figures 5.7–5.9 present the same graphics for the RG-failure rate estimator.

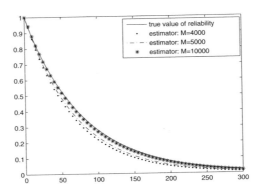

Fig. 5.1. Consistency of reliability estimator

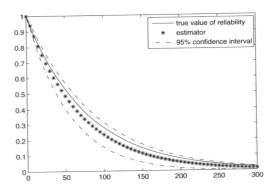

Fig. 5.2. Confidence interval of reliability

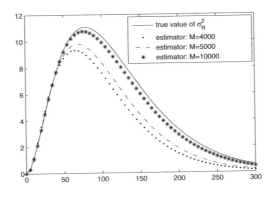

Fig. 5.3. Consistency of $\sigma_R^2(k)$ estimator

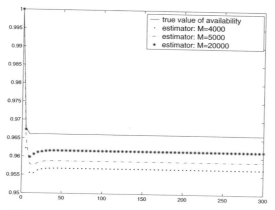

Fig. 5.4. Consistency of availability estimator

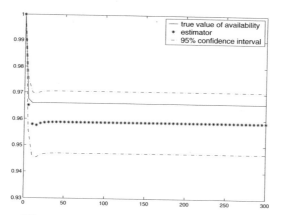

Fig. 5.5. Confidence interval of availability

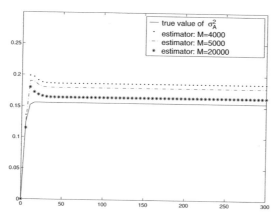

Fig. 5.6. Consistency of $\sigma_A^2(k)$ estimator

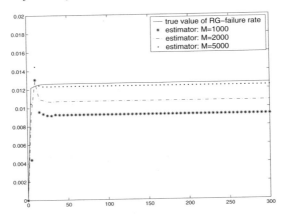

Fig. 5.7. Consistency of RG-failure rate estimator

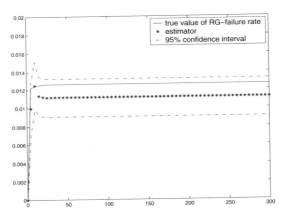

Fig. 5.8. Confidence interval of RG-failure rate

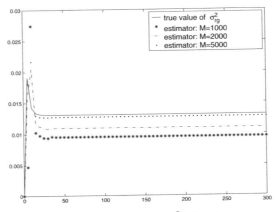

Fig. 5.9. Consistency of $\sigma_r^2(k)$ estimator

Exercises

Exercise 5.1. Show that $\mathbf{P}_{11} \in \mathcal{M}_U(\mathbb{N})$ defined by Equation (5.1) can be expressed as

$$\mathbf{P}_{11}(k) = (\boldsymbol{\delta}\mathbf{I} - \mathbf{q}_{11})^{(-1)} * (\mathbf{I} - \mathbf{H}_1)(k)$$

and

$$(\mathbf{P}_{11})_{ij}(k) = \mathbb{P}(Z_k = j, Z_l \in U, l = 1, \ldots, k-1 \mid Z_0 = i), i, j \in U.$$

Hint: Show first that $\mathbf{P}_{11}$ defined in Equation (5.1) is the solution of the Markov renewal equation

$$\mathbf{P}_{11} = I - \mathbf{H}_1 + \mathbf{q}_{11} * \mathbf{P}_{11} \tag{5.38}$$

(the proof is similar to that of Proposition 3.2) and use Theorem 3.1 to obtain the first expression.

Exercise 5.2. Show that the chain $Y = (Y_k)_{k \in \mathbb{N}}$ defined in the second proof of Proposition 5.1 by

$$Y_k = \begin{cases} Z_k, & \text{if } k < T_D, \\ \Delta, & \text{if } k \geq T_D, \end{cases}$$

is a semi-Markov chain with the semi-Markov kernel

$$\mathbf{q}_Y(k) = \begin{pmatrix} \mathbf{q}_{11}(k) & \mathbf{q}_{12}(k)\,\mathbf{1}_{s-s_1} \\ \mathbf{0}_{1s_1} & 0 \end{pmatrix}, \ k \in \mathbb{N}.$$

Exercise 5.3. Show that, for a nonergodic discrete-time semi-Markov system (cf. Remark 5.3), the availability of the system is identical to the reliability and it is computed using Equation (5.3),

$$A(k) = R(k) = \boldsymbol{\alpha}_1\,\mathbf{P}_{11}(k)\mathbf{1}_{s_1}, \ k \in \mathbb{N}.$$

Exercise 5.4. Derive the Markov renewal equation of maintainability of a discrete-time semi-Markov system.

Hint: Follow the same steps as in the proof of Proposition 5.1 and obtain the analogous of Equation (5.8).

Exercise 5.5. Show that in a discrete-time setting, we have the following relationship between the two failure rate averages Λ and $\widetilde{\Lambda}$ (see Section 5.1.4):

$$\Lambda(k) < \widetilde{\Lambda}(k), \ \text{for all } k \in \mathbb{N}^*.$$

Exercise 5.6 (alternative proof of Proposition 5.5). Let $(Z_n)_{n \in \mathbb{N}}$ be a SMC of state space $E = U \cup D$, and let $(S_n^j)_{n \in \mathbb{N}}$ be the successive passage times in a fixed state $j \in U$, N_j the number of visits to state j before hitting set D, and X_{jk} the kth sojourn time of the SMC in state j, $0 \leq k \leq N_j$.

(1) Express the MTTF of the system as a function of $(X_{jk})_k$ and N_j, $j \in U$.

(2) Use Wald Lemma (Theorem E.8) and the expression of the mean waiting time of a Markov chain before entering a subspace $D \subset E$ in order to derive from (1) the expression of the MTTF given in Equation (5.19).

Exercise 5.7. Prove that, if there exists an $i, 1 \leq i \leq s_1$, such that $\sum_{j=1}^{s_1} p_{ij} < 1$, then $\mathbf{I} - \mathbf{p}_{11}$ is nonsingular. Symmetrically, if there exists an $l, s_1 + 1 \leq l \leq s$, such that $\sum_{j=s_1+1}^{s} p_{lj} < 1$, then $\mathbf{I} - \mathbf{p}_{22}$ is nonsingular.

Thus under these conditions, the hypotheses of Proposition (5.5) are fulfilled and we have the expressions of MTTF and MTTR given in Equations (5.19), respectively (5.20).

Exercise 5.8. Consider a discrete-time semi-Markov system.

1. Prove the consistency of the availability estimator given by Equation (5.33).
2. Give an estimator of the maintainability and prove its consistency and asymptotic normality.

Exercise 5.9. Consider a discrete-time semi-Markov system.

1. If the subset U of up states has only one element, show that MUT=MTTF and give an explicit expression of MUT.
2. Similarly, if the subset D of the down states has only one element, show that MDT=MTTR and give an explicit expression of MDT.

Exercise 5.10. Consider the three-state Markov chain defined in Exercise 3.3. Give explicit theoretical expressions of the reliability, availability, and mean time to failure, when $U = \{1, 2\}$ and $D = \{3\}$. Compute also numerically the values of these reliability indicators.

Exercise 5.11. Let us consider a system observed on the time axis $\mathbb{N}$, with geometric lifetime distribution $T \sim G(p)$ and geometric repair time distribution $V \sim G(q)$, $0 < p, q < 1$. Give its reliability, availability, maintainability, MTTF, MTTR, MUT, MDT, and MTBF.

Exercise 5.12. Prove the asymptotic normality of the reliability estimator given in (5.28) by means of Pyke and Schaufele CLT for Markov renewal chains (Theorem 3.5).

Exercise 5.13. Let $\mathcal{S}$ be a binary reparable system (see Section 2.4 and Example 3.3), with lifetime distribution $f(n), n \in \mathbb{N}$, and repair time distribution $g(n), n \in \mathbb{N}$. Give the estimator of its availability and derive its consistency and asymptotic normality.

Exercise 5.14. Let $\mathcal{S}$ be a multistate SM system, with lifetime distribution T. Define by $\mu(n) := \mathbb{E}(T - n \mid T > n)$ its residual mean lifetime. Give an expression of $\mu(n)$ in terms of the semi-Markov kernel $\mathbf{q}$.

Exercise 5.15. Give an estimator of $\mu(n)$ (Exercise 5.14) and derive its asymptotic properties.

6

Hidden Semi-Markov Model and Estimation

In this chapter, we introduce the discrete-time hidden semi-Markov model, we investigate the asymptotic properties of the nonparametric maximum-likelihood estimators of the basic quantities of the model, namely, consistency and asymptotic normality, and we derive an EM algorithm, for practically obtaining the maximum-likelihood estimators.

Hidden semi-Markov models (HSMMs) are an extension of hidden Markov models (HMMs). Since being introduced by Baum and Petrie (1966), the HMMs have become very popular in a wide range of applications like biology (Churchill, 1992; Durbin et al., 1998), speech recognition (Rabiner, 1989), image processing, and text recognition. A recent book that covers much work on the field is Cappé et al. (2005).

In parallel with the extensive use of HMMs in applications and with the related statistical inference work (see, e.g., Baum and Petrie, 1966; Leroux, 1992; Bickel et al., 1998; Jensen and Petersen, 1999; Douc and Matias, 2001; Douc et al., 2004; Koski, 2001), a new type of model was derived, initially in the domain of speech recognition. In fact, the hidden Markov processes present a serious disadvantage, namely, the fact that the sojourn times of the hidden process are geometrically or exponentially distributed. Such a model was found to be inappropriate for some speech recognition applications (see Levinson, 1986, for a detailed discussion). To solve this problem, Ferguson (1980) proposed a model that allows arbitrary sojourn time distributions for the hidden process. As the hidden process becomes semi-Markovian, this model is called a hidden semi-Markov model (also known as a variable-duration HMM or explicit-duration HMM).

Different variations of the HSMM have been proposed, modeling the duration times by a parametric family of continuous distributions (see Levinson, 1986, with an application to the two-parameter gamma distributions), by an exponential family of distributions (Mitchell and Jamieson, 1993) or by the negative binomial distribution (Durbin et al., 1998).

V.S. Barbu, N. Limnios, *Semi-Markov Chains and Hidden Semi-Markov Models toward Applications*, DOI: 10.1007/978-0-387-73173-5_6,

More recent papers develop computational aspects for HSMMs, with various applications: Guédon (2003) (EM algorithm, Derin's scheme, applications in speech processing and plant structure analysis), Guédon (1999) (algorithms for computing characteristic distributions), Sansom and Thomson (2001) (EM algorithm, discussion of nonparametric and parametric cases for the sojourn time distributions, applications in climatology), Sansom and Thompson (2003); Sansom and Thomson (2007) (rainfall modeling), Bulla and Bulla (2006) (EM algorithm, financial applications), and Yu and Kobayashi (2003) (EM algorithm for HSMMs with missing observation data or multiple observation sequences, applications in mobility modeling).

To the best of our knowledge, most of the existing works on HSMMs consist only in the construction of different types of models and in the associated algorithmic studies. In the literature of HSMMs there are no studies on the asymptotic properties of the MLE, as is the case in the HMM context. The aim of the first part of this chapter is to investigate the asymptotic properties of the MLE for a HSMM, following the lines of Baum and Petrie (1966) and Bickel et al. (1998). Afterward, we derive an EM algorithm (Baum et al., 1970; Dempster et al., 1977; Devijver, 1985) for the computation of the maximum-likelihood estimators.

Combining the flexibility of the semi-Markov processes with the proved advantages of HMMs, we obtain HSMMs, which are a powerful tool for applications and offer a rich statistical framework. Our principal motivation for undertaking this work is the importance of such models in DNA analysis.

6.1 Hidden Semi-Markov Model

Before giving any formal definition of the hidden semi-Markov model, let us first see a concrete application in genetics.

Example 6.1 (CpG islands in a DNA sequence). Consider a DNA sequence, that is, a sequence of the four nucleotides A, C, G, and T, i.e., an element of the space $\{A, C, G, T\}^{\mathbb{N}}$,

$$\{TAGTGGAACGACCGGATCC\ldots\}.$$

It is known that the presence of the pairs C–G is relatively rare in the genome. Nevertheless, there are some regions within a DNA sequence where the frequency of C–G pairs, as well as the frequency of nucleotides C and G themselves, is more important. It has been proved that these regions, called CpG islands, play a key role in the coding mechanism, so finding them is of great importance in genetic research.

Several mathematical models have been proposed for detecting CpG islands (see Durbin et al., 1998, for the use of Markov and hidden Markov models). We will present in the sequel the use of the hidden Markov model for detection of CpG islands and we will also see why we think that it is more natural to use a hidden semi-Markov model instead.

Suppose that the DNA sequence is modeled by a sequence of conditionally independent random variables Y, with state space $D = \{A, C, G, T\}$. Suppose also that the possible presence of a CpG island is modeled by a Markov chain Z with state space $E = \{0, 1\}$. Having $(y_0, \ldots, y_M)$ a truncated sample path of Y, we set $Z_n = 1$ if y_n is a nucleotide inside a CpG island and $Z_n = 0$ otherwise:

- $Y : \underbrace{TAGTGGAATG}\underbrace{CGACG} \cdots$ - DNA sequence

- $Z : 0\,0\,0\,0\,0\,0\,0\,0\,0\,0\,1\,1\,1\,1\,1 \cdots$ - CpG islands indicators.

Suppose that $(Z_n)_{n \in \mathbb{N}}$ is a Markov chain and that the observed nucleotides Y_n are generated according to the corresponding Z_n. This is a typical example of a hidden Markov model. From a practical point of view, the main drawback of this type of approach is that we suppose that the length of windows of 0s and 1s follows geometric distributions since we impose a Markovian evolution on process Z. For this reason it is more natural to let Z be a semi-Markov chain, allowing a more realistic behavior of the model, as the length of windows of 0s and 1s can follow any probability distribution on $\mathbb{N}$, instead of a geometric one in the Markov case. In this way, we obtain what is called a hidden semi-Markov model. Obviously, the HMM is a particular case of the HSMM.

Example 6.2 (hidden Markov chains for detecting an unfair die). Consider two dice, a fair one and an unfair one. When rolling the unfair die, there is a 1/2 probability of getting a 6 and a 1/10 probability of getting 1, 2, 3, 4, or 5. After rolling the fair die, the probability that the next game will be done with the unfair die is 1/10. On the other hand, after rolling the unfair die, the probability that the next game will be done with the fair die is 1/2.

Let $Z_0, Z_1, \ldots$ be the random variable sequence of successively used dice, with value 0 for the fair die and 1 for unfair one. Consider also $Y_0, Y_1, \ldots$ the random variable sequence, with values in $\{1, 2, 3, 4, 5, 6\}$ representing the successive values of the rolled dice. In practical terms, only sequence Y is observed, whereas chain Z is "hidden" (unobserved, unknown). The couple (Z, Y) is a hidden Markov chain, that is, Z is an unobserved ergodic Markov chain and Y is a sequence of conditional independent random variables, in the sense that the distribution of Y_n depends only on $Z_n, n \in \mathbb{N}$.

Let us compute:

1. The probability $\mathbb{P}(Y_n = i \mid Z_0 = 1)$, $1 \leq i \leq 6$;
2. The limit $\lim_{n \to \infty} \mathbb{P}(Y_n = i \mid Z_0 = 1)$, $1 \leq i \leq 6$.

Set $E = \{0, 1\}$ for the state space of the Markov chain Z and note that its associated transition matrix is

$$\mathbf{p} = \begin{pmatrix} 9/10 & 1/10 \\ 1/2 & 1/2 \end{pmatrix}.$$

The conditional distributions of Y_n, given the state of Z_n, $n \in \mathbb{N}$, are as follows:

$\mathbb{P}(Y_n = i \mid Z_n = 1) = 1/10$, for all $i = 1, \ldots, 5$, $\mathbb{P}(Y_n = 6 \mid Z_n = 1) = 1/2$;
$\mathbb{P}(Y_n = i \mid Z_n = 0) = 1/6$, for all $i = 1, \ldots, 6$.

Let us now compute the probabilities of interest.

1. We have

$$\mathbb{P}(Y_n = i \mid Z_0 = 1) = \sum_{l \in E} \mathbb{P}(Y_n = i, Z_n = l \mid Z_0 = 1)$$

$$= \sum_{l \in E} \mathbb{P}(Y_n = i \mid Z_n = l, Z_0 = 1) \mathbb{P}(Z_n = l \mid Z_0 = 1)$$

$$= \sum_{l \in E} \mathbb{P}(Y_n = i \mid Z_n = l) p_{1l}^n,$$

where p_{1l}^n is the element $(1, l)$ of $\mathbf{p}^n$, the n-fold matrix product of $\mathbf{p}$ (Appendix D). Using the previous computations, given a state i and a positive integer n, one can immediately obtain the values of $\mathbb{P}(Y_n = i \mid Z_0 = 1)$.

2. To obtain the limit $\lim_{n \to \infty} \mathbb{P}(Y_n = i \mid Z_0 = 1)$, $1 \le i \le 6$, we start with the relation obtained above,

$$\mathbb{P}(Y_n = i \mid Z_0 = 1) = \sum_{l \in E} \mathbb{P}(Y_n = i \mid Z_n = l) p_{1l}^n.$$

First, note that the probabilities $\mathbb{P}(Y_n = i \mid Z_n = l)$ do not depend on $n \in \mathbb{N}$, so the limit as n tends to infinity concerns only p_{1l}^n. Second, from Proposition D.7 we know that

$$\lim_{n \to \infty} p_{1l}^n = \nu(l),$$

where $\boldsymbol{\nu} = (\nu(0) \ \nu(1))$ is the stationary (invariant) distribution of the Markov chain $(Z_n)_{n \in \mathbb{N}}$. We compute the stationary distribution $\boldsymbol{\nu}$ by solving the system $\boldsymbol{\nu} \, \mathbf{p} = \boldsymbol{\nu}$, with the additional condition $\nu(0) + \nu(1) = 1$, and we get $\nu(0) = 5/6$, $\nu(1) = 1/6$. Consequently, we obtain

$$\lim_{n \to \infty} \mathbb{P}(Y_n = i \mid Z_0 = 1)$$
$$= \mathbb{P}(Y_n = i \mid Z_n = 0)\nu(0) + \mathbb{P}(Y_n = i \mid Z_n = 1)\nu(1).$$

Finally, for $i = 1, \ldots, 5$, we get

$$\lim_{n \to \infty} \mathbb{P}(Y_n = i \mid Z_0 = 1) = \frac{1}{6} \times \frac{5}{6} + \frac{1}{10} \times \frac{1}{6} = \frac{7}{45},$$

whereas for $i = 6$ the limit is

$$\lim_{n \to \infty} \mathbb{P}(Y_n = 6 \mid Z_0 = 1) = \frac{1}{6} \times \frac{5}{6} + \frac{1}{2} \times \frac{1}{6} = \frac{2}{9}.$$

Let us now formally define a hidden semi-Markov model. We will take into account two different types of such models, the so-called hidden SM-M0 model and the hidden SM-Mk, $k \geq 1$ model. Let $Z = (Z_n)_{n \in \mathbb{N}}$ be a semi-Markov chain with finite state space $E = \{1, \ldots, s\}$ and $Y = (Y_n)_{n \in \mathbb{N}}$ be a stationary sequence of random variables with finite state space $A = \{1, \ldots, d\}$, i.e., for any $n \in \mathbb{N}$, $a \in A$, and $i \in E$ we have that $\mathbb{P}(Y_n = a \mid Z_n = i)$ is independent of n.

Before giving the definitions, let us introduce some notation. Let $l, k \in \mathbb{N}$ be two nonnegative integers such that $l \leq k$, and let $a_l, \ldots, a_k \in A$. We will denote by Y_l^k the vector $Y_l^k := (Y_l, \ldots, Y_k)$ and we will write $\{Y_l^k = a_l^k\}$ for the event $\{Y_l = a_l, \ldots, Y_k = a_k\}$. When all these states represent the same state, say $a \in A$, we simply denote by $\{Y_l^k = a\}$ the event $\{Y_l = a, \ldots, Y_k = a\}$. We also denote by $\{Y_l^k = \cdot\}$ the event $\{Y_l^k = \cdot\} := \{Y_l = \cdot, \ldots, Y_k = \cdot\}$. We gave all this notation in terms of chain Y, but it can be obviously used for chain Z.

Definition 6.1 (hidden semi-Markov chain of type SM-M0).

1. *Let $Y = (Y_n)_{n \in \mathbb{N}}$ be conditionally independent random variables, given a sample path of the SMC Z, i.e., for all $a \in A, j \in E, n \in \mathbb{N}^*$, the following relation holds true:*

$$\mathbb{P}(Y_n = a \mid Y_0^{n-1} = \cdot, Z_n = i, Z_0^{n-1} = \cdot) = \mathbb{P}(Y_n = a \mid Z_n = i). \, (6.1)$$

   The chain $(Z, Y) = (Z_n, Y_n)_{n \in \mathbb{N}}$ is called a hidden semi-Markov chain of type SM-M0, where the index 0 stands for the order of Y regarded as a conditional Markov chain.
2. *For (Z, Y) a hidden semi-Markov chain of type SM-M0, let us define $R = (R_{i;a}; i \in E, a \in A) \in \mathcal{M}_{E \times A}$ as the conditional distribution of chain Y*

$$R_{i;a} := \mathbb{P}(Y_n = a \mid Z_n = i), \tag{6.2}$$

   called the emission probability matrix.

Definition 6.2 (hidden semi-Markov chain of type SM-Mk).

1. *Let $Y = (Y_n)_{n \in \mathbb{N}}$ be a homogeneous Markov chain of order k, $k \geq 1$, conditioned on the SMC Z, i.e., for all $a_0, \ldots, a_k \in A$, $i \in E$, $n \in \mathbb{N}^*$, the following relation holds true:*

$$\mathbb{P}(Y_{n+1} = a_k \mid Y_{n-k+1}^n = a_0^{k-1}, Y_0^{n-k} = \cdot, Z_{n+1} = i, Z_0^n = \cdot)$$
$$= \mathbb{P}(Y_{n+1} = a_k \mid Y_{n-k+1}^n = a_0^{k-1}, Z_{n+1} = i). \tag{6.3}$$

   The chain $(Z, Y) = (Z_n, Y_n)_{n \in \mathbb{N}}$ is called a hidden semi-Markov chain of type SM-Mk, where the index k stands for the order of the conditional Markov chain Y.

2. For (Z, Y) a hidden semi-Markov chain of type SM-Mk, let us define $R = (R_{i;a_0,\ldots,a_k}; \; i \in E, a_0, \ldots, a_k \in A) \in \mathcal{M}_{E \times A \times \ldots \times A}$ as the transition matrix of the conditional Markov chain Y

$$R_{i;a_0,\ldots,a_k} := \mathbb{P}(Y_{n+1} = a_k \mid Y_{n-k+1}^n = a_0^{k-1}, Z_{n+1} = i), \qquad (6.4)$$

called the emission probability matrix of the conditional Markov chain Y.

Example 6.3 (hidden semi-Markov chains for detecting an unfair die). Let us consider the problem of an unfair die detection presented in Example 6.2 and see how we can propose a hidden semi-Markov modeling instead of a hidden Markov one.

As before, we have two dice, an unfair one and a fair one. When rolling the unfair die, there is a $1/2$ probability of getting a 6 and a $1/10$ probability of getting $1, 2, 3, 4$, or 5. After rolling the fair die n times, the probability that the next roll will be done with the unfair die is $f(n)$, where $\mathbf{f} := (f(n))_{n \in \mathbb{N}^*}$ is a distribution on $\mathbb{N}^*$. On the other hand, after rolling the unfair die n times, the probability that the next roll will be done with the fair die is $g(n)$, where $\mathbf{g} := (g(n))_{n \in \mathbb{N}^*}$ is a distribution on $\mathbb{N}^*$.

Let $Z_0, Z_1, \ldots$ be the random sequence of successively used dice, with value 0 for the fair die and 1 for the fake die. Consider also $Y_0, Y_1, \ldots$ the random variable sequence, with values in $\{1, 2, 3, 4, 5, 6\}$ representing the successive values of the rolled dice. The couple $(Z, Y) = (Z_n, Y_n)_{n \in \mathbb{N}}$ is a hidden semi-Markov chain of type SM-M0.

A schematic representation of a hidden semi-Markov model is given in Figure 1.3. Note that in the Y line we either have k dependence arrows (the SM-Mk model) or we do not have dependence arrows at all (the SM-M0 model).

Many types of concrete applications can be investigated using this type of model. Some examples: Y can be a received signal and Z the emitted signal; see Y as the GPS position of a car, whereas Z represents the real (and unknown) position of the car; in reliability and maintenance, Y can be the noise (or any other indicator) made by an engine, whereas Z is the real state of the engine; Y is an observed DNA sequence and Z is an unknown chain related to the DNA coding mechanism (for example, Z can be the sequence of indicators for CpG islands, as shown in Example 6.1).

6.2 Estimation of a Hidden Semi-Markov Model

Let $(Z, Y) = (Z_n, Y_n)_{n \in \mathbb{N}}$ be a hidden SM-M0 chain with finite state space $E \times A$. We suppose that the semi-Markov chain Z is not directly observed and that the observations are described by the sequence of conditionally independent random variables $Y = (Y_n)_{n \in \mathbb{N}}$. Starting from a sample path $y = y_0^M =$

$(y_0, \ldots, y_M)$ of observations, we want to estimate the characteristics of the underlying semi-Markov chain, as well as the conditional distribution of $Y = (Y_n)_{n \in \mathbb{N}}$. All the results of this chapter will be obtained under Assumption A1 of Chapter 3, page 61, that the SMC $(Z_n)_{n \in \mathbb{N}}$ is irreducible.

6.2.1 Consistency of Maximum-Likelihood Estimator

Let $U = (U_n)_{n \in \mathbb{N}}$ be the backward-recurrence times of the semi-Markov chain $(Z_n)_{n \in \mathbb{N}}$, that is,

$$U_n := n - S_{N(n)}. \tag{6.5}$$

One can check that the chain $(Z, U) = (Z_n, U_n)_{n \in \mathbb{N}}$ is a Markov chain with state space $E \times \mathbb{N}$ (see, e.g., Anselone, 1960; Limnios and Oprişan, 2001). Let us denote by $\widetilde{\mathbf{p}} := (p_{(i,t_1)(j,t_2)})_{i,j \in E, t_1, t_2 \in \mathbb{N}}$ its transition matrix. We can easily prove the following result, which gives the transition matrix $\widetilde{\mathbf{p}}$ in terms of the semi-Markov kernel $\mathbf{q}$.

Proposition 6.1. *For all $i, j \in E, t_1, t_2 \in \mathbb{N}$, the transition probabilities of the Markov chain (Z, U) are given by:*

$$p_{(i,t_1)(j,t_2)} = \begin{cases} q_{ij}(t_1 + 1)/\overline{H_i}(t_1), & \text{if } i \neq j \text{ and } t_2 = 0, \\ \overline{H_i}(t_1 + 1)/\overline{H_i}(t_1), & \text{if } i = j \text{ and } t_2 - t_1 = 1, \\ 0, & \text{otherwise}, \end{cases} \tag{6.6}$$

where $\overline{H_i}(\cdot)$ is the survival function of sojourn time in state i (see Equation (4.4)).

The following result concerns the stationary distribution of the Markov chain (Z, U). See Chryssaphinou et al. (2008) for the proof.

Proposition 6.2 (stationary distribution of the MC (Z, U)). *Consider an aperiodic MRC $(J_n, S_n)_{n \in \mathbb{N}}$ that satisfies Assumptions A1 and A2. Then the stationary probability distribution $\widetilde{\pi} = (\pi_{i,u})_{i \in E, u \in \mathbb{N}}$ of the Markov chain (Z, U) is given by*

$$\pi_{i,u} := \frac{1 - H_i(u)}{\mu_{ii}}. \tag{6.7}$$

Reduced state space. We shall consider that the conditional distributions of sojourn times, $f_{ij}(\cdot), i, j \in E, i \neq j$, have the same bounded support, $\text{supp} f_{ij}(\cdot) = D := \{1, \ldots, \widetilde{n}\}$ for all $i, j \in E, i \neq j$. From a practical point of view there is only little loss of generality that follows from this assumption, since in applications we always take into account just a finite support for a distribution, provided that the neglected part observes an error condition.

Remark 6.1. First, we mention that choosing the same support for all the distributions is only a question of computational convenience. Few changes need to be done in order to apply the method presented here to the case of different finite supports for sojourn time distribution.

Second, we will consider that D is also the support of all $q_{ij}(\cdot), i, j \in E, i \neq j$, that is, any transition from an arbitrary state i to an arbitrary state $j, i \neq j$, is allowed. If one needs a model where transitions from some states i to some states j are not allowed, then the parameters corresponding to these transitions need to be eliminated from the parameter space and all the computations we do here will apply. In conclusion, the only real limitation of the model we present here comes from the fact that we ask the sojourn time distributions to have finite support. The readers interested in exact computations for the case of different finite supports for sojourn time distributions can find them in Trevezas and Limnios (2008b). This paper considers observations Y with values in a subset of a Euclidean space and also takes into account backward recurrence time dependence for the observed process.

In conclusion, for the reasons discussed above, we shall suppose that the Markov chain $(Z_n, U_n)_{n \in \mathbb{N}}$ has the finite state space $E \times D$ and the transition matrix $\widetilde{\mathbf{p}} := (p_{(i,t_1)(j,t_2)})_{i,j \in E, t_1, t_2 \in D}$. All the work in the rest of this chapter will be done under the assumption:

A3 The conditional sojourn time distributions have finite support D.

Taking into account the conditional independence, Relation (6.1), for all $a \in A, j \in E$, and $t \in D$, we have

$$R_{i;a} = \mathbb{P}(Y_n = a \mid Z_n = i) = \mathbb{P}(Y_n = a \mid Z_n = i, U_n = t). \tag{6.8}$$

Consequently, starting from the initial hidden semi-Markov chain $(Z_n, Y_n)_{n \in \mathbb{N}}$, we have an associated hidden Markov chain $((Z, U), Y) = ((Z_n, U_n), Y_n)_{n \in \mathbb{N}}$, with $(Z_n, U_n)_{n \in \mathbb{N}}$ a Markov chain and $(Y_n)_{n \in \mathbb{N}}$ a sequence of conditionally independent random variables. This new hidden Markov model is defined by:

- The transition matrix $\widetilde{\mathbf{p}} = (p_{(i,t_1)(j,t_2)})_{i,j \in E, t_1, t_2 \in D}$ of the Markov chain (Z, U), with $p_{(i,t_1)(j,t_2)}$ given by Equation (6.6);
- The conditional distribution $\mathbf{R}$ of the sequence Y, given by Equation (6.8);
- The initial distribution of the hidden Markov chain $((Z, U), Y)$.

As was done previously in Chapters 4 and 5, we will not estimate the initial distribution, since we are concerned only with one trajectory. Further, we consider that the HSMM is stationary. Consequently, we will not take into account the initial distribution in the parameter space.

In order to obtain the parameter space of the hidden Markov model, note first that

- $q_{ij}(t_1 + 1) = 0$ for $t_1 + 1 > \widetilde{n}$,
- $H_i(t_1 + 1) = \sum_{k=t_1+2}^{\infty} \sum_{j \in E} q_{i,j}(k) = 0$ for $t_1 + 2 > \widetilde{n}$.

Thus, for all $i, j \in E, t_1, t_2 \in \mathbb{N}$, the transition probabilities of the Markov chain (Z, U) given in Proposition 6.1 can be written for our model as follows:

$$p_{(i,t_1)(j,t_2)} = \begin{cases} q_{ij}(t_1+1)/\overline{H_i}(t_1), & \text{if } i \neq j, t_2 = 0, \text{ and } 0 \leq t_1 \leq \tilde{n}-1, \\ \overline{H_i}(t_1+1)/\overline{H_i}(t_1), & \text{if } i = j, t_2 - t_1 = 1, \text{ and } 0 \leq t_1 \leq \tilde{n}-2, \\ 0, & \text{otherwise.} \end{cases}$$

(6.9)

Second, for any state $i \in E$ and duration t_1, $0 \leq t_1 \leq \tilde{n}-1$, we have

$$\sum_{j \in E} \sum_{t_2=0}^{\tilde{n}-1} p_{(i,t_1)(j,t_2)} = 1,$$

which, taking into account (6.9), can be written

$$p_{(i,t_1)(i,t_1+1)} \mathbf{1}_{\{0,1,\dots,\tilde{n}-2\}}(t_1) + \sum_{j \neq i} p_{(i,t_1)(j,0)} = 1 \tag{6.10}$$

Thus, we need to eliminate $s\tilde{n}$ parameters, by expressing them in terms of the others (cf. Trevezas and Limnios, 2008b). For instance, for any $i \in E$ and t_1, $0 \leq t_1 \leq \tilde{n}-1$, we will express $p_{(i,t_1)(j,0)}$ (with a certain choice for j) as a function of the other parameters as follows:

$$p_{(i,t_1)(s,0)} = 1 - p_{(i,t_1)(i,t_1+1)} - \sum_{l \neq i,s} p_{(i,t_1)(l,0)}, \text{ if } 0 \leq t_1 \leq \tilde{n}-2, i \neq s,$$

$$p_{(s,t_1)(s-1,0)} = 1 - p_{(s,t_1)(s,t_1+1)} - \sum_{l \neq s-1,s} p_{(s,t_1)(l,0)}, \text{ if } 0 \leq t_1 \leq \tilde{n}-2,$$

$$p_{(i,\tilde{n}-1)(s,0)} = 1 - \sum_{l \neq i,s} p_{(i,\tilde{n}-1)(l,0)}, \text{ if } i \neq s,$$

$$p_{(s,\tilde{n}-1)(s-1,0)} = 1 - \sum_{l \neq s-1,s} p_{(s,\tilde{n}-1)(l,0)}.$$

Note that we had $(s^2\tilde{n} - s)$ parameters non identically zero for the transition matrix of the MC (Z, U), from which we removed $s\tilde{n}$ parameters. Thus, the minimal number of non identically zero parameters that describe the behavior of (Z, U) is $b_1 := s^2\tilde{n} - s\tilde{n} - s$.

Similarly, for the conditional distributions of Y given by Equation (6.8) we have s linear relations,

$$\sum_{a=1}^{d} R_{i;a} = 1, i \in E,$$

so we have to express s parameters as functions of the others. For example, we set

$$R_{i;d} = 1 - \sum_{a=1}^{d-1} R_{i;a}, \text{ for any state } i \in E. \tag{6.11}$$

Thus, the minimal number of parameters for the conditional distribution of Y is $b_2 := sd - s$.

Consequently, we have the parameter space $\Theta := \Theta_1 \times \Theta_2$ for the hidden Markov chain $((Z,U),Y)$, where

$$\Theta_1 \subset \{\boldsymbol{\theta}_1 = (\theta_{1;1}, \theta_{1;2}, \ldots, \theta_{1;b_1}) \mid 0 \le \theta_{1;r} \le 1, r = 1, \ldots, b_1\} \subset \mathbb{R}^{b_1}$$

is the parameter space corresponding to the Markov chain (Z,U), and

$$\Theta_2 \subset \{\boldsymbol{\theta}_2 = (\theta_{2;1}, \theta_{2;2}, \ldots, \theta_{2;b_2}) \mid 0 \le \theta_{2;r} \le 1, r = 1, \ldots, b_2\} \subset \mathbb{R}^{b_2}$$

is the parameter space corresponding to the conditional distribution of Y. So $\Theta := \Theta_1 \times \Theta_2 \subset \mathbb{R}^b$, where $b := b_1 + b_2 = s^2 \widetilde{n} + sd - s\widetilde{n} - 2s$. A generic element of Θ will be denoted by

$$\boldsymbol{\theta} = (\boldsymbol{\theta}_1, \boldsymbol{\theta}_2) = (\theta_1, \ldots, \theta_b) = \left((p_{(i,t_1)(j,t_2)})_{i,j,t_1,t_2}, (R_{ia})_{i,a}\right).$$

Note that in the description of the parameter $\boldsymbol{\theta}$ in terms of $p_{(i,t_1)(j,t_2)}$ and R_{ia} we consider only the non identically zero parameters, and all the dependent parameters have been removed, as described above. When we will need to consider also the dependent parameters $R_{i;d}, i \in E$, (as in Theorem 6.5) we will denote the entire matrix of the conditional distribution of Y by $(R_{ia})_{i\in E, a\in A}$ instead of $(R_{ia})_{i,a}$. Let us also denote by $\boldsymbol{\theta}^0 = (\boldsymbol{\theta}_1^0, \boldsymbol{\theta}_2^0) = \left((p_{(i,t_1)(j,t_2)}^0)_{i,j,t_1,t_2}, (R_{ia}^0)_{i,a}\right)$ the true value of the parameter.

For $(Y_0, \ldots, Y_M)$ a sample path of observations, the likelihood function for an observation of the hidden Markov chain $((Z,U),Y)$ is given by

$$p_{\boldsymbol{\theta}}(Y_0^n) = \sum_{z_0^M, u_0^M} \pi_{z_0, u_0} \prod_{k=1}^{M} p_{(z_{k-1}, u_{k-1})(z_k, u_k)} \prod_{k=0}^{M} R_{z_k; Y_k}, \tag{6.12}$$

where $(\pi_{i,u})_{i,u}$ is the stationary distribution of the Markov chain (Z,U) defined in Equation (6.7). We also consider the likelihood function for an observation of the initial hidden semi-Markov chain in Section 6.4.2 (for the hidden SM-M0 model) and in Section 6.4.3 (for the hidden SM-M1 model).

As all the chains are assumed to be stationary, we consider that the time scale of all the processes is $\mathbb{Z}$ instead of $\mathbb{N}$, so we shall work with $(Z,U) = (Z_n, U_n)_{n\in\mathbb{Z}}$ and $Y = (Y_n)_{n\in\mathbb{Z}}$ instead of $(Z,U) = (Z_n, U_n)_{n\in\mathbb{N}}$ and $Y = (Y_n)_{n\in\mathbb{N}}$.

In conclusion, what follows is carried out within the framework of Baum and Petrie (1966) and Petrie (1969) and we have the following result.

Theorem 6.1. *Under assumptions A1 and A3, given a sample of observations Y_0^M, the maximum-likelihood estimator $\widehat{\boldsymbol{\theta}}(M)$ of $\boldsymbol{\theta}^0$ is strongly consistent as M tends to infinity.*

Proof. Note that we can show that the Markov chain (Z, U) is aperiodic. On the other hand, from Assumption A1 we obtain that the chain (Z, U) is also irreducible. For these reasons, one can always find a positive integer k such that the kth power of the transition matrix of the Markov chain (Z, U) has the following property:

$$\widetilde{\mathbf{p}}^{\,k} = \left(p_{(i,t_1)(j,t_2)}^{(k)} \right)_{i,j \in E, t_1, t_2 \in D}, \text{ with } p_{(i,t_1)(j,t_2)}^{(k)} > 0 \text{ for all } i, j \in E, t_1, t_2 \in D.$$

See Bickel and Ritov (1996) for a similar discussion. Applying Theorem 3.4. of Baum and Petrie (1966), we get the desired result. $\qquad\square$

Remark 6.2. We stress the fact that the convergence in the preceding theorem is in the quotient topology, in order to render the model identifiable. To be more specific, we define an equivalence relation on the parameter space Θ, by setting $\boldsymbol{\theta} \sim \widetilde{\boldsymbol{\theta}}$ if $\boldsymbol{\theta}$ and $\widetilde{\boldsymbol{\theta}}$ define the same finite-dimensional distributions for the chain Y. We consider the quotient set $\Theta/\sim$ with respect to this equivalence relation. The convergence of $\widehat{\boldsymbol{\theta}}(M)$ to $\boldsymbol{\theta}^0$ means in fact convergence between the corresponding equivalence classes. See Baum and Petrie (1966), Petrie (1969), or Leroux (1992) for more details.

Thus, we have the consistency of the MLE of $\boldsymbol{\theta}_1^0 = (p_{(i,t_1)(j,t_2)}^0)_{i,j,t_1,t_2}$ and $\boldsymbol{\theta}_2^0 = (R_{ia}^0)_{i,a}$, denoted by $\widehat{\boldsymbol{\theta}}_1(M) = (\widehat{p}_{(i,t_1)(j,t_2)}(M))_{i,j,t_1,t_2}$ and $\widehat{\boldsymbol{\theta}}_2(M) = (\widehat{R}_{ia}(M))_{i,a}$. Note that the consistency of the maximum-likelihood estimator of $(R_{ia}^0)_{i \in E, a \in A}$, (i.e., taking also into account the s parameters expressed in Equations (6.11) as functions of the others) can be immediately obtained using the continuous mapping theorem (Theorem E.9), the consistency of $\widehat{\boldsymbol{\theta}}_2(M) = (\widehat{R}_{ia}(M))_{i,a}$, and Equations (6.11).

The following two theorems use these results in order to prove the consistency of the maximum-likelihood estimators of the true value of the semi-Markov kernel $(q_{ij}^0(k))_{i,j \in E, i \neq j, k \in D}$ and of the true value of the transition matrix of the embedded Markov chain $(p_{i,j}^0)_{i,j \in E, i \neq j}$.

Theorem 6.2. *Under assumptions A1 and A3, given a sample of observations Y_0^M, the maximum-likelihood estimator $(\widehat{q}_{ij}(k, M))_{i,j \in E, i \neq j, k \in D}$ of $(q_{ij}^0(k))_{i,j \in E, i \neq j, k \in D}$ is strongly consistent as M tends to infinity.*

Proof. Using the expression of $(p_{(i,t_1)(j,t_2)})_{i,j \in E, t_1, t_2 \in D}$ given in Equation (6.9) and the $s\widetilde{n}$ linear relations between $p_{(i,t_1)(j,t_2)}$ described above, we will write $(q_{ij}(k))_{i,j \in E, i \neq j, k \in D}$ as a function Φ of $\boldsymbol{\theta}_1 = (p_{(i,t_1)(j,t_2)})_{i,j,t_1,t_2}$.

Let $i, j \in E, i \neq j$, be arbitrarily fixed states. First, we have

$$q_{ij}(1) = p_{(i,0)(j,0)}$$

and taking into account the dependence relations between $p_{(i,t_1)(j,t_2)}$ we obtain

$$q_{ij}(1) = \begin{cases} 1 - p_{(i,0)(i,1)} - \sum_{l \neq i,j} p_{(i,0)(l,0)}, \\ \qquad \text{if } i = s, j = s - 1 \text{ or } i \neq s, j = s, \\ p_{(i,0)(j,0)}, \text{ otherwise.} \end{cases} \tag{6.13}$$

For $k \geq 2, k \in D$ arbitrarily fixed, we have

$$\overline{H_i}(t + 1)/\overline{H_i}(t) = p_{(i,t)(i,t+1)}, \quad t = 0, \ldots, k - 2,$$

and taking the product of these equalities we obtain

$$\prod_{t=0}^{k-2} \overline{H_i}(t + 1)/\overline{H_i}(t) = \prod_{t=0}^{k-2} p_{(i,t)(i,t+1)}, \text{ so } \overline{H_i}(k - 1)/\overline{H_i}(0) = \prod_{t=0}^{k-2} p_{(i,t)(i,t+1)}.$$

Consequently, we get

$$\overline{H_i}(k - 1) = \prod_{t=0}^{k-2} p_{(i,t)(i,t+1)}, \quad k \geq 2. \tag{6.14}$$

On the other hand, from Equation (6.9) we have $p_{(i,k-1)(j,0)} = q_{ij}(k)/\overline{H_i}(k-1)$ and from Equation (6.14) we get

$$q_{ij}(k) = p_{(i,k-1)(j,0)}\overline{H_i}(k - 1) = p_{(i,k-1)(j,0)} \prod_{t=0}^{k-2} p_{(i,t)(i,t+1)}.$$

Taking into account the dependence relations between $p_{(i,t_1)(j,t_2)}$ we obtain for $2 \leq k \leq \widetilde{n} - 1$

$$q_{ij}(k) = \begin{cases} \left(1 - p_{(i,k-1)(i,k)} - \sum_{l \neq i,j} p_{(i,k-1)(l,0)}\right) \prod_{t=0}^{k-2} p_{(i,t)(i,t+1)}, \\ \qquad \text{if } i = s, j = s - 1 \text{ or } i \neq s, j = s, \\ p_{(i,k-1)(j,0)} \prod_{t=0}^{k-2} p_{(i,t)(i,t+1)}, \text{ otherwise,} \end{cases} \tag{6.15}$$

and, for $k = \widetilde{n}$, we have

$$q_{ij}(\widetilde{n}) = \begin{cases} \left(1 - \sum_{l \neq i,j} p_{(i,\widetilde{n}-1)(l,0)}\right) \prod_{t=0}^{\widetilde{n}-2} p_{(i,t)(i,t+1)}, \\ \qquad \text{if } i = s, j = s - 1 \text{ or } i \neq s, j = s, \\ p_{(i,\widetilde{n}-1)(j,0)} \prod_{t=0}^{\widetilde{n}-2} p_{(i,t)(i,t+1)}, \text{ otherwise.} \end{cases} \tag{6.16}$$

In conclusion, we can define the function Φ

$$\Phi : \Theta_1 \to \mathbb{R}^{s(s-1)\widetilde{n}}$$

by setting

$$\Phi\left((p_{(i,t_1)(j,t_2)})_{i,j,t_1,t_2}\right) := (q_{ij}(k))_{i,j\in E, i\neq j, k\in D}, \tag{6.17}$$

where $q_{ij}(k)$ are defined by Relations (6.13), (6.15), and (6.16). The consistency of the estimator

$$\widehat{\boldsymbol{\theta}_1}(M) = (\widehat{p}_{(i,t_1)(j,t_2)}(M))_{i,j,t_1,t_2}$$

obtained in Theorem 6.1 and the continuous mapping theorem (Theorem E.9) applied to the function Φ give

$$(\widehat{q}_{ij}(k, M))_{i,j\in E, i\neq j, k\in D} \xrightarrow[M\to\infty]{a.s.} (q_{ij}^0(k))_{i,j\in E, i\neq j, k\in D}.$$

$\square$

Theorem 6.3. *Under Assumptions A1 and A3, given a sample of observations Y_0^M, the maximum-likelihood estimator $(\widehat{p}_{i,j}(M))_{i,j\in E, i\neq j}$ of $(p_{i,j}^0)_{i,j\in E, i\neq j}$ is strongly consistent as M tends to infinity.*

Proof. As we have considered finite supports for the sojourn time distributions,

$$\operatorname{supp} f_{ij}(\cdot) = D = \{1, \dots, \widetilde{n}\} \subset \mathbb{N}, \text{ for all } i, j \in E,$$

for all $k > \widetilde{n}$ we have $q_{ij}(k) = p_{ij} f_{ij}(k) = 0$. For this reason, we can write the transition probabilities of the embedded Markov chain J as

$$p_{ij} = \sum_{k=0}^{\infty} q_{ij}(k) = \sum_{k=1}^{\widetilde{n}} q_{ij}(k). \tag{6.18}$$

Define the function

$$\varphi : \mathbb{R}^{s\times(s-1)\times\widetilde{n}} \to \mathbb{R}^{s\times(s-1)}$$

by

$$\varphi\left((x_{ijk})_{i,j\in E, i\neq j, k\in D}\right) := \Big(\sum_{k=1}^{\widetilde{n}} x_{ijk}\Big)_{i,j\in E, i\neq j}.$$

Consequently, the matrix $\mathbf{p}$ can be written as a function of $q_{ij}(k)$,

$$(p_{ij})_{i,j\in E, i\neq j} = \varphi\left((q_{ij}(k))_{i,j\in E, i\neq j, k\in D}\right). \tag{6.19}$$

Using the consistency of the estimator $(\widehat{q}_{ij}(k, M))_{i,j\in E, i\neq j, k\in D}$ obtained in Theorem 6.2 and the continuous mapping theorem (Theorem E.9) applied to function φ, we obtain

$$(\widehat{p}_{ij}(M))_{i,j\in E, i\neq j} \xrightarrow[M\to\infty]{a.s.} (p_{ij}^0)_{i,j\in E, i\neq j}.$$

$\square$

6.2.2 Asymptotic Normality of Maximum-Likelihood Estimator

For $(Y_0, \ldots, Y_n)$ a sample path of observations we denote by $\sigma_{Y_0^n}(\boldsymbol{\theta}^0) :=$ $-\mathbb{E}_{\boldsymbol{\theta}^0}\left(\frac{\partial^2 \log p(Y_0^n)}{\partial \theta_u \partial \theta_v}\Big|_{\boldsymbol{\theta}=\boldsymbol{\theta}^0}\right)_{u,v}$ the Fisher information matrix computed in $\boldsymbol{\theta}^0$, where $p_{\boldsymbol{\theta}}(Y_0^n)$ is the associated likelihood function (Equation 6.12).

Let

$$\sigma(\boldsymbol{\theta}^0) = \left(\sigma_{u,v}(\boldsymbol{\theta}^0)\right)_{u,v} := -\mathbb{E}_{\boldsymbol{\theta}^0}\left(\frac{\partial^2 \log \mathbb{P}_{\boldsymbol{\theta}}(Y_0 \mid Y_{-1}, Y_{-2}, \ldots)}{\partial \theta_u \partial \theta_v}\Big|_{\boldsymbol{\theta}=\boldsymbol{\theta}^0}\right)_{u,v} \tag{6.20}$$

be the asymptotic Fisher information matrix computed in $\boldsymbol{\theta}^0$ (see Baum and Petrie (1966) or Douc et al. (2004) for the definition of $\sigma(\boldsymbol{\theta}^0)$ as a limiting matrix of Fisher information matrices).

From Theorem 3 of Douc (2005) we know that $\sigma(\boldsymbol{\theta}^0)$ is nonsingular if and only if there exists an integer $n \in \mathbb{N}$ such that $\sigma_{Y_0^n}(\boldsymbol{\theta}^0)$ is nonsingular. Consequently, all our work will be done under the following assumption.

A4   There exists an integer $n \in \mathbb{N}$ such that the matrix $\sigma_{Y_0^n}(\boldsymbol{\theta}^0)$ is nonsingular.

The following result is a direct application of Theorem 5.3 in Baum and Petrie (1966) (see also Bickel et al. (1998)).

Theorem 6.4. *Under Assumptions A1, A3, and A4, the random vector*

$$\sqrt{M}\left[\widehat{\boldsymbol{\theta}}(M) - \boldsymbol{\theta}^0\right] = \sqrt{M}\Big[\left((\widehat{p}_{(i,t_1)(j,t_2)}(M))_{i,j,t_1,t_2}, (\widehat{R}_{ia}(M))_{i,a}\right)$$
$$- \left((p^0_{(i,t_1)(j,t_2)})_{i,j,t_1,t_2}, (R^0_{ia})_{i,a}\right)\Big]$$

is asymptotically normal, as $M \to \infty$, with zero mean and covariance matrix $\sigma(\boldsymbol{\theta}^0)^{-1}$.

From this theorem we immediately obtain the asymptotic normality of the conditioned transition matrix R of chain Y.

Theorem 6.5. *Under Assumptions A1, A3, and A4, the random vector*

$$\sqrt{M}\Big[(\widehat{R}_{ia}(M))_{i\in E, a\in A} - (R^0_{ia})_{i\in E, a\in A}\Big]$$

is asymptotically normal, as $M \to \infty$, with zero mean and covariance matrix

$$\Sigma_R = \phi' \cdot \sigma(\boldsymbol{\theta}^0)_{22}^{-1} \cdot \phi'^{\top}$$

where ϕ' is given in Equation (6.22) bellow and we have considered the following partition of the matrix $\sigma(\boldsymbol{\theta}^0)^{-1}$:

$$\sigma(\boldsymbol{\theta}^0)^{-1} = \begin{pmatrix} \overbrace{\sigma(\boldsymbol{\theta}^0)_{11}^{-1}}^{b_1} & \overbrace{\sigma(\boldsymbol{\theta}^0)_{12}^{-1}}^{b_2} \\ \sigma(\boldsymbol{\theta}^0)_{21}^{-1} & \sigma(\boldsymbol{\theta}^0)_{22}^{-1} \end{pmatrix} \begin{matrix} \} \ b_1 \\ \} \ b_2 \end{matrix}. \tag{6.21}$$

Proof. First, note that the asymptotic normality obtained in Theorem 6.4 immediately yields that

$$\sqrt{M}\left[\widehat{\boldsymbol{\theta}}_2(M) - \boldsymbol{\theta}_2^0\right] = \sqrt{M}\left[\left(\widehat{R}_{ia}(M)\right)_{i,a} - \left(R_{ia}^0\right)_{i,a}\right]$$

is asymptotically normal, as $M \to \infty$, with zero mean and covariance matrix $\sigma(\boldsymbol{\theta}^0)_{22}^{-1}$.

Let us arrange the vector $\boldsymbol{\theta}_2$ as follows

$$\boldsymbol{\theta}_2 = (\theta_{2;1}, \theta_{2;2}, \ldots, \theta_{2;b_2}) = (R_{1;1}, \ldots, R_{s;1}, \ldots, R_{1;d-1}, \ldots, R_{s;d-1}).$$

We will write $(R_{ia})_{i\in E, a\in A}$ as a function of $\boldsymbol{\theta}_2 = (R_{ia})_{i,a}$, i.e., we include also the terms excluded by Relations 6.11. Let

$$\phi : \Theta_2 \to \mathbb{R}^{sd}$$

be defined by

$$\phi(\boldsymbol{\theta}_2) := (R_{1;1}, \ldots, R_{s;1}, \ldots, R_{1;d-1}, \ldots, R_{s;d-1}, 1 - \sum_{a=1}^{d-1} R_{1;a}, \ldots, 1 - \sum_{a=1}^{d-1} R_{s;a}).$$

We have $(R_{ia})_{i\in E, a\in A} = \phi(\boldsymbol{\theta}_2)$, where the matrix $(R_{ia})_{i\in E, a\in A}$ has been expressed as a vector, with the elements taken column-wise. Applying delta method we obtain that the vector

$$\sqrt{M}\left[(\widehat{R}_{ia}(M))_{i\in E, a\in A} - (R_{ia}^0)_{i\in E, a\in A}\right]$$

is asymptotically normal, as $M \to \infty$, with zero mean and covariance matrix

$$\Sigma_R = \phi' \, \sigma(\boldsymbol{\theta}^0)_{22}^{-1} \, \phi'^{\top}$$

whith $\phi' \in \mathcal{M}_{sd \times b_2}$ the derivative matrix of ϕ given by

$$\phi' = \begin{pmatrix} \mathbf{I}_{b_2} \\ \mathbf{A} \end{pmatrix}, \tag{6.22}$$

where $\mathbf{I}_{b_2} \in \mathcal{M}_{b_2 \times b_2}$ is the identity matrix and $\mathbf{A} \in \mathcal{M}_{s \times b_2}$ is defined by $\mathbf{A} = -(\underbrace{\mathbf{I}_s \ldots \mathbf{I}_s}_{d-1 \text{ blocks}})$, with $\mathbf{I}_s \in \mathcal{M}_{s \times s}$ identity matrix. Thus, we obtain the desired result. $\square$

The following result concerns the asymptotic normality of the semi-Markov kernel estimator.

Theorem 6.6. *Under Assumptions A1, A3, and A4, the random vector*

$$\sqrt{M}\left[(\widehat{q}_{ij}(k,M))_{i,j\in E, i\neq j, k\in D} - (q_{ij}^0(k))_{i,j\in E, i\neq j, k\in D}\right] \qquad (6.23)$$

is asymptotically normal, as $M \to \infty$, with zero mean and covariance matrix

$$\Sigma_q = \Phi' \, \sigma(\boldsymbol{\theta}^0)_{11}^{-1} \, \Phi'^{\top}, \qquad (6.24)$$

where $\Phi' \in \mathcal{M}_{s(s-1)\tilde{n}\times b_1}$ is defined in (6.26) bellow and $\sigma(\boldsymbol{\theta}^0)_{11}^{-1}\mathcal{M}_{b_1\times b_1}$ was introduced in (6.21) as the sub-matrix of $\sigma(\boldsymbol{\theta}^0)^{-1}$ corresponding to the parameter space Θ_1.

Proof. First, the same type of remark as done at the beginning of the proof of Theorem 6.5 holds true also here. Indeed, the asymptotic normality obtained in Theorem 6.4 immediately yields that

$$\sqrt{M}\left[\widehat{\boldsymbol{\theta}}_1(M) - \boldsymbol{\theta}_1^0\right] = \sqrt{M}\left[(\widehat{p}_{(i,t_1)(j,t_2)}(M))_{i,j,t_1,t_2} - \left(p_{(i,t_1)(j,t_2)}^0\right)_{i,j,t_1,t_2}\right]$$

is asymptotically normal, as $M \to \infty$, with zero mean and covariance matrix $\sigma(\boldsymbol{\theta}^0)_{11}^{-1}$.

As we saw in the proof of Theorem 6.2 (Equation 6.17), we can express $(q_{ij}(k))_{i,j\in E, i\neq j, k\in D}$ as a function of $(p_{(i,t_1)(j,t_2)})_{i,j,t_1,t_2}$. Thus we can write

$$\sqrt{M}\left[(\widehat{q}_{ij}(k,M))_{i,j\in E, i\neq j, k\in D} - (q_{ij}^0(k))_{i,j\in E, i\neq j, k\in D}\right]$$
$$= \sqrt{M}\left[\Phi\left((\widehat{p}_{(i,t_1)(j,t_2)}(M))_{i,j,t_1,t_2}\right) - \Phi\left((\widehat{p}_{(i,t_1)(j,t_2)}^0)_{i,j,t_1,t_2}\right)\right],$$

with the function Φ defined in Equations (6.13), (6.15), and (6.16). Applying the delta method, we obtain from Theorem 6.4 that the random vector

$$\sqrt{M}\left[(\widehat{q}_{ij}(k,M))_{i,j\in E, k\in D} - (q_{ij}^0(k))_{i,j\in E, k\in D}\right]$$

is asymptotically normal, as $M \to \infty$, with zero mean and covariance matrix $\Sigma_q = \Phi' \, \sigma(\boldsymbol{\theta}^0)_{11}^{-1} \, \Phi'^{\top}$, where Φ' is the derivative matrix of Φ.

In order to compute explicitly Φ' and to have a block-matrix representation of it, we need to arrange in a convenient way the vectors $\boldsymbol{\theta}_1 = (p_{(i,t_1)(j,t_2)})_{i,j,t_1,t_2} \in \Theta_1$ and $(q_{ij}(k))_{i,j\in E, i\neq j, k\in D} \in \mathbb{R}^{s(s-1)\tilde{n}}$. We have

$$\boldsymbol{\theta}_1 = (p_{(i,t_1)(j,t_2)})_{i,j,t_1,t_2}$$

$$= \Big(\underbrace{\underbrace{p_{(1,0)(2,0)}, \ldots, p_{(1,0)(s-1,0)}}_{s-2}, \ldots, \underbrace{p_{(s,0)(1,0)}, \ldots, p_{(s,0)(s-2,0)}}_{s-2}}_{s \text{ blocks}},$$

$$\vdots$$

$$\underbrace{\underbrace{p_{(1,\widetilde{n}-1)(2,0)}, \ldots, p_{(1,\widetilde{n}-1)(s-1,0)}}_{s-2}, \ldots, \underbrace{p_{(s,\widetilde{n}-1)(1,0)}, \ldots, p_{(s,\widetilde{n}-1)(s-2,0)}}_{s-2}}_{s \text{ blocks}},$$

$$\underbrace{\underbrace{p_{(1,0)(1,1)}, \ldots, p_{(1,\widetilde{n}-2)(1,\widetilde{n}-1)}}_{\widetilde{n}-1}, \ldots, \underbrace{p_{(s,0)(s,1)}, \ldots, p_{(s,\widetilde{n}-2)(s,\widetilde{n}-1)}}_{\widetilde{n}-1}}_{s \text{ blocks}} \Big),$$

$$(q_{ij}(k))_{i,j\in E, i\neq j, k\in D} = \Big(\underbrace{\underbrace{q_{12}(1), \ldots, q_{1s}(1)}_{s-1}, \ldots, \underbrace{q_{s1}(1), \ldots, q_{s(s-1)}(1)}_{s-1}}_{s \text{ blocks}},$$

$$\vdots$$

$$\underbrace{\underbrace{q_{12}(\widetilde{n}), \ldots, q_{1s}(\widetilde{n})}_{s-1}, \ldots, \underbrace{q_{s1}(\widetilde{n}), \ldots, q_{s(s-1)}(\widetilde{n})}_{s-1}}_{s \text{ blocks}} \Big).$$

We obtain $\Phi' \in \mathcal{M}_{s(s-1)\widetilde{n} \times b_1}$, the jacobian matrix of Φ, given by

$$\Phi' = (q'_{ij}(k))_{i,j\in E, i\neq j, k\in D}$$

$$= \Big(\underbrace{\underbrace{q'_{12}(1), \ldots, q'_{1s}(1)}_{s-1}, \ldots, \underbrace{q'_{s1}(1), \ldots, q'_{s(s-1)}(1)}_{s-1}}_{s \text{ blocks}}, \tag{6.25}$$

$$\underbrace{\underbrace{q'_{12}(2), \ldots, q'_{1s}(2)}_{s-1}, \ldots, \underbrace{q'_{s1}(2), \ldots, q'_{s(s-1)}(2)}_{s-1}}_{s \text{ blocks}},$$

$$\vdots$$

$$\underbrace{\underbrace{q'_{12}(\widetilde{n}), \ldots, q'_{1s}(\widetilde{n})}_{s-1}, \ldots, \underbrace{q'_{s1}(\widetilde{n}), \ldots, q'_{s(s-1)}(\widetilde{n})}_{s-1}}_{s \text{ blocks}} \Big)^{\top},$$

where $q'_{ij}(k)$ is the row vector defined by

$$q'_{ij}(k) = \left(\frac{\partial q_{ij}}{\partial \theta_{1;m}}\right)_{m=1,\ldots,b_1} = \left(\frac{\partial q_{ij}}{\partial p_{(l,t_1)(r,t_2)}}\right)_{l,r,t_1,t_2} \in \mathbb{R}^{b_1} = \mathbb{R}^{s^2\tilde{n}-s\tilde{n}-s}.$$

Let us now compute derivatives $\frac{\partial q_{ij}(k)}{\partial p_{(l,t_1)(r,t_2)}}$ for all $i,j \in E, i \neq j, k \in D = \{1,\ldots,\tilde{n}\}$, using the expressions of $q_{ij}(k)$ in terms of $p_{(l,t_1)(r,t_2)}$ given in Equations (6.13), (6.15), and (6.16). We will give only the non identical zero derivatives.

(i) For $k = 1$:

- If $i = s, j = s - 1$ or $i \neq s, j = s$

$$\frac{\partial q_{ij}(1)}{\partial p_{(i,0)(i,1)}} = -1, \quad \frac{\partial q_{ij}(1)}{\partial p_{(i,0)(l,0)}} = -1, \, l \neq i,j.$$

- Otherwise

$$\frac{\partial q_{ij}(1)}{\partial p_{(i,0)(j,0)}} = 1.$$

(ii) For $k = 2,\ldots,\tilde{n} - 1$, $t = 0,\ldots,k - 2$:

- If $i = s, j = s - 1$ or $i \neq s, j = s$

$$\frac{\partial q_{ij}(k)}{\partial p_{(i,k-1)(i,k)}} = -\prod_{m=0}^{k-2} p_{(i,m)(i,m+1)},$$

$$\frac{\partial q_{ij}(k)}{\partial p_{(i,k-1)(l,0)}} = -\prod_{m=0}^{k-2} p_{(i,m)(i,m+1)}, \, l \neq i,j,$$

$$\frac{\partial q_{ij}(k)}{\partial p_{(i,t)(i,t+1)}} = \left(1 - p_{(i,k-1)(i,k)} - \sum_{l\neq i,j} p_{(i,k-1)(l,0)}\right) \prod_{\substack{m=0 \\ m\neq t}}^{k-2} p_{(i,m)(i,m+1)}.$$

- Otherwise

$$\frac{\partial q_{ij}(k)}{\partial p_{(i,k-1)(j,0)}} = \prod_{m=0}^{k-2} p_{(i,m)(i,m+1)},$$

$$\frac{\partial q_{ij}(k)}{\partial p_{(i,t)(i,t+1)}} = p_{(i,k-1)(j,0)} \prod_{\substack{m=0 \\ m\neq t}}^{k-2} p_{(i,m)(i,m+1)}.$$

(iii) For $k = \tilde{n}$, $t = 0,\ldots,\tilde{n} - 2$:

- If $i = s, j = s - 1$ or $i \neq s, j = s$

$$\frac{\partial q_{ij}(\widetilde{n})}{\partial p_{(i,\widetilde{n}-1)(l,0)}} = -\prod_{m=0}^{\widetilde{n}-2} p_{(i,m)(i,m+1)}, \ l \neq i, j,$$

$$\frac{\partial q_{ij}(\widetilde{n})}{\partial p_{(i,t)(i,t+1)}} = \left(1 - \sum_{l \neq i,j} p_{(i,\widetilde{n}-1)(l,0)}\right) \prod_{\substack{m=0 \\ m \neq t}}^{\widetilde{n}-2} p_{(i,m)(i,m+1)}.$$

- Otherwise

$$\frac{\partial q_{ij}(\widetilde{n})}{\partial p_{(i,\widetilde{n}-1)(j,0)}} = \prod_{m=0}^{\widetilde{n}-2} p_{(i,m)(i,m+1)},$$

$$\frac{\partial q_{ij}(\widetilde{n})}{\partial p_{(i,t)(i,t+1)}} = p_{(i,\widetilde{n}-1)(j,0)} \prod_{\substack{m=0 \\ m \neq t}}^{\widetilde{n}-2} p_{(i,m)(i,m+1)}.$$

In order to have a compact block-matrix representation of the jacobian Φ', we need to introduce some additional notation. Let $\mathbf{C}$ et $\mathbf{D}$ be matrices defined by blocks as follows

$$\mathbf{C} = \begin{pmatrix} \mathbf{I}_{s-2} \\ -\mathbf{1}_{s-2}^\top \end{pmatrix} \in \mathcal{M}_{(s-1)\times(s-2)}, \ \mathbf{D} = \begin{pmatrix} \mathbf{1}_{s-2}^\top \\ \mathbf{I}_{s-2} \end{pmatrix} \in \mathcal{M}_{(s-1)\times(s-2)}$$

where $\mathbf{I}_{s-2} \in \mathcal{M}_{(s-2)\times(s-2)}$ is the identity matrix and $\mathbf{1}_{s-1}^\top$ the $(s-1)$-row vector whose elements are all 1. Let also $\mathbf{E} \in \mathcal{M}_{(s-1)\times(s-2)}$ be defined by $a_{11} = -1$, $a_{ij} = 0$ for all $(i,j) \neq (1,1)$.

For all $k = 2, \ldots, \widetilde{n}$ and $i \in E$, let us define:

- The scalars $\alpha_{i;k} := \prod_{t=0}^{k-2} p_{(i,t)(i,t+1)}$,
- The matrices $\mathbf{B}_{i;k} \in \mathcal{M}_{(s-1)\times(\widetilde{n}-1)}$, whose elements $B_{i;k}(u,v)$ are given below ($u = 1, \ldots, s-1$, $v = 1, \ldots, \widetilde{n}-1$).

(i) For $k = 2, \ldots, \widetilde{n}-1$:

$$B_{i;k}(u,v) := \begin{cases} -\prod_{m=0}^{k-2} p_{(i,m)(i,m+1)}, \text{ if } u = s-1, v = k, \\ \left(1 - p_{(i,k-1)(i,k)} - \sum_{l \neq i} p_{(i,k-1)(l,0)}\right) \prod_{\substack{m=0 \\ m \neq v-1}}^{k-2} p_{(i,m)(i,m+1)}, \\ \quad \text{if } u = s-1, v \leq k-1, \\ p_{(i,k-1)(u,0)} \prod_{\substack{m=0 \\ m \neq v-1}}^{k-2} p_{(i,t)(i,t+1)}/p_{(i,v-1)(i,v)}, \\ \quad \text{if } v \leq k-1, u < i, u \neq s-1, \\ p_{(i,k-1)(u+1,0)} \prod_{\substack{m=0 \\ m \neq v-1}}^{k-2} p_{(i,t)(i,t+1)}/p_{(i,v-1)(i,v)}, \\ \quad \text{if } v \leq k-1, u \geq i, u \neq s-1, \\ 0, \text{ otherwise.} \end{cases}$$

(ii) For $k = \tilde{n}$:

$$
B_{i;\tilde{n}}(u,v) := \begin{cases}
\left(1 - \sum_{l \neq i} P_{(i,\tilde{n}-1)(l,0)}\right) \prod_{\substack{m=0 \\ m \neq v-1}}^{\tilde{n}-2} P_{(i,m)(i,m+1)}, \\
\quad \text{if } u = s-1, v \leq \tilde{n}-1, \\
P_{(i,\tilde{n}-1)(u,0)} \prod_{\substack{m=0 \\ m \neq v-1}}^{\tilde{n}-2} P_{(i,t)(i,t+1)} \big/ P_{(i,v-1)(i,v)}, \\
\quad \text{if } v \leq \tilde{n}-1, u < i, u \neq s-1, \\
P_{(i,\tilde{n}-1)(u+1,0)} \prod_{\substack{m=0 \\ m \neq v-1}}^{\tilde{n}-2} P_{(i,t)(i,t+1)} \big/ P_{(i,v-1)(i,v)}, \\
\quad \text{if } v \leq \tilde{n}-1, u \geq i, u \neq s-1, \\
0, \quad \text{otherwise.}
\end{cases}
$$

Using the derivatives of q_{ij}, we can rewrite the jacobian of Φ in terms of scalars $\alpha_{i;k}$ and matrices $\mathbf{C}$, $\mathbf{D}$, $\mathbf{E}$, and $\mathbf{B}_{i;k}$ as follows:

$$
\Phi' = \begin{pmatrix}
\mathbf{C} & & & & & & \mathbf{E} & & & \\
& \ddots & & & & & & \ddots & & \\
& & \mathbf{C} & & & & & & \mathbf{E} & \\
& & \alpha_{1;2}\mathbf{D} & & & & & & B_{1;2} & \\
& & & \ddots & & & & & & \ddots \\
& & & & \alpha_{s;2}\mathbf{D} & & & & \vdots & B_{s;2} \\
& & & & & \alpha_{1;\tilde{n}}\mathbf{D} & & & & B_{1;\tilde{n}} & \vdots \\
& & & & & & \ddots & & & \ddots \\
& & & & & & \alpha_{s;\tilde{n}}\mathbf{D} & & & B_{s;\tilde{n}}
\end{pmatrix}. \tag{6.26}
$$

Note that in this representation of Φ' we have made appear only the non-null blocks.   □

6.3 Monte Carlo Algorithm

We want to present here a Monte Carlo algorithm for realizing a trajectory in the time interval $[0, M]$, $M \in \mathbb{N}$, of a given hidden semi-Markov chain $(Z, Y) = (Z_n, Y_n)_{n \in \mathbb{N}}$ of type SM-M0. It is the direct analog of the Monte Carlo algorithm proposed in Section 3.5 for generating a sample path of a given semi-Markov chain in the time interval $[0, M]$. We only need to have an additional step for generating the observed chain Y, according to the emission probability matrix $\mathbf{R} = (R_{i;a}; i \in E, a \in A) \in \mathcal{M}_{E \times A}$.

The output of the algorithm consists in:

- The successive visited states of the semi-Markov chain Z, up to time M, $(J_0, \ldots, J_k)$, with $k = N(M)$ the number of jumps of the semi-Markov chain up to time M;
- The successive jump times of the semi-Markov chain Z, up to time M, $(S_0, \ldots, S_k)$, where $S_k \leq M < S_{k+1}$;
- The successive visited states of the conditionally independent random variables (Y_n), $Y_0, \ldots, Y_M$.

Algorithm

1. Set $k = 0, S_0 = 0$ and sample J_0 from the initial distribution α;
2. Sample the random variable $J \sim p_{J_k,\cdot}$ and set $J_{k+1} = J(\omega)$;
3. Sample the random variable $X \sim F_{J_k J_{k+1}}(\cdot)$;
4. Set $S_{k+1} = S_k + X$;
5. For $m = S_k$ to $\max(M, S_{k+1} - 1)$ do
   sample the random variable $Y \sim R_{J_k,\cdot}$ and set $Y_m = Y(\omega)$;
6. If $S_{k+1} \geq M$, then end;
7. Else, set $k = k + 1$ and continue to step 2.

6.4 EM Algorithm for a Hidden Semi-Markov Model

We want now to present an EM algorithm, adapted from Guédon and Cocozza-Thivent (1990), allowing one to obtain the MLEs of the hidden semi-Markov models we consider. The existing literature on EM algorithms for hidden semi-Markov models (Ferguson, 1980; Levinson, 1986; Sansom and Thomson, 2001; Guédon, 2003; Bulla and Bulla, 2006) consider, generally, parametric families for the sojourn time distributions. On the other hand, most of the hidden semi-Markov models we meet in the literature suppose particular cases of semi-Markov processes.

In the sequel, we present two different versions of the algorithm, one for the SM-M0 model and the other for the SM-M1 model. We want to stress the fact that we consider a general semi-Markov chain, i.e., the kernel is given by $q_{ij}(k) = p_{ij} f_{ij}(k)$.

6.4.1 Preliminaries

Let (Z, Y) be a hidden semi-Markov chain of type SM-M0 or SM-M1. For a sample path of length $M + 1$ of the observed chain Y, $y = y_0^M = (y_0, \ldots, y_M)$, we denote by $\mathbf{y}_l^k := \{Y_l^k = y_l^k\}, l, k \in \mathbb{N}, 0 \leq l \leq k \leq M$ and $\mathbf{y} := \mathbf{y}_0^M = \{Y = y\}$.

Put $Z = Z_0^M = (Z_0, \ldots, Z_M)$ for the semi-Markov chain Z, or, in an equivalent way, $(J, X, U_M) = (J_0^{N(M)}, X_1^{N(M)}, U_M)$. We recall that $U_M := M - S_{N(M)}$ represents the censored sojourn time of the semi-Markov chain Z. For all $n \in \mathbb{N}, j \in E$, we define:

$$F_n(j) := \mathbb{P}(Z_n = j, Z_{n-1} \neq j \mid \mathbf{y}_0^n), n \neq 0,$$
$$F_0(j) := \mathbb{P}(Z_0 = j \mid \mathbf{y}_0) = \mu(j, y_0) / \sum_{i \in E} \mu(i, y_0),$$
$$L_n(j) := \mathbb{P}(Z_n = j \mid \mathbf{y}_0^M)$$
$$= \mathbb{P}(Z_n = j, Z_{n-1} \neq j \mid \mathbf{y}_0^M) + \mathbb{P}(Z_n = j, Z_{n-1} = j \mid \mathbf{y}_0^M), n \neq 0,$$
$$L_0(j) := \mathbb{P}(Z_0 = j \mid \mathbf{y}_0^M),$$
$$L_{1;n}(j) := \mathbb{P}(Z_n = j, Z_{n-1} \neq j \mid \mathbf{y}_0^M), n \neq 0,$$
$$L_{2;n}(j) := \mathbb{P}(Z_n = j, Z_{n-1} = j \mid \mathbf{y}_0^M), n \neq 0,$$
$$P_l := \mathbb{P}(\mathbf{y}_l \mid \mathbf{y}_0^{l-1}), l \neq 0, \quad P_0 := \mathbb{P}(\mathbf{y}_0) = \sum_{i \in E} \mu(i, y_0),$$

where, for all $i \in E, a \in A$, $\mu(i, a) := \mathbb{P}(Z_0 = i, Y_0 = a)$ is the initial distribution of the couple chain (Z, Y).

We have

$$\mathbb{P}(\mathbf{y}_0^n) = \mathbb{P}(\mathbf{y}_0)\mathbb{P}(\mathbf{y}_1 \mid \mathbf{y}_0) \ldots \mathbb{P}(\mathbf{y}_n \mid \mathbf{y}_0^{n-1}) = \prod_{l=0}^{n} P_l$$

and

$$\mathbb{P}(\mathbf{y}_n^{n+t} \mid \mathbf{y}_0^{n-1}) = \prod_{l=n}^{n+t} P_l,$$

with the convention

$$\sum_{t=t_1}^{t_2} g(t) := 0, \qquad \prod_{t=t_1}^{t_2} g(t) := 1, \text{ for } t_1 > t_2,$$

for $g(t)$ an arbitrary function of t.

For technical reasons, we will express the number of transitions $N_i(M)$, $N_{ij}(M)$, and $N_{ij}(k, M)$, $i, j \in E, k \leq M$, in terms of the semi-Markov chain Z, instead of the embedded Markov chain J. Thus, we can write

$$N_i(M) := \sum_{n=0}^{N(M)-1} \mathbf{1}_{\{J_n=i\}} = \sum_{n=1}^{M} \mathbf{1}_{\{Z_n=i;Z_{n-1}\neq i\}} - \mathbf{1}_{\{Z_M=i\}} + \mathbf{1}_{\{Z_0=i\}};$$

$$N_{ij}(M) := \sum_{n=1}^{N(M)} \mathbf{1}_{\{J_{n-1}=i,J_n=j\}} = \sum_{n=1}^{M} \mathbf{1}_{\{Z_{n-1}=i,Z_n\neq i,Z_n=j\}};$$

$$N_{ij}(k, M) := \sum_{n=1}^{N(M)} \mathbf{1}_{\{J_{n-1}=i,J_n=j,X_n=k\}}$$

$$= \sum_{n=1}^{M-k} \mathbf{1}_{\{Z_{n+k}=j,Z_n^{n+k-1}=i,Z_{n-1}\neq i\}} + \mathbf{1}_{\{Z_k=j,Z_0^{k-1}=i\}}.$$

6.4.2 EM Algorithm for Hidden Model SM-M0

Let (Z, Y) be a hidden model of the SM-M0 type. In the sequel, we will see how we can apply the principle of the classical EM algorithm in order to find the nonparametric MLEs.

The Likelihood Function

Let $\boldsymbol{\theta} := (\mathbf{p}, \mathbf{f}, \mathbf{R})$ be the parameter of the model, or, equivalently, $\boldsymbol{\theta} := (\mathbf{q}, \mathbf{R})$. Note that here we consider directly the parameter space of the hidden semi-Markov model (Z, Y), instead of the parameter space of the associated hidden Markov model $((Z, U), Y)$, as was done in Section 6.2. Denote by $g_M(Y_0^M \mid \boldsymbol{\theta})$ the likelihood function of the incomplete data (the known data) $Y = Y_0^M$ and by $f_M(Y_0^M, Z_0^M \mid \boldsymbol{\theta})$ the likelihood function of the complete data $(Y, Z) = (Y_0^M, Z_0^M)$. The following relationship holds true between the two likelihood functions:

$$g_M(Y_0^M \mid \boldsymbol{\theta}) = \sum_{(Z_0,\ldots,Z_M) \in E^{M+1}} f_M(Y_0^M, Z_0^M \mid \boldsymbol{\theta}).$$

The likelihood of complete data can be expressed as

$$
\begin{aligned}
&f_M(Y_0^M, Z_0^M \mid \boldsymbol{\theta}) \\
&= \prod_{k=1}^{N(M)} \left[p_{J_{k-1}J_k} f_{J_{k-1}J_k}(X_k) \prod_{l=S_{k-1}}^{S_k-1} R_{J_{k-1},Y_l} \right] \overline{H}_{J_{N(M)}}(U_M) \prod_{l=S_{N(M)}}^{M} R_{J_{N(M)},Y_l},
\end{aligned}
$$

where $\overline{H}_i(\cdot)$ is the survival function in state i for the semi-Markov chain. As we are concerned only with one sample path of the chains, we have not considered the initial distribution in the expression of the likelihood function.

For all $i \in E$ and $b \in A$, we denote by $N_{i;b}(M) := \sum_{n=0}^{M} \mathbf{1}_{\{Y_n=b, Z_n=i\}}$ the number of visits of state b by the chain Y, up to time M, with the semi-Markov chain Z visiting state i. The log-likelihood of the complete data is given by

$$
\begin{aligned}
&\log(f_M(Y_0^M, Z_0^M \mid \boldsymbol{\theta})) \\
&= \log \left[\prod_{i,j \in E} p_{ij}^{N_{ij}(M)} \prod_{i,j \in E} \prod_{k \in \mathbb{N}^*} f_{ij}(k)^{N_{ij}(k,M)} \right. \\
&\quad \left. \times \prod_{i \in E} \prod_{b \in A} R_{i;b}^{N_{i;b}(M)} (1 - \sum_{l=1}^{k} \sum_{j \in E} q_{ij}(l)) \right] \\
&= \sum_{i,j \in E} N_{ij}(M) \log p_{ij} + \sum_{i,j \in E} \sum_{k \in \mathbb{N}^*} N_{ij}(k,M) \log f_{ij}(k) \\
&\quad + \sum_{i \in E} \sum_{b \in A} N_{i;b}(M) \log R_{i;b} + \log \left(\sum_{k \geq M - S_{N(M)}} (1 - \sum_{l=1}^{k} \sum_{j \in E} q_{ij}(l)) \right).
\end{aligned}
$$

Roughly speaking, the EM algorithm is based on the idea (cf. Dempster et al., 1977; Baum et al., 1970) of maximizing the expectation of the log-likelihood function of the complete data, conditioned by the observation $\mathbf{y}$ and by a known value $\boldsymbol{\theta}^{(m)}$ of the parameter, instead of maximizing the likelihood function of observed data $g_M(Y_0^M \mid \boldsymbol{\theta})$. Consequently, the function to be maximized is

$$Q(\boldsymbol{\theta} \mid \boldsymbol{\theta}^{(m)}) := \mathbb{E}_{\boldsymbol{\theta}^{(m)}}[\log(f_M(Y_0^M, Z_0^M \mid \boldsymbol{\theta})) \mid \mathbf{y}].$$

Taking into account the expression of the log-likelihood $\log(f_M(Y_0^M, Z_0^M \mid \boldsymbol{\theta}))$ we can write the conditional expectation $Q(\boldsymbol{\theta} \mid \boldsymbol{\theta}^{(m)})$ as

$$Q(\boldsymbol{\theta} \mid \boldsymbol{\theta}^{(m)}) := \mathbb{E}_{\boldsymbol{\theta}^{(m)}}[\log(f_M(Y_0^M, Z_0^M \mid \boldsymbol{\theta})) \mid \mathbf{y}]$$

$$= \sum_{i,j \in E} \log p_{ij} \sum_{n=1}^{M} \mathbb{P}_{\boldsymbol{\theta}^{(m)}}(Z_{n-1} = i, Z_n = j \mid \mathbf{y})$$

$$+ \sum_{i,j \in E} \sum_{k \in \mathbb{N}^*} \log f_{ij}(k) \sum_{n=0}^{M-k} \mathbb{P}_{\boldsymbol{\theta}^{(m)}}(Z_{n+k} = j, Z_n^{n+k-1} = i, Z_{n-1} \neq i \mid \mathbf{y})$$

$$+ \sum_{i \in E} \sum_{b \in A} \log R_{i;b} \sum_{n=0}^{M} \mathbb{P}_{\boldsymbol{\theta}^{(m)}}(Z_n = i \mid \mathbf{y}) \mathbf{1}_{\{Y_n = b\}}$$

$$+ \mathbb{E}_{\boldsymbol{\theta}^{(m)}} \left[\log \left(\sum_{k \geq M - S_{N(M)}} \left(1 - \sum_{l=1}^{k} \sum_{j \in E} q_{ij}(l) \right) \right) \mid \mathbf{y} \right],$$

where we used the relations

$$\mathbb{E}_{\boldsymbol{\theta}^{(m)}}[\mathbf{1}_{\{Z_{n-1}=i, Z_n=j\}} \mid \mathbf{y}] = \mathbb{P}_{\boldsymbol{\theta}^{(m)}}(Z_{n-1} = i, Z_n = j \mid \mathbf{y}),$$

$$\mathbb{E}_{\boldsymbol{\theta}^{(m)}}[\mathbf{1}_{\{Z_n=i, Y_n=b\}} \mid \mathbf{y}] = \mathbf{1}_{\{Y_n=b\}} \mathbb{P}_{\boldsymbol{\theta}^{(m)}}(Z_n = i \mid \mathbf{y}).$$

In what follows, the last term of $Q(\boldsymbol{\theta} \mid \boldsymbol{\theta}^{(m)})$ will be neglected; the information provided by this term is not important asymptotically, because this term corresponds to $u_M = M - S_{N(M)}$ and we know that $u_M/M \xrightarrow[M \to \infty]{a.s.} 0$ (Lemma 4.1).

The EM algorithm works as follows. We start with an arbitrary chosen value $\boldsymbol{\theta}^{(0)}$ of parameter $\boldsymbol{\theta}$. By a forward-backard algorithm (Section 6.4.2) we obtain the quantities $L_n(j), F_n(j)$ and P_n previously defined. By maximizing $Q(\boldsymbol{\theta} \mid \boldsymbol{\theta}^{(0)})$ with respect to $\boldsymbol{\theta}$ (Section 6.4.2), we will see that the solutions of the maximum-likelihood equations can be expressed in terms of $L_n(j), F_n(j)$, and P_n. For this reason, we obtain an update $\boldsymbol{\theta}^{(1)}$ of parameter $\boldsymbol{\theta}$. We continue recurrently, obtaining a sequence $(\boldsymbol{\theta}^{(m)})_{m \in \mathbb{N}}$ of updates of the parameter. This sequence verifies

$$g_M(Y_0^M \mid \boldsymbol{\theta}^{(m+1)}) \geq g_M(Y_0^M \mid \boldsymbol{\theta}^{(m)}), \ m \in \mathbb{N},$$

so the EM algorithm provides updates of the parameter that increase the likelihood function of the incomplete data. For this reason, the maximizing of $Q(\boldsymbol{\theta} \mid \boldsymbol{\theta}^{(m)})$ with respect to $\boldsymbol{\theta}$ provides a maximum of $g_M(Y_0^M \mid \boldsymbol{\theta})$ (which can be a local one).

We run the algorithm up to the moment when a particular stopping condition is fulfilled (maximal number of iterations, stationarity of the obtained estimates, stationarity of the likelihood, etc.). Choosing this stopping condition is one of the problems of the EM algorithm. We will not discuss here this topic, but the interested reader can see, e.g., Biernacki et al. (2003).

We will now present the EM algorithm obtained for an SM-M0 model. The proofs of the relations are postponed to the end of the chapter, where they are provided for the more general model of the SM-M1 type.

Estimation Step

Forward

- INITIALIZATION: For $n = 0$

$$P_0 = \sum_{i \in E} \mu(i, y_0), \quad F_0(j) = \frac{\mu(j, y_0)}{P_0}.$$

- For $n = 1, \ldots, M$

$$
\begin{aligned}
P_n = & \sum_{i \in E} \sum_{j \neq i} \sum_{t=1}^{n-1} \sum_{u=1}^{M-n} \frac{q_{ij}(u+t) F_{n-t}(i) \prod_{p=n-t+1}^{n} R_{i;y_p}}{\prod_{p=n-t+1}^{n-1} P_p} \\
& + \sum_{i \in E} \sum_{j \neq i} \sum_{t=1}^{n-1} \frac{q_{ij}(t) F_{n-t}(i) R_{j;y_n} \prod_{p=n-t+1}^{n-1} R_{i;y_p}}{\prod_{p=n-t+1}^{n-1} P_p} \\
& + \sum_{i \in E} \sum_{t=1}^{n-1} \frac{[1 - \sum_{k=1}^{M-n+t} \sum_{j \in E} q_{ij}(k)] F_{n-t}(i) \prod_{p=n-t+1}^{n} R_{i;y_p}}{\prod_{p=n-t+1}^{n-1} P_p} \\
& + \sum_{i \in E} \sum_{j \neq i} \sum_{u=1}^{M-n+1} \frac{q_{ij}(n+u) \mu(i, y_0) \prod_{p=1}^{n} R_{i;y_p}}{\prod_{p=0}^{n-1} P_p} \\
& + \sum_{i \in E} \sum_{j \neq i} \frac{q_{ij}(n) \mu(i, y_0) R_{j;y_n} \prod_{p=1}^{n-1} R_{i;y_p}}{\prod_{p=0}^{n-1} P_p} \\
& + \sum_{i \in E} \frac{[1 - \sum_{k=1}^{M} \sum_{j \in E} q_{ij}(k)] \mu(i, y_0) \prod_{p=1}^{n} R_{i;y_p}}{\prod_{p=0}^{n-1} P_p},
\end{aligned}
\tag{6.27}
$$

$$F_n(j) = \sum_{t=1}^{n-1} \sum_{i \neq j} \frac{q_{ij}(t) F_{n-t}(i) R_{j;y_n} \prod_{p=n-t+1}^{n-1} R_{i;y_p}}{\prod_{p=n-t+1}^{n} P_p}$$

$$+ \sum_{i \neq j} \frac{q_{ij}(n) \mu(i, y_0) R_{j;y_n} \prod_{p=1}^{n-1} R_{i;y_p}}{\prod_{p=0}^{n} P_p}. \tag{6.28}$$

Backward

- For $n = M$

$$L_{1;M}(i) := \mathbb{P}(Z_M = i, Z_{M-1} \neq i \mid \mathbf{y}_0^M) = F_M(i), \tag{6.29}$$

$$L_{2;M}(i) = \sum_{u=2}^{M} \frac{[1 - \sum_{l=1}^{u-1} \sum_{j \in E} q_{ij}(l)] F_{M-u+1}(i) \prod_{p=M-u+2}^{M} R_{i;y_p}}{\prod_{l=M-u+2}^{M} P_l}$$

$$+ \frac{\mu(i, y_0)[1 - \sum_{l=1}^{M} \sum_{j \in E} q_{ij}(l)] \prod_{p=1}^{M} R_{i;y_p}}{\prod_{l=0}^{M} P_l}. \tag{6.30}$$

- For $n = M - 1, \ldots, 1$

$$L_{1;n}(i) = \sum_{t=1}^{M-n} \sum_{j \neq i} \frac{q_{ij}(t) F_n(i) L_{1;n+t}(j) R_{j;y_{n+t}} \prod_{p=n+1}^{n+t-1} R_{i;y_p}}{F_{n+t}(j) \prod_{p=n+1}^{n+t} P_p}$$

$$+ \frac{[1 - \sum_{l=1}^{M-n} \sum_{j \in E} q_{ij}(l)] F_n(i) \prod_{p=n+1}^{M} R_{i;y_p}}{\prod_{p=n+1}^{M} P_p}, \tag{6.31}$$

$$L_{2;n}(i) = \sum_{t=0}^{M-n-1} \sum_{u=2}^{n} \sum_{j \neq i} \frac{q_{ij}(t+u) F_{n-u+1}(i) L_{1;n+t+1}(j) R_{j;y_{n+t+1}} \prod_{p=n-u+2}^{n+t} R_{i;y_p}}{F_{n+t+1}(j) \prod_{p=n-u+2}^{n+t+1} P_p}$$

$$+ \sum_{u=2}^{n} \frac{[1 - \sum_{l=1}^{M-n+u-1} \sum_{j \in E} q_{ij}(l)] F_{n-u+1}(i) \prod_{p=n-u+2}^{M} R_{i;y_p}}{\prod_{l=n-u+2}^{M} P_l}$$

$$+ \sum_{t=0}^{M-n-1} \sum_{j \neq i} \frac{q_{ij}(n+t+1) L_{1;n+t+1}(j) \mu(i, y_0) R_{j;y_{n+t+1}} \prod_{p=1}^{n+t} R_{i;y_p}}{F_{n+t+1}(j) \prod_{p=0}^{n+t+1} P_p}$$

$$+ \frac{\mu(i, y_0)[1 - \sum_{l=1}^{M} \sum_{j \in E} q_{ij}(l)] \prod_{p=1}^{M} R_{i;y_p}}{\prod_{l=0}^{M} P_l}. \tag{6.32}$$

- For $n = 0$

$$L_0(i) = \sum_{t=1}^{M} \sum_{j \neq i} \frac{L_{1;t}(j)\mu(i, y_0)q_{ij}(t)R_{j;y_t} \prod_{p=1}^{t-1} R_{i;y_p}}{F_t(j) \prod_{p=0}^{t} P_p}$$
$$+ \frac{\mu(i, y_0)[1 - \sum_{l=1}^{M} \sum_{j \in E} q_{ij}(l)] \prod_{p=1}^{M} R_{i;y_p}}{\prod_{p=0}^{M} P_p}. \tag{6.33}$$

Maximization Step

Maximizing $Q(\boldsymbol{\theta} \mid \boldsymbol{\theta}^{(m)})$ with respect to $\boldsymbol{\theta}$, for all $i, j \in E, b \in A$, we get

$$p_{ij}^{(m+1)} = \frac{\sum_{n=1}^{M} \mathbb{P}_{\boldsymbol{\theta}^{(m)}}(Z_n = j, Z_{n-1} = i \mid \mathbf{y})}{\sum_{n=0}^{M-1} \mathbb{P}_{\boldsymbol{\theta}^{(m)}}(Z_n = i, Z_{n-1} \neq i \mid \mathbf{y}) - \mathbb{P}_{\boldsymbol{\theta}^{(m)}}(Z_M = i \mid \mathbf{y})}, \tag{6.34}$$

where, for $n = 0$, we have set

$$\mathbb{P}_{\boldsymbol{\theta}^{(m)}}(Z_n = i, Z_{n-1} \neq i \mid \mathbf{y}) := \mathbb{P}_{\boldsymbol{\theta}^{(m)}}(Z_0 = i \mid \mathbf{y}) = L_0^{(m)}(i);$$

$$f_{ij}^{(m+1)}(k) = \frac{\sum_{n=0}^{M-k} \mathbb{P}_{\boldsymbol{\theta}^{(m)}}(Z_{n+k} = j, Z_n^{n+k-1} = i, Z_{n-1} \neq i \mid \mathbf{y})}{\sum_{n=1}^{M} \mathbb{P}_{\boldsymbol{\theta}^{(m)}}(Z_n = j, Z_{n-1} = i \mid \mathbf{y})}, \tag{6.35}$$

where, for $n = 0$, we have set

$$\mathbb{P}_{\boldsymbol{\theta}^{(m)}}(Z_{n+k} = j, Z_n^{n+k-1} = i, Z_{n-1} \neq i \mid \mathbf{y})$$
$$:= \mathbb{P}_{\boldsymbol{\theta}^{(m)}}(Z_k = j, Z_0^{k-1} = i \mid \mathbf{y});$$

$$q_{ij}^{(m+1)}(k) = f_{ij}^{(m+1)}(k)p_{ij}^{(m+1)} \tag{6.36}$$
$$= \frac{\sum_{n=0}^{M-k} \mathbb{P}_{\boldsymbol{\theta}^{(m)}}(Z_{n+k} = j, Z_n^{n+k-1} = i, Z_{n-1} \neq i \mid \mathbf{y})}{\sum_{n=0}^{M-1} \mathbb{P}_{\boldsymbol{\theta}^{(m)}}(Z_n = i, Z_{n-1} \neq i \mid \mathbf{y}) - \mathbb{P}_{\boldsymbol{\theta}^{(m)}}(Z_M = i \mid \mathbf{y})};$$

$$R_{i;b}^{(m+1)} = \frac{\sum_{n=1}^{M} \mathbb{P}_{\boldsymbol{\theta}^{(m)}}(Z_n = i \mid \mathbf{y})\mathbf{1}_{\{Y_n = b\}}}{\sum_{n=1}^{M} \mathbb{P}_{\boldsymbol{\theta}^{(m)}}(Z_n = i \mid \mathbf{y})}. \tag{6.37}$$

We compute $p_{ij}^{(m+1)}, f_{ij}^{(m+1)}(k), q_{ij}^{(m+1)}(k)$ and $R_{i;b}^{(m+1)}$ using the mth step of the estimation part and we obtain:

$$p_{ij}^{(m+1)} = \sum_{n=1}^{M} \Big[\sum_{k=1}^{n-1} \frac{q_{ij}(k)F_{n-k}(i)L_{1;n}(j)R_{j;y_n} \prod_{p=n-k+1}^{n-1} R_{i;y_p}}{F_n(j) \prod_{p=n-k+1}^{n} P_p} \tag{6.38}$$
$$+ \frac{q_{ij}(n)\mu(i, y_0)L_{1;n}(j)R_{j;y_n} \prod_{p=1}^{n-1} R_{i;y_p}}{F_n(j) \prod_{p=0}^{n} P_p} \Big] \Big/ \Big[\sum_{n=0}^{M-1} L_{1;n}^{(m)}(i) - L_{2;M}^{(m)}(i) \Big],$$

$$f_{ij}^{(m+1)}(k) = \sum_{n=0}^{M-k} \left[\frac{q_{ij}(k)F_n(i)L_{1;n+k}(j)R_{j;y_{n+k}} \prod_{p=n+1}^{n+k-1} R_{i;y_p}}{F_{n+k}(j) \prod_{p=n+1}^{n+k} P_p} \right]$$

$$\Big/ \sum_{n=1}^{M} \left[\sum_{l=1}^{n-1} \frac{q_{ij}(l)F_{n-l}(i)L_{1;n}(j)R_{j;y_n} \prod_{p=n-l+1}^{n-1} R_{i;y_p}}{F_n(j) \prod_{p=n-l+1}^{n} P_p} \right.$$

$$\left. + \frac{q_{ij}(n)\mu(i,y_0)L_{1;n}(j)R_{j;y_n} \prod_{p=1}^{n-1} R_{i;y_p}}{F_n(j) \prod_{p=0}^{n} P_p} \right], \tag{6.39}$$

$$q_{ij}^{(m+1)}(k) = \frac{1}{\sum_{n=0}^{M-1} L_{1;n}^{(m)}(i) - L_{2;M}^{(m)}(i)}$$

$$\cdot \sum_{n=0}^{M-k} \left[\frac{q_{ij}(k)F_n(i)L_{1;n+k}(j)R_{j;y_{n+k}} \prod_{p=n+1}^{n+k-1} R_{i;y_p}}{F_{n+k}(j) \prod_{p=n+1}^{n+k} P_p} \right], \tag{6.40}$$

$$R_{i;b}^{(m+1)} = \frac{\sum_{n=0}^{M} L_n^{(m)}(i) \mathbf{1}_{\{Y_n=b\}}}{\sum_{n=0}^{M} L_n^{(m)}(i)}, \tag{6.41}$$

where we have set

$$L_{1;0}^{(m)}(i) := L_0^{(m)}(i).$$

6.4.3 EM Algorithm for Hidden Model SM-M1

Let (Z, Y) be a hidden model of the SM-M1 type. In the sequel, we will present the analog of the EM algorithm constructed for the SM-M0 model.

The Likelihood Function

The likelihood function of the complete data for a hidden system SM-M1 is given by

$$f_M(Y_0^M, Z_0^M \mid \boldsymbol{\theta}) = \prod_{k=1}^{N(M)} \left[p_{J_{k-1}J_k} f_{J_{k-1}J_k}(X_k) \prod_{l=S_{k-1}}^{S_k-1} R_{J_{k-1},Y_{l-1},Y_l} \right]$$

$$\times \overline{H}_{J_{N(M)}}(U_M) \prod_{l=S_{N(M)}}^{M} R_{J_{N(M)},Y_{l-1},Y_l}.$$

As we did before, we have not considered the initial distribution in the expression of the likelihood function. For any $i \in E$ and any $a, b \in A$, let

$$N_{i;ab}(M) := \sum_{n=1}^{N(M)} \mathbf{1}_{\{Y_{n-1}=a, Y_n=b, Z_n=i\}},$$

be the number of transitions of chain Y from a to b, up to time M, with the semi-Markov chain Z visiting state i when Y is visiting state b.

The log-likelihood of the complete data can be written as

$$\log(f_M(Y_0^M, Z_0^M \mid \boldsymbol{\theta}))$$

$$= \log\Big[\prod_{i,j\in E} p_{ij}^{N_{ij}(M)} \prod_{i,j\in E} \prod_{k\in\mathbb{N}^*} f_{ij}(k)^{N_{ij}(k,M)}$$

$$\times \prod_{i\in E} \prod_{a,b\in A} R_{i;a,b}^{N_{i;ab}(M)} (1 - \sum_{l=1}^{k}\sum_{j\in E} q_{ij}(l))\Big]$$

$$= \sum_{i,j\in E} N_{ij}(M)\log p_{ij} + \sum_{i,j\in E}\sum_{k\in\mathbb{N}^*} N_{ij}(k,M)\log f_{ij}(k)$$

$$+ \sum_{i\in E}\sum_{a,b\in A} N_{i;ab}(M)\log R_{i;a,b} + \log\Big(\sum_{k\geq M-S_{N(M)}} (1 - \sum_{l=1}^{k}\sum_{j\in E} q_{ij}(l))\Big).$$

Consequently, the conditional expectation $Q(\boldsymbol{\theta}\mid\boldsymbol{\theta}^{(m)})$ can be written as

$$Q(\boldsymbol{\theta}\mid\boldsymbol{\theta}^{(m)}) := \mathbb{E}_{\boldsymbol{\theta}^{(m)}}[\log(f_M(Y_0^M, Z_0^M \mid \boldsymbol{\theta}))\mid \mathbf{y}]$$

$$= \sum_{i,j\in E}\log p_{ij}\sum_{n=1}^{M}\mathbb{P}_{\boldsymbol{\theta}^{(m)}}(Z_{n-1}=i, Z_n=j\mid \mathbf{y})$$

$$+ \sum_{i,j\in E}\sum_{k\in\mathbb{N}^*}\log f_{ij}(k)\sum_{n=0}^{M-k}\mathbb{P}_{\boldsymbol{\theta}^{(m)}}(Z_{n+k}=j, Z_n^{n+k-1}=i, Z_{n-1}\neq i\mid \mathbf{y})$$

$$+ \sum_{i\in E}\sum_{a,b\in A}\sum_{n=1}^{M}\mathbb{P}_{\boldsymbol{\theta}^{(m)}}(Z_n=i\mid \mathbf{y})\mathbf{1}_{\{Y_{n-1}=a,Y_n=b\}}\log R_{i;a,b}$$

$$+ \mathbb{E}_{\boldsymbol{\theta}^{(m)}}\Big[\log\Big(\sum_{k\geq M-S_{N(M)}} (1 - \sum_{l=1}^{k}\sum_{j\in E} q_{ij}(l))\Big)\mid \mathbf{y}\Big].$$

As we already did for the SM-M0 model, the last term of the right-hand side will be neglected. For the hidden SM-M1 model, the algorithm takes the following form.

Estimation Step

Forward

- INITIALIZATION: For $n = 0$

$$P_0 = \sum_{i\in E}\mu(i, y_0), \quad F_0(j) = \frac{\mu(j, y_0)}{\sum_{i\in E}\mu(i, y_0)}.$$

- For $n = 1, \ldots, M$

$$P_n = \sum_{i \in E} \sum_{j \neq i} \sum_{t=1}^{n-1} \sum_{u=1}^{M-n} \frac{q_{ij}(u+t) F_{n-t}(i) \prod_{p=n-t+1}^{n} R_{i;y_{p-1},y_p}}{\prod_{p=n-t+1}^{n-1} P_p}$$

$$+ \sum_{i \in E} \sum_{j \neq i} \sum_{t=1}^{n-1} \frac{q_{ij}(t) F_{n-t}(i) R_{j;y_{n-1},y_n} \prod_{p=n-t+1}^{n-1} R_{i;y_{p-1},y_p}}{\prod_{p=n-t+1}^{n-1} P_p}$$

$$+ \sum_{i \in E} \sum_{t=1}^{n-1} \frac{[1 - \sum_{k=1}^{M-n+t} \sum_{j \in E} q_{ij}(k)] F_{n-t}(i) \prod_{p=n-t+1}^{n} R_{i;y_{p-1},y_p}}{\prod_{p=n-t+1}^{n-1} P_p}$$

$$+ \sum_{i \in E} \sum_{j \neq i} \sum_{u=1}^{M-n+1} \frac{q_{ij}(n+u) \mu(i,y_0) \prod_{p=1}^{n} R_{i;y_{p-1},y_p}}{\prod_{p=0}^{n-1} P_p}$$

$$+ \sum_{i \in E} \sum_{j \neq i} \frac{q_{ij}(n) \mu(i,y_0) R_{j;y_{n-1},y_n} \prod_{p=1}^{n-1} R_{i;y_{p-1},y_p}}{\prod_{p=0}^{n-1} P_p}$$

$$+ \sum_{i \in E} \frac{[1 - \sum_{k=1}^{M} \sum_{j \in E} q_{ij}(k)] \mu(i,y_0) \prod_{p=1}^{n} R_{i;y_{p-1},y_p}}{\prod_{p=0}^{n-1} P_p}, \tag{6.42}$$

$$F_n(j) = \sum_{t=1}^{n-1} \sum_{i \neq j} \frac{q_{ij}(t) F_{n-t}(i) R_{j;y_{n-1},y_n} \prod_{p=n-t+1}^{n-1} R_{i;y_{p-1},y_p}}{\prod_{p=n-t+1}^{n} P_p}$$

$$+ \sum_{i \neq j} \frac{q_{ij}(n) \mu(i,y_0) R_{j;y_{n-1},y_n} \prod_{p=1}^{n-1} R_{i;y_{p-1},y_p}}{\prod_{p=0}^{n} P_p}. \tag{6.43}$$

Backward

- For $n = M$

$$L_{1;M}(i) := \mathbb{P}(Z_M = i, Z_{M-1} \neq i \mid \mathbf{y}_0^M) = F_M(i), \tag{6.44}$$

$$L_{2;M}(i) = \sum_{u=2}^{M} \frac{[1 - \sum_{l=1}^{u-1} \sum_{j \in E} q_{ij}(l)] F_{M-u+1}(i) \prod_{p=M-u+2}^{M} R_{i;y_{p-1},y_p}}{\prod_{l=M-u+2}^{M} P_l}$$

$$+ \frac{\mu(i,y_0)[1 - \sum_{l=1}^{M} \sum_{j \in E} q_{ij}(l)] \prod_{p=1}^{M} R_{i;y_{p-1},y_p}}{\prod_{l=0}^{M} P_l}. \tag{6.45}$$

- For $n = M-1, \ldots, 1$

$$L_{1;n}(i) = \sum_{t=1}^{M-n} \sum_{j \neq i} \frac{q_{ij}(t) F_n(i) L_{1;n+t}(j) R_{j;y_{n+t-1},y_{n+t}} \prod_{p=n+1}^{n+t-1} R_{i;y_{p-1},y_p}}{F_{n+t}(j) \prod_{p=n+1}^{n+t} P_p}$$

$$+\frac{[1-\sum_{l=1}^{M-n}\sum_{j\in E}q_{ij}(l)]F_n(i)\prod_{p=n+1}^{M}R_{i;y_{p-1},y_p}}{\prod_{p=n+1}^{M}P_p},\qquad(6.46)$$

$$L_{2;n}(i)=\sum_{t=0}^{M-n-1}\sum_{u=2}^{n}\sum_{j\neq i}\frac{q_{ij}(t+u)F_{n-u+1}(i)L_{1;n+t+1}(j)}{F_{n+t+1}(j)\prod_{p=n-u+2}^{n+t+1}P_p}$$

$$\times R_{j;y_{n+t},y_{n+t+1}}\prod_{p=n-u+2}^{n+t}R_{i;y_{p-1},y_p}$$

$$+\sum_{u=2}^{n}\frac{[1-\sum_{l=1}^{M-n+u-1}\sum_{j\in E}q_{ij}(l)]F_{n-u+1}(i)\prod_{p=n-u+2}^{M}R_{i;y_{p-1},y_p}}{\prod_{l=n-u+2}^{M}P_l}$$

$$+\sum_{t=0}^{M-n-1}\sum_{j\neq i}\frac{q_{ij}(n+t+1)L_{1;n+t+1}(j)\mu(i,y_0)}{F_{n+t+1}(j)\prod_{p=0}^{n+t+1}P_p}$$

$$\times R_{j;y_{n+t},y_{n+t+1}}\prod_{p=1}^{n+t}R_{i;y_{p-1},y_p}$$

$$+\frac{\mu(i,y_0)[1-\sum_{l=1}^{M}\sum_{j\in E}q_{ij}(l)]\prod_{p=1}^{M}R_{i;y_{p-1},y_p}}{\prod_{l=0}^{M}P_l}.\qquad(6.47)$$

- For $n=0$

$$L_0(i)=\sum_{t=1}^{M}\sum_{j\neq i}\frac{L_{1;t}(j)\mu(i,y_0)q_{ij}(t)R_{j;y_{t-1},y_t}\prod_{p=1}^{t-1}R_{i;y_{p-1},y_p}}{F_t(j)\prod_{p=0}^{t}P_p}$$

$$+\frac{\mu(i,y_0)[1-\sum_{l=1}^{M}\sum_{j\in E}q_{ij}(l)]\prod_{p=1}^{M}R_{i;y_{p-1},y_p}}{\prod_{p=0}^{M}P_p}.\qquad(6.48)$$

Maximization Step

Maximizing $Q(\boldsymbol{\theta}\mid\boldsymbol{\theta}^{(m)})$ with respect to $\boldsymbol{\theta}$, for all $i,j\in E, a,b\in A$, we get

$$p_{ij}^{(m+1)}=\frac{\sum_{n=1}^{M}\mathbb{P}_{\boldsymbol{\theta}^{(m)}}(Z_n=j,Z_{n-1}=i\mid\mathbf{y})}{\sum_{n=0}^{M-1}\mathbb{P}_{\boldsymbol{\theta}^{(m)}}(Z_n=i,Z_{n-1}\neq i\mid\mathbf{y})-\mathbb{P}_{\boldsymbol{\theta}^{(m)}}(Z_M=i\mid\mathbf{y})},\;(6.49)$$

where, for $n=0$, we set

$$\mathbb{P}_{\boldsymbol{\theta}^{(m)}}(Z_n=i,Z_{n-1}\neq i\mid\mathbf{y}):=\mathbb{P}_{\boldsymbol{\theta}^{(m)}}(Z_0=i\mid\mathbf{y})=L_0^{(m)}(i);$$

$$f_{ij}^{(m+1)}(k)=\frac{\sum_{n=0}^{M-k}\mathbb{P}_{\boldsymbol{\theta}^{(m)}}(Z_{n+k}=j,Z_n^{n+k-1}=i,Z_{n-1}\neq i\mid\mathbf{y})}{\sum_{n=1}^{M}\mathbb{P}_{\boldsymbol{\theta}^{(m)}}(Z_n=j,Z_{n-1}=i\mid\mathbf{y})},\;(6.50)$$

where, for $n = 0$, we have set

$$\mathbb{P}_{\boldsymbol{\theta}^{(m)}}(Z_{n+k} = j, Z_n^{n+k-1} = i, Z_{n-1} \neq i \mid \mathbf{y}) := \mathbb{P}_{\boldsymbol{\theta}^{(m)}}(Z_k = j, Z_0^{k-1} = i, \mid \mathbf{y});$$

$$
\begin{aligned}
q_{ij}^{(m+1)}(k) &= f_{ij}^{(m+1)}(k)p_{ij}^{(m+1)} \\
&= \frac{\sum_{n=0}^{M-k} \mathbb{P}_{\boldsymbol{\theta}^{(m)}}(Z_{n+k} = j, Z_n^{n+k-1} = i, Z_{n-1} \neq i \mid \mathbf{y})}{\sum_{n=0}^{M-1} \mathbb{P}_{\boldsymbol{\theta}^{(m)}}(Z_n = i, Z_{n-1} \neq i \mid \mathbf{y}) - \mathbb{P}_{\boldsymbol{\theta}^{(m)}}(Z_M = i \mid \mathbf{y})};
\end{aligned}
\tag{6.51}
$$

$$R_{i;a,b}^{(m+1)} = \frac{\sum_{n=1}^{M} \mathbb{P}_{\boldsymbol{\theta}^{(m)}}(Z_n = i \mid \mathbf{y})\mathbf{1}_{\{Y_{n-1}=a,Y_n=b\}}}{\sum_{n=1}^{M} \mathbb{P}_{\boldsymbol{\theta}^{(m)}}(Z_n = i \mid \mathbf{y})\mathbf{1}_{\{Y_{n-1}=a\}}}. \tag{6.52}$$

Using the mth step of the estimation part, we get

$$
\begin{aligned}
p_{ij}^{(m+1)} = \sum_{n=1}^{M} \Bigg[&\sum_{k=1}^{n-1} \frac{q_{ij}(k)F_{n-k}(i)L_{1;n}(j)R_{j;y_{n-1},y_n} \prod_{p=n-k+1}^{n-1} R_{i;y_{p-1},y_p}}{F_n(j)\prod_{p=n-k+1}^{n} P_p} \\
&+ \frac{q_{ij}(n)\mu(i,y_0)L_{1;n}(j)R_{j;y_{n-1},y_n} \prod_{p=1}^{n-1} R_{i;y_{p-1},y_p}}{F_n(j)\prod_{p=0}^{n} P_p} \Bigg] \\
&\Bigg/ \Bigg[\sum_{n=0}^{M-1} L_{1;n}^{(m)}(i) - L_{2;M}^{(m)}(i) \Bigg],
\end{aligned}
\tag{6.53}
$$

$$
\begin{aligned}
f_{ij}^{(m+1)}(k) = &\sum_{n=0}^{M-k} \Bigg[\frac{q_{ij}(k)F_n(i)L_{1;n+k}(j)R_{j;y_{n+k-1},y_{n+k}} \prod_{p=n+1}^{n+k-1} R_{i;y_{p-1},y_p}}{F_{n+k}(j)\prod_{p=n+1}^{n+k} P_p} \Bigg] \\
&\Bigg/ \sum_{n=1}^{M} \Bigg[\sum_{l=1}^{n-1} \frac{q_{ij}(l)F_{n-l}(i)L_{1;n}(j)R_{j;y_{n-1},y_n} \prod_{p=n-l+1}^{n-1} R_{i;y_{p-1},y_p}}{F_n(j)\prod_{p=n-l+1}^{n} P_p} \\
&+ \frac{q_{ij}(n)\mu(i,y_0)L_{1;n}(j)R_{j;y_{n-1},y_n} \prod_{p=1}^{n-1} R_{i;y_{p-1},y_p}}{F_n(j)\prod_{p=0}^{n} P_p} \Bigg],
\end{aligned}
\tag{6.54}
$$

$$
\begin{aligned}
q_{ij}^{(m+1)}(k) = &\frac{1}{\sum_{n=0}^{M-1} L_{1;n}^{(m)}(i) - L_{2;M}^{(m)}(i)} \\
&\times \sum_{n=0}^{M-k} \Bigg[\frac{q_{ij}(k)F_n(i)L_{1;n+k}(j)R_{j;y_{n+k-1},y_{n+k}} \prod_{p=n+1}^{n+k-1} R_{i;y_{p-1},y_p}}{F_{n+k}(j)\prod_{p=n+1}^{n+k} P_p} \Bigg],
\end{aligned}
\tag{6.55}
$$

$$R_{i;a,b}^{(m+1)} = \frac{\sum_{n=1}^{M} L_n^{(m)}(i)\mathbf{1}_{\{Y_{n-1}=a,Y_n=b\}}}{\sum_{n=1}^{M} L_n^{(m)}(i)\mathbf{1}_{\{Y_{n-1}=a\}}}, \tag{6.56}$$

with

$$L_{1;0}^{(m)}(i) := L_0^{(m)}(i).$$

6.4.4 Proofs

We give here the proofs of the relationships of the algorithm, in the case of a hidden SM-M1 model. From a technical point of view, the results are based on the conditional independence results given in Lemmas (B.4)–(B.6) from Appendix B.

The relationships of the EM algorithm for the hidden SM-M0 model can be proved in the same manner, with minor modifications, using Lemmas (B.1)–(B.3) from Appendix B.

Estimation-Forward Step

We shall justify the expression of $F_n(j)$ given in Equation (6.43). For $n = 1, \ldots, M$, and $j \in E$, we have

$$F_n(j) = \mathbb{P}(Z_n = j, Z_{n-1} \neq j \mid \mathbf{y}_0^n) = \sum_{i \neq j} \mathbb{P}(Z_n = j, Z_{n-1} = i \mid \mathbf{y}_0^n)$$

$$= \sum_{i \neq j} \sum_{t=1}^{n-1} \mathbb{P}(Z_n = j, Z_{n-t}^{n-1} = i, Z_{n-t-1} \neq i \mid \mathbf{y}_0^n)$$

$$+ \sum_{i \neq j} \mathbb{P}(Z_n = j, Z_0^{n-1} = i \mid \mathbf{y}_0^n).$$

Let us express the two terms of $F_n(j)$ by means of the semi-Markov kernel and of the conditional transition matrix of Y.

- Computing the first term of $F_n(j)$

$$\mathbb{P}(Z_n = j, Z_{n-t}^{n-1} = i, Z_{n-t-1} \neq i \mid \mathbf{y}_0^n)$$

$$= \mathbb{P}(Z_n = j, Z_{n-t}^{n-1} = i, Z_{n-t-1} \neq i, \mathbf{y}_{n-t+1}^n \mid \mathbf{y}_0^{n-t}) \big/ \mathbb{P}(\mathbf{y}_{n-t+1}^n \mid \mathbf{y}_0^{n-t})$$

$$= \mathbb{P}(\mathbf{y}_{n-t+1}^n \mid Z_n = j, Z_{n-t}^{n-1} = i, Z_{n-t-1} \neq i, \mathbf{y}_0^{n-t})$$

$$\times \mathbb{P}(Z_n = j, Z_{n-t}^{n-1} = i \mid Z_{n-t} = i, Z_{n-t-1} \neq i, \mathbf{y}_0^{n-t})$$

$$\times \mathbb{P}(Z_{n-t} = i, Z_{n-t-1} \neq i \mid \mathbf{y}_0^{n-t}) / \mathbb{P}(\mathbf{y}_{n-t+1}^n \mid \mathbf{y}_0^{n-t})$$

$$= q_{ij}(t) F_{n-t}(i) R_{j;y_{n-1},y_n} \prod_{p=n-t+1}^{n-1} R_{i;y_{p-1},y_p} \Big/ \prod_{p=n-t+1}^{n} P_p.$$

- Computing the second term of $F_n(j)$

Similarly, we have:

$$\mathbb{P}(Z_n = j, Z_0^{n-1} = i \mid \mathbf{y}_0^n) = \mathbb{P}(Z_n = j, Z_0^{n-1} = i, \mathbf{y}_0^n)/\mathbb{P}(\mathbf{y}_0^n)$$

$$= \mathbb{P}(\mathbf{y}_1^n \mid Z_n = j, Z_0^{n-1} = i, \mathbf{y}_0)\mathbb{P}(Z_n = j, Z_1^{n-1} = i \mid Z_0 = i, \mathbf{y}_0)$$

$$\times \mathbb{P}(Z_0 = i, \mathbf{y}_0)\Big/ \prod_{p=0}^{n} P_p$$

$$= q_{ij}(n)\mu(i, \mathbf{y}_0)R_{j;y_{n-1},y_n} \prod_{p=1}^{n-1} R_{i;y_{p-1},y_p}\Big/ \prod_{p=0}^{n} P_p.$$

Let us prove the expression of P_l given in Equation (6.42). For $l = 1, \ldots, M$, we have

$$P_l = \sum_{i \in E} \mathbb{P}(Z_{l-1} = i, \mathbf{y}_l \mid \mathbf{y}_0^{l-1})$$

$$= \sum_{i \in E} \sum_{j \neq i} \sum_{t=1}^{l-1} \sum_{u=0}^{M-l} \mathbb{P}(Z_{l+u} = j, Z_{l-t}^{l+u-1} = i, Z_{l-t-1} \neq i, \mathbf{y}_l \mid \mathbf{y}_0^{l-1})$$

$$+ \sum_{i \in E} \sum_{t=1}^{l-1} \mathbb{P}(Z_{l-t}^M = i, Z_{l-t-1} \neq i, \mathbf{y}_l \mid \mathbf{y}_0^{l-1})$$

$$+ \sum_{i \in E} \sum_{j \neq i} \sum_{u=0}^{M-l} \mathbb{P}(Z_{l+u} = j, Z_0^{l+u-1} = i, \mathbf{y}_l \mid \mathbf{y}_0^{l-1})$$

$$+ \sum_{i \in E} \mathbb{P}(Z_0^M = i, \mathbf{y}_l \mid \mathbf{y}_0^{l-1}).$$

We shall write each of the four terms of P_l in terms of the semi-Markov kernel and of the conditional transition matrix of Y.

- Computing the first term of P_l for $u = 0$

$$\mathbb{P}(Z_l = j, Z_{l-t}^{l-1} = i, Z_{l-t-1} \neq i, \mathbf{y}_l \mid \mathbf{y}_0^{l-1})$$

$$= \mathbb{P}(Z_l = j, Z_{l-t}^{l-1} = i, Z_{l-t-1} \neq i, \mathbf{y}_{l-t+1}^l \mid \mathbf{y}_0^{l-t})/\mathbb{P}(\mathbf{y}_{l-t+1}^{l-1} \mid \mathbf{y}_0^{l-t})$$

$$= \mathbb{P}(\mathbf{y}_{l-t+1}^l \mid Z_l = j, Z_{l-t}^{l-1} = i, Z_{l-t-1} \neq i, \mathbf{y}_0^{l-t})$$

$$\times \mathbb{P}(Z_l = j, Z_{l-t+1}^{l-1} = i \mid Z_{l-t} = i, Z_{l-t-1} \neq i, \mathbf{y}_0^{l-t})$$

$$\times \mathbb{P}(Z_{l-t} = i, Z_{l-t-1} \neq i \mid \mathbf{y}_0^{l-t})\Big/ \prod_{p=l-t+1}^{l-1} P_p$$

$$= q_{ij}(t)F_{l-t}(i)R_{j;y_{l-1},y_l} \prod_{p=l-t+1}^{l-1} R_{i;y_{p-1},y_p}\Big/ \prod_{p=l-t+1}^{l-1} P_p.$$

- Computing the first term of P_l for $u \neq 0$

A similar computation yields

$$q_{ij}(t+u)F_{l-t}(i) \prod_{p=l-t+1}^{l} R_{i;y_{p-1},y_p} \Big/ \prod_{p=l-t+1}^{l-1} P_p.$$

- Computing the second term of P_l

We have

$$\mathbb{P}(Z_{l-t}^M = i, Z_{l-t-1} \neq i, \mathbf{y}_l \mid \mathbf{y}_0^{l-1})$$
$$= \mathbb{P}(Z_{l-t}^M = i, Z_{l-t-1} \neq i, \mathbf{y}_{l-t+1}^l \mid \mathbf{y}_0^{l-t})/\mathbb{P}(\mathbf{y}_{l-t+1}^l \mid \mathbf{y}_0^{l-t})$$
$$= \mathbb{P}(\mathbf{y}_{l-t+1}^l \mid Z_{l-t}^M = i, Z_{l-t-1} \neq i, \mathbf{y}_0^{l-t})$$
$$\times \mathbb{P}(Z_{l-t+1}^M = i \mid Z_{l-t} = i, Z_{l-t-1} \neq i, \mathbf{y}_0^{l-t})$$
$$\times \mathbb{P}(Z_{l-t} = i, Z_{l-t-1} \neq i \mid \mathbf{y}_0^{l-t}) \Big/ \prod_{p=l-t+1}^{l-1} P_p$$
$$= \left[1 - \sum_{k=1}^{M-l+t} \sum_{j \in E} q_{ij}(k) \right] F_{l-t}(i) \prod_{p=l-t+1}^{l} R_{i;y_{p-1},y_p} \Big/ \prod_{p=l-t+1}^{l-1} P_p,$$

where we used the fact that

$$\mathbb{P}(Z_{l-t+1}^M = i \mid Z_{l-t} = i, Z_{l-t-1} \neq i)$$
$$= \mathbb{P}(X_{N(t)+1} \geq M - l + t + 1 \mid J_{N(t)} = i)$$
$$= \overline{H}_i(M - l + t)$$
$$= 1 - \sum_{k=1}^{M-l+t} \sum_{j \in E} q_{ij}(k).$$

- Computing the third term of P_l for $u = 0$

We have

$$\mathbb{P}(Z_l = j, Z_0^{l-1} = i, y_l \mid \mathbf{y}_0^{l-1})$$
$$= \mathbb{P}(Z_l = j, Z_0^{l-1} = i, \mathbf{y}_1^l \mid \mathbf{y}_0)/\mathbb{P}(\mathbf{y}_1^{l-1} \mid \mathbf{y}_0)$$
$$= \mathbb{P}(\mathbf{y}_1^l \mid Z_l = j, Z_0^{l-1} = i, \mathbf{y}_0)\mathbb{P}(Z_l = j, Z_1^{l-1} = i \mid Z_0 = i, \mathbf{y}_0)$$
$$\times \mathbb{P}(Z_0 = i \mid \mathbf{y}_0) \Big/ \prod_{p=1}^{l-1} P_p$$
$$= q_{ij}(l)\mu(i, y_0)R_{j;y_{l-1},y_l} \prod_{p=1}^{l-1} R_{i;y_{p-1},y_p} \Big/ \prod_{p=0}^{l-1} P_p.$$

- Computing the third term of P_l for $u \neq 0$

Similarly, for $u \neq 0$ we get the expression

$$q_{ij}(l+u)\mu(i,y_0) \prod_{p=1}^{l} R_{i;y_{p-1},y_p} \Big/ \prod_{p=0}^{l-1} P_p$$

for the third term of P_l.

- Computing the fourth term of P_l

An analogous computation yields

$$\Big[1 - \sum_{k=1}^{M} \sum_{j \in E} q_{ij}(k)\Big] \mu(i,y_0) \prod_{p=1}^{l} R_{i;y_{p-1},y_p} \Big/ \prod_{p=0}^{l-1} P_p.$$

Estimation-Backward Step

In the sequel, we prove the expressions of $L_{1;n}(i)$ and $L_{2;n}(i)$ given in Section 6.4.3.

Proof of Relation (6.46)

For $n = 1, \ldots, M$, and $i \in E$, we have

$$L_{1;n}(i) = \mathbb{P}(Z_n = i, Z_{n-1} \neq i \mid \mathbf{y}_0^M)$$

$$= \sum_{j \neq i} \sum_{t=1}^{M-n} \mathbb{P}(Z_{n+t} = j, Z_n^{n+t-1} = i, Z_{n-1} \neq i, \mathbf{y}_0^M) / \mathbb{P}(\mathbf{y}_0^M)$$

$$+ \mathbb{P}(Z_n^M = i, Z_{n-1} \neq i, \mathbf{y}_0^M) / \mathbb{P}(\mathbf{y}_0^M).$$

We have

$$\mathbb{P}(Z_{n+t} = j, Z_n^{n+t-1} = i, Z_{n-1} \neq i, \mathbf{y}_0^M)$$
$$= \mathbb{P}(\mathbf{y}_{n+t+1}^M \mid Z_{n+t} = j, Z_n^{n+t-1} = i, Z_{n-1} \neq i, \mathbf{y}_0^{n+t})$$
$$\times \mathbb{P}(\mathbf{y}_{n+1}^{n+t} \mid Z_{n+t} = j, Z_n^{n+t-1} = i, Z_{n-1} \neq i, \mathbf{y}_0^n)$$
$$\times \mathbb{P}(Z_{n+t} = j, Z_{n+1}^{n+t-1} = i \mid Z_n = i, Z_{n-1} \neq i, \mathbf{y}_0^n)$$
$$\times \mathbb{P}(Z_n = i, Z_{n-1} \neq i \mid \mathbf{y}_0^n)\mathbb{P}(\mathbf{y}_0^n)$$

$$= q_{ij}(t)F_n(i)R_{j;y_{n+t-1},y_{n+t}} \prod_{p=n+1}^{n+t-1} R_{i;y_{p-1},y_p}$$
$$\times \mathbb{P}(\mathbf{y}_{n+t+1}^M \mid Z_{n+t} = j, Z_{n+t-1} \neq j, \mathbf{y}_{n+t})\mathbb{P}(\mathbf{y}_0^n),$$

$$\mathbb{P}(\mathbf{y}_0^M) = \frac{\mathbb{P}(Z_{n+t} = j, Z_{n+t-1} \neq j, \mathbf{y}_0^M)}{\mathbb{P}(Z_{n+t} = j, Z_{n+t-1} \neq j \mid \mathbf{y}_0^M)}$$

$$= \frac{\mathbb{P}(\mathbf{y}_{n+t+1}^M \mid Z_{n+t} = j, Z_{n+t-1} \neq j, \mathbf{y}_0^{n+t})}{L_{1;n+t}(j)}$$
$$\times \mathbb{P}(Z_{n+t} = j, Z_{n+t-1} \neq j \mid \mathbf{y}_0^{n+t})\mathbb{P}(\mathbf{y}_0^{n+t})$$

$$= \frac{\mathbb{P}(\mathbf{y}_{n+t+1}^M \mid Z_{n+t} = j, Z_{n+t-1} \neq j, \mathbf{y}_{n+t}) F_{n+t}(j) \mathbb{P}(\mathbf{y}_0^{n+t})}{L_{1;n+t}(j)},$$

and

$$\mathbb{P}(Z_n^M = i, Z_{n-1} \neq i, \mathbf{y}_0^M)$$
$$= \mathbb{P}(\mathbf{y}_{n+1}^M \mid Z_n^M = i, Z_{n-1} \neq i, \mathbf{y}_0^n)$$
$$\times \mathbb{P}(Z_{n+1}^M = i \mid Z_n = i, Z_{n-1} \neq i, \mathbf{y}_0^n)$$
$$\times \mathbb{P}(Z_n = i, Z_{n-1} \neq i \mid \mathbf{y}_0^n) \mathbb{P}(\mathbf{y}_0^n)$$
$$= \left[1 - \sum_{l=1}^{M-n} \sum_{j \in E} q_{ij}(l) \right] F_n(i) \prod_{p=n+1}^M R_{i;y_{p-1},y_p} \mathbb{P}(\mathbf{y}_0^n).$$

Finally, we get

$$L_{1;n}(i) = \sum_{t=1}^{M-n} \sum_{j \neq i} \frac{q_{ij}(t) F_n(i) L_{1;n+t}(j) R_{j;y_{n+t-1},y_{n+t}} \prod_{p=n+1}^{n+t-1} R_{i;y_{p-1},y_p}}{F_{n+t}(j) \prod_{p=n+1}^{n+t} P_p}$$
$$+ \frac{[1 - \sum_{l=1}^{M-n} \sum_{j \in E} q_{ij}(l)] F_n(i) \prod_{p=n+1}^M R_{i;y_{p-1},y_p}}{\prod_{p=n+1}^M P_p}.$$

Proof of Relation (6.47)

For $n = 1, \ldots, M$, and $i \in E$, we have

$$L_{2;n}(i) := \mathbb{P}(Z_n = i, Z_{n-1} = i \mid \mathbf{y}_0^M)$$
$$= \sum_{j \neq i} \sum_{t=0}^{M-n-1} \sum_{u=2}^n \mathbb{P}(Z_{n+t+1} = j, Z_{n-u+1}^{n+t} = i, Z_{n-u} \neq i \mid \mathbf{y}_0^M)$$
$$+ \sum_{u=2}^n \mathbb{P}(Z_{n-u+1}^M = i, Z_{n-u} \neq i \mid \mathbf{y}_0^M)$$
$$+ \sum_{j \neq i} \sum_{t=0}^{M-n-1} \mathbb{P}(Z_{n+t+1} = j, Z_0^{n+t} = i, \mid \mathbf{y}_0^M)$$
$$+ \mathbb{P}(Z_0^M = i \mid \mathbf{y}_0^M).$$

We shall compute the four terms that appear in the expression of $L_{2;n}(i)$.

- Computing the first term of $L_{2;n}(i)$

$$\mathbb{P}(Z_{n+t+1} = j, Z_{n-u+1}^{n+t} = i, Z_{n-u} \neq i, \mathbf{y}_0^M)$$
$$= q_{ij}(t+u) F_{n-u+1}(i) R_{j;y_{n+t},y_{n+t+1}} \prod_{p=n-u+2}^{n+t} R_{i;y_{p-1},y_p}$$
$$\times \mathbb{P}(\mathbf{y}_{n+t+2}^M \mid Z_{n+t+1} = j, Z_{n+t} \neq j, \mathbf{y}_{n+t+1}) \mathbb{P}(\mathbf{y}_0^{n-u+1}).$$

On the other hand, we have

$$
\begin{aligned}
\mathbb{P}(\mathbf{y}_0^M) &= \frac{\mathbb{P}(Z_{n+t+1} = j, Z_{n+t} \neq j, \mathbf{y}_0^M)}{\mathbb{P}(Z_{n+t+1} = j, Z_{n+t} \neq j \mid \mathbf{y}_0^M)} \\
&= \frac{\mathbb{P}(\mathbf{y}_{n+t+2}^M \mid Z_{n+t+1} = j, Z_{n+t} \neq j, \mathbf{y}_0^{n+t+1})}{L_{1;n+t+1}(j)} \\
&\quad \times \mathbb{P}(Z_{n+t+1} = j, Z_{n+t} \neq j \mid \mathbf{y}_0^{n+t+1})\mathbb{P}(\mathbf{y}_0^{n+t+1}) \\
&= \frac{\mathbb{P}(\mathbf{y}_{n+t+2}^M \mid Z_{n+t+1} = j, Z_{n+t} \neq j, \mathbf{y}_{n+t+1})F_{n+t+1}(j)\mathbb{P}(\mathbf{y}_0^{n+t+1})}{L_{1;n+t+1}(j)}.
\end{aligned}
$$

Consequently, the first term of $L_{2;n}(i)$ is

$$
\sum_{j \neq i} \sum_{t=0}^{M-n-1} \sum_{u=2}^{n} \frac{q_{ij}(t+u)F_{n-u+1}(i)L_{1;n+t+1}(j)R_{j;y_{n+t},y_{n+t+1}} \prod_{p=n-u+2}^{n+t} R_{i;y_{p-1},y_p}}{F_{n+t+1}(j) \prod_{p=n-u+2}^{n+t+1} P_p}.
$$

- Computing the second term of $L_{2;n}(i)$

$$
\begin{aligned}
&\mathbb{P}(Z_{n-u+1}^M = i, Z_{n-u} \neq i, \mathbf{y}_0^M) \\
&= \mathbb{P}(\mathbf{y}_{n-u+2}^M \mid Z_{n-u+1}^M = i, Z_{n-u} \neq i, \mathbf{y}_0^{n-u+1}) \\
&\quad \times \mathbb{P}(Z_{n-u+2}^M = i \mid Z_{n-u+1} = i, Z_{n-u} \neq i, \mathbf{y}_0^{n-u+1}) \\
&\quad \times \mathbb{P}(Z_{n-u+1} = i, Z_{n-u} \neq i \mid \mathbf{y}_0^{n-u+1}) \times \mathbb{P}(\mathbf{y}_0^{n-u+1}) \\
&= \Big[1 - \sum_{l=1}^{M-n+u-1} \sum_{j \in E} q_{ij}(l)\Big] F_{n-u+1}(i) \prod_{p=n-u+2}^{M} R_{i;y_{p-1},y_p} \mathbb{P}(\mathbf{y}_0^{n-u+1}).
\end{aligned}
$$

We obtain the second term of $L_{2;n}(i)$ as

$$
\sum_{u=2}^{n} \frac{[1 - \sum_{l=1}^{M-n+u-1} \sum_{j \in E} q_{ij}(l)] F_{n-u+1}(i) \prod_{p=n-u+2}^{M} R_{i;y_{p-1},y_p}}{\prod_{l=n-u+2}^{M} P_l}.
$$

- Computing the third term of $L_{2;n}(i)$

$$
\begin{aligned}
&\mathbb{P}(Z_{n+t+1} = j, Z_0^{n+t} = i, \mathbf{y}_0^M) \\
&= \mathbb{P}(\mathbf{y}_{n+t+2}^M \mid Z_{n+t+1} = j, Z_0^{n+t} = i, \mathbf{y}_0^{n+t+1}) \\
&\quad \times \mathbb{P}(\mathbf{y}_1^{n+t+1} \mid Z_{n+t+1} = j, Z_0^{n+t} = i, \mathbf{y}_0) \\
&\quad \times \mathbb{P}(Z_{n+t+1} = j, Z_1^{n+t} = i \mid Z_0 = i, \mathbf{y}_0) \cdot \mathbb{P}(Z_0 = i, \mathbf{y}_0) \\
&= q_{ij}(n+t+1)\mu(i,y_0)R_{j;y_{n+t},y_{n+t+1}} \prod_{p=1}^{n+t} R_{i;y_{p-1},y_p} \\
&\quad \times \mathbb{P}(\mathbf{y}_{n+t+2}^M \mid Z_{n+t+1} = j, Z_{n+t} \neq j, \mathbf{y}_{n+t+1}) \cdot \mu(i,y_0).
\end{aligned}
$$

Using

$$\mathbb{P}(\mathbf{y}_0^M) = \frac{\mathbb{P}(\mathbf{y}_{n+t+2}^M \mid Z_{n+t+1} = j, Z_{n+t} \neq j, \mathbf{y}_{n+t+1}) F_{n+t+1}(j) \mathbb{P}(\mathbf{y}_0^{n+t+1})}{L_{1;n+t+1}(j)}$$

we obtain the third term of $L_{2;n}(i)$

$$\sum_{t=0}^{M-n-1} \sum_{j \neq i} \frac{q_{ij}(n+t+1) L_{1;n+t+1}(j) \mu(i, y_0) R_{j;y_{n+t}, y_{n+t+1}} \prod_{p=1}^{n+t} R_{i;y_{p-1}, y_p}}{F_{n+t+1}(j) \prod_{l=0}^{n+t+1} P_l}.$$

• Computing the fourth term of $L_{2;n}(i)$

$$\mathbb{P}(Z_0^M = i, \mathbf{y}_0^M)$$
$$= \mathbb{P}(\mathbf{y}_1^M \mid Z_0^M = i, \mathbf{y}_0) \mathbb{P}(Z_1^M = i \mid Z_0 = i, \mathbf{y}_0) \cdot \mathbb{P}(Z_0 = i, \mathbf{y}_0)$$
$$= \mu(i, y_0)[1 - \sum_{l=1}^{M} \sum_{j \in E} q_{ij}(l)] \prod_{p=1}^{M} R_{i;y_{p-1}, y_p}.$$

We obtain the fourth term of $L_{2;n}(i)$

$$\frac{\mu(i, y_0) \left[1 - \sum_{l=1}^{M} \sum_{j \in E} q_{ij}(l) \right] \prod_{p=1}^{M} R_{i;y_{p-1}, y_p}}{\prod_{l=0}^{M} P_l}.$$

Proof of Relation (6.45)

The proof is similar to that of Relation (6.47).

Proof of Relation (6.48)

$$L_0(i) := \mathbb{P}(Z_0 = i \mid \mathbf{y}_0^M)$$
$$= \sum_{j \neq i} \sum_{t=1}^{M} \frac{\mathbb{P}(Z_t = j, Z_0^{t-1} = i, \mathbf{y}_0^M)}{\mathbb{P}(\mathbf{y}_0^M)} + \frac{\mathbb{P}(Z_0^M = i, \mathbf{y}_0^M)}{\mathbb{P}(\mathbf{y}_0^M)}.$$

We have

$$\mathbb{P}(Z_t = j, Z_0^{t-1} = i, \mathbf{y}_0^M)$$
$$= \mathbb{P}(\mathbf{y}_{t+1}^M \mid Z_t = j, Z_0^{t-1} = i, \mathbf{y}_0^t)$$
$$\times \mathbb{P}(\mathbf{y}_1^t \mid Z_t = j, Z_0^{t-1} = i, \mathbf{y}_0)$$
$$\times \mathbb{P}(Z_t = j, Z_1^{t-1} = i \mid Z_0 = i, \mathbf{y}_0) \cdot \mathbb{P}(Z_0 = i, \mathbf{y}_0)$$
$$= q_{ij}(t) R_{j;y_{t-1}, y_t} \prod_{p=1}^{t-1} R_{i;y_{p-1}, y_p}$$
$$\times \mu(i, y_0) \mathbb{P}(\mathbf{y}_{t+1}^M \mid Z_t = j, Z_{t-1} \neq j, \mathbf{y}_t),$$

$$\mathbb{P}(\mathbf{y}_0^M) = \frac{\mathbb{P}(Z_t = j, Z_{t-1} \neq j, \mathbf{y}_0^M)}{\mathbb{P}(Z_t = j, Z_{t-1} \neq j \mid \mathbf{y}_0^M)}$$

$$= \frac{\mathbb{P}(\mathbf{y}_{t+1}^M \mid Z_t = j, Z_{t-1} \neq j, \mathbf{y}_0^t)\mathbb{P}(Z_t = j, Z_{t-1} \neq j \mid \mathbf{y}_0^t)\mathbb{P}(\mathbf{y}_0^t)}{L_{1;t}(j)}$$

$$= \frac{\mathbb{P}(\mathbf{y}_{t+1}^M \mid Z_t = j, Z_{t-1} \neq j, \mathbf{y}_t)F_t(j)\mathbb{P}(\mathbf{y}_0^t)}{L_{1;t}(j)},$$

and

$$\mathbb{P}(Z_0^M = i, \mathbf{y}_0^M)$$
$$= \mathbb{P}(\mathbf{y}_1^M \mid Z_0^M = i, \mathbf{y}_0)\mathbb{P}(Z_1^M = i \mid Z_0 = i, \mathbf{y}_0)\mu(i, y_0)$$
$$= \mu(i, y_0)\left[1 - \sum_{l=1}^{M}\sum_{j \in E} q_{ij}(l)\right]\prod_{p=1}^{M} R_{i;y_{p-1},y_p}.$$

Finally, we get

$$L_0(i) = \sum_{t=1}^{M}\sum_{j \neq i} \frac{L_{1;t}(j)\mu(i, y_0)q_{ij}(t)R_{j;y_{t-1},y_t}\prod_{p=1}^{t-1} R_{i;y_{p-1},y_p}}{F_t(j)\prod_{p=0}^{t} P_p}$$

$$+ \frac{\mu(i, y_0)[1 - \sum_{l=1}^{M}\sum_{j \in E} q_{ij}(l)]\prod_{p=1}^{M} R_{i;y_{p-1},y_p}}{\prod_{p=0}^{M} P_p}.$$

Maximization Step

Proof of Relations (6.49)–(6.52).

Using the relationship

$$\sum_{k=1}^{\infty} f_{ij}(k) = 1, \tag{6.57}$$

we can write $Q(\boldsymbol{\theta} \mid \boldsymbol{\theta}^{(m)})$ as

$$Q(\boldsymbol{\theta} \mid \boldsymbol{\theta}^{(m)}) \tag{6.58}$$

$$= \sum_{i,j \in E} \log p_{ij} \sum_{n=1}^{M} \mathbb{P}_{\boldsymbol{\theta}^{(m)}}(Z_{n-1} = i, Z_n = j \mid \mathbf{y})$$

$$+ \sum_{i,j \in E}\sum_{k \in \mathbb{N}^*} \log f_{ij}(k) \sum_{n=0}^{M-k} \mathbb{P}_{\boldsymbol{\theta}^{(m)}}(Z_{n+k} = j, Z_n^{n+k-1} = i, Z_{n-1} \neq i \mid \mathbf{y})$$

$$+ \sum_{i \in E}\sum_{a,b \in A}\sum_{n=1}^{N(M)} \mathbb{P}_{\boldsymbol{\theta}^{(m)}}(Z_n = i \mid \mathbf{y})\mathbf{1}_{\{Y_{n-1}=a,Y_n=b\}} \log R_{i;a,b}$$

$$+ \sum_{i,j \in E}\sum_{k \in \mathbb{N}^*} \lambda_{ij}(1 - f_{ij}(k)),$$

with λ_{ij} arbitrarily chosen. In order to obtain the maximum-likelihood estimator of $f_{ij}(k)$, we maximize (6.58) with respect to $f_{ij}(k)$ and we get

$$f_{ij}^{(m+1)}(k, M) = \frac{\sum_{n=0}^{M-k} \mathbb{P}_{\boldsymbol{\theta}^{(m)}}(Z_{n+k} = j, Z_n^{n+k-1} = i, Z_{n-1} \neq i \mid \mathbf{y})}{\lambda_{ij}}.$$

Equation (6.57) becomes

$$1 = \frac{\sum_{k=0}^{M} \sum_{n=0}^{M-k} \mathbb{P}_{\boldsymbol{\theta}^{(m)}}(Z_{n+k} = j, Z_n^{n+k-1} = i, Z_{n-1} \neq i \mid \mathbf{y})}{\lambda_{ij}}$$

$$= \frac{\sum_{n=1}^{M} \mathbb{P}_{\boldsymbol{\theta}^{(m)}}(Z_{n-1} = i, Z_n = j \mid \mathbf{y})}{\lambda_{ij}}.$$

Finally, we infer that the values λ_{ij} that maximize (6.58) with respect to $f_{ij}(k)$ are

$$\widehat{\lambda}_{ij}(M) = \sum_{n=1}^{M} \mathbb{P}_{\boldsymbol{\theta}^{(m)}}(Z_{n-1} = i, Z_n = j \mid \mathbf{y}),$$

and we get

$$f_{ij}^{(m+1)}(k, M) = \frac{\sum_{n=0}^{M-k} \mathbb{P}_{\boldsymbol{\theta}^{(m)}}(Z_{n+k} = j, Z_n^{n+k-1} = i, Z_{n-1} \neq i \mid \mathbf{y})}{\sum_{n=1}^{M} \mathbb{P}_{\boldsymbol{\theta}^{(m)}}(Z_{n-1} = i, Z_n = j \mid \mathbf{y})}.$$

The expression of $p_{ij}^{(m+1)}(M)$ can be obtained by the same method. Indeed, starting from the equality $\sum_{j \in E} p_{ij} = 1$, and introducing in (6.58) the term $\sum_{i \in E} \lambda_i(1 - \sum_{j \in E} p_{ij})$, with λ_i arbitrarily chosen, we maximize (6.58) with respect to p_{ij} and we obtain Relation (6.49).

The same method is used for obtaining $R_M^{(m+1)}(i; a, b)$. Indeed, starting from the equality $\sum_{b \in A} R(i; a, b) = 1$, and introducing in (6.58) the term $\sum_{i \in E} \sum_{a \in A} \lambda_{ia}(1 - \sum_{b \in A} R(i; a, b))$, with λ_{ia} arbitrarily chosen, we maximize (6.58) with respect to $R(i; a, b)$ and we obtain Relation (6.52).

Proof of Relations (6.53)–(6.56).

For all $i, j \in E, a, b \in A, m \in \mathbb{N}$, we shall compute $p_{ij}^{(m+1)}, f_{ij}^{(m+1)}(k), q_{ij}^{(m+1)}(k)$ and $R^{(m+1)}(i; a, b)$ using the mth step of the Forward-Backward algorithm.

The probabilities

$$\mathbb{P}_{\boldsymbol{\theta}^{(m)}}(Z_n = i, Z_{n-1} \neq i \mid \mathbf{y}_0^M) = L_{1;n}^{(m)}(i)$$

and

$$\mathbb{P}_{\boldsymbol{\theta}^{(m)}}(Z_n = j \mid \mathbf{y}_0^M) = L_n^{(m)}(i), n \neq 0,$$

have already been obtained during the mth step. All we need is to compute

$$\mathbb{P}_{\boldsymbol{\theta}^{(m)}}(Z_{n+k} = j, Z_n^{n+k-1} = i, Z_{n-1} \neq i \mid \mathbf{y})$$

and
$$\mathbb{P}_{\boldsymbol{\theta}^{(m)}}(Z_n = j, Z_{n-1} = i \mid \mathbf{y}).$$

From the proof of the expression of $L_{1;n}(i)$ we have

$$\mathbb{P}_{\boldsymbol{\theta}^{(m)}}(Z_{n+k} = j, Z_n^{n+k-1} = i, Z_{n-1} \neq i \mid \mathbf{y})$$
$$= \frac{q_{ij}(k)F_n(i)L_{1;n+k}(j)R_{j;y_{n+k-1},y_{n+k}} \prod_{p=n+1}^{n+k-1} R_{i;y_{p-1},y_p}}{F_{n+k}(j) \prod_{p=n+1}^{n+k} P_p}. \qquad (6.59)$$

For $n = 1, \ldots, M$, and $i, j \in E$, we have

$$\mathbb{P}_{\boldsymbol{\theta}^{(m)}}(Z_n = j, Z_{n-1} = i \mid \mathbf{y})$$
$$= \sum_{l=1}^{n-1} \mathbb{P}(Z_n = j, Z_{n-1} = i, \ldots, Z_{n-l} = i, Z_{n-l-1} \neq i \mid \mathbf{y}_0^M)$$
$$+ \mathbb{P}(Z_n = j, Z_{n-1} = i, \ldots, Z_0 = i \mid \mathbf{y}_0^M).$$

- Computing the first term of $\mathbb{P}_{\boldsymbol{\theta}^{(m)}}(Z_n = j, Z_{n-1} = i \mid \mathbf{y})$

  We have

$$\mathbb{P}(Z_n = j, Z_{n-l}^{n-1} = i, Z_{n-l-1} \neq i, \mathbf{y}_0^M)$$
$$= \mathbb{P}(\mathbf{y}_{n+1}^M \mid Z_n = j, Z_{n-l}^{n-1} = i, Z_{n-l-1} \neq i, \mathbf{y}_0^n)$$
$$\times \mathbb{P}(\mathbf{y}_{n-l+1}^n \mid Z_n = j, Z_{n-l}^{n-1} = i, Z_{n-l-1} \neq i, \mathbf{y}_0^{n-l})$$
$$\times \mathbb{P}(Z_n = j, Z_{n-l+1}^{n-1} = i \mid Z_{n-l} = i, Z_{n-l-1} \neq i, \mathbf{y}_0^{n-l})$$
$$\times \mathbb{P}(Z_{n-l} = i, Z_{n-l-1} \neq i \mid \mathbf{y}_0^{n-l}) \cdot \mathbb{P}(\mathbf{y}_0^{n-l})$$
$$= q_{ij}(l)F_{n-l}(i)R_{j;y_{n-1},y_n} \prod_{p=n-l+1}^{n-1} R_{i;y_{p-1},y_p}$$
$$\times \mathbb{P}(\mathbf{y}_{n+1}^M \mid Z_n = j, Z_{n-1} \neq j, \mathbf{y}_n)\mathbb{P}(\mathbf{y}_0^n).$$

On the other hand, we have

$$\mathbb{P}(\mathbf{y}_0^M) = \frac{\mathbb{P}(Z_n = j, Z_{n-1} \neq j, \mathbf{y}_0^M)}{\mathbb{P}(Z_n = j, Z_{n-1} \neq j \mid \mathbf{y}_0^M)}$$
$$= \frac{\mathbb{P}(\mathbf{y}_{n+1}^M \mid Z_n = j, Z_{n-1} \neq j, \mathbf{y}_0^n)\mathbb{P}(Z_n = j, Z_{n-1} \neq j \mid \mathbf{y}_0^n)\mathbb{P}(\mathbf{y}_0^n)}{L_{1;n}(j)}$$
$$= \frac{\mathbb{P}(\mathbf{y}_{n+1}^M \mid Z_n = j, Z_{n-1} \neq j, \mathbf{y}_n)F_n(j)\mathbb{P}(\mathbf{y}_0^n)}{L_{1;n}(j)}.$$

Consequently, the first term of $\mathbb{P}_{\boldsymbol{\theta}^{(m)}}(Z_n = j, Z_{n-1} = i \mid \mathbf{y})$ is

$$\sum_{l=1}^{n-1} \frac{q_{ij}(l)F_{n-l}(i)L_{1;n}(j)R_{j;y_{n-1},y_n} \prod_{p=n-l+1}^{n-1} R_{i;y_{p-1},y_p}}{F_n(j) \prod_{p=n-l+1}^{n} P_p}.$$

- Computing the second term of $\mathbb{P}_{\boldsymbol{\theta}^{(m)}}(Z_n = j, Z_{n-1} = i \mid \mathbf{y})$

  We have

$$
\begin{aligned}
&\mathbb{P}(Z_n = j, Z_0^{n-1} = i, \mathbf{y}_0^M) \\
&= \mathbb{P}(\mathbf{y}_{n+1}^M \mid Z_n = j, Z_0^{n-1} = i, \mathbf{y}_0^n) \\
&\quad \times \mathbb{P}(\mathbf{y}_1^n \mid Z_n = j, Z_0^{n-1} = i, \mathbf{y}_0) \\
&\quad \times \mathbb{P}(Z_n = j, Z_1^{n-1} = i \mid Z_0 = i, \mathbf{y}_0) \cdot \mathbb{P}(Z_0 = i, \mathbf{y}_0) \\
&= q_{ij}(n)\mu(i, y_0)R_{j;y_{n-1},y_n} \prod_{p=1}^{n-1} R_{i;y_{p-1},y_p} \\
&\quad \times \mathbb{P}(\mathbf{y}_{n+1}^M \mid Z_n = j, Z_{n-1} \neq j, \mathbf{y}_n) \cdot \mu(i, y_0).
\end{aligned}
$$

Using the relation

$$
\mathbb{P}(\mathbf{y}_0^M) = \frac{\mathbb{P}(\mathbf{y}_{n+1}^M \mid Z_n = j, Z_{n-1} \neq j, \mathbf{y}_n)F_n(j)\mathbb{P}(\mathbf{y}_0^n)}{L_{1;n}(j)}
$$

we obtain that the second term of $\mathbb{P}_{\boldsymbol{\theta}^{(m)}}(Z_n = j, Z_{n-1} = i \mid \mathbf{y})$ is

$$
\frac{q_{ij}(n)\mu(i, y_0)L_{1;n}(j)R_{j;y_{n-1},y_n} \prod_{p=1}^{n-1} R_{i;y_{p-1},y_p}}{F_n(j) \prod_{p=0}^{n} P_p}
$$

and

$$
\begin{aligned}
&\mathbb{P}_{\boldsymbol{\theta}^{(m)}}(Z_n = j, Z_{n-1} = i \mid \mathbf{y}) \\
&= \sum_{l=1}^{n-1} \frac{q_{ij}(l)F_{n-l}(i)L_{1;n}(j)R_{j;y_{n-1},y_n} \prod_{p=n-l+1}^{n-1} R_{i;y_{p-1},y_p}}{F_n(j) \prod_{p=n-l+1}^{n} P_p} \\
&\quad + \frac{q_{ij}(n)\mu(i, y_0)L_{1;n}(j)R_{j;y_{n-1},y_n} \prod_{p=1}^{n-1} R_{i;y_{p-1},y_p}}{F_n(j) \prod_{p=0}^{n} P_p}.
\end{aligned}
\tag{6.60}
$$

Equations (6.59) and (6.60), together with the expressions of $F_n(\cdot)$, $L_{1;n}(\cdot)$, $L_{2;n}(\cdot)$, $L_n(\cdot)$ obtained in the forward-backward algorithm, provide the expressions of $p_{ij}^{(m+1)}$, $f_{ij}^{(m+1)}(k)$, $q_{ij}^{(m+1)}(k)$, and $R^{(m+1)}(i; a, b)$ given in Equations (6.53)–(6.56).

6.4.5 Example: CpG Island Detection

Let us continue Example 6.1 of CpG island detection. Suppose that the observed DNA sequence is modeled by a sequence of conditionally independent random variables Y, with state space $D = \{0, 1, 2, 3\}$ (we replace the set $\{A, C, G, T\}$ by the set $\{0, 1, 2, 3\}$). Suppose that the possible presence of a CpG island is modeled by a semi-Markov chain Z with state space $E = \{0, 1\}$.

Having $(y_0, \ldots, y_M)$ as a sample path of Y, we set $Z_n = 1$ if y_n is a nucleotide inside from a CpG island and $Z_n = 0$ otherwise:

$$Z : 0011100001110000011 \ldots.$$

We have here an SM-M0 hidden model. In this context, finding CpG islands from an observed DNA sequence corresponds to estimating the kernel of semi-Markov chain Z from an observed sample path $(y_0, \ldots, y_M)$. In order to obtain the maximum-likelihood estimators, we apply the EM algorithm presented above.

In a first step, we generate a sample path of length m of an SM-M0 hidden model with the following characteristics:

- The kernel, $q_{0,1} := W_{0.7, \, 0.9}$, $q_{1,0} := W_{0.5, \, 0.7}$, discrete-time Weibull random variables;
- The conditional distribution of Y:
  - $(R_{0;a})_{a \in A} = (0.25 \; 0.25 \; 0.25 \; 0.25)$,
  - $(R_{1;a})_{a \in A} = (0.1 \; 0.4 \; 0.4 \; 0.1)$.

Remark 6.3. Since Z has only two states and the transition matrix of the embedded Markov chain has null diagonal ($p_{ii} = 0$ for all $i \in E$), the semi-Markov kernel is the same as the conditional distribution of the sojourn time, $q_{0,1} = f_{0,1}$ and $q_{1,0} = f_{1,0}$. Moreover, as the sojourn time in a state is independent on the next visited state (there is only one state to be visited at each jump), we have $f_{0,1} \equiv h_0$ and $f_{1,0} \equiv h_1$, where h_i is the unconditional sojourn time distribution in state i (Definition 3.4). As mentioned in Remark 3.3, we have here a particular semi-Markov kernel of the type $q_{ij}(k) = p_{ij} f_i(k)$.

In a second step of our example, the semi-Markov kernel q is estimated from a sample path of the sequence Y obtained from simulation. In the following figures, the kernel estimators are represented when considering different initial values for our EM algorithm. In each figure we have: the initial value for the algorithm, the true value and the estimator of the kernel, obtained performing 100 iterations of the algorithm.

Results in Figure 6.1 have been obtained with the following initial values for the EM algorithm:

- $q_{0,1}^{(0)} = U(1, \ldots, 15)$: uniform discrete-time distribution on $[1, 15]$;
- $q_{1,0}^{(0)} = U(1, \ldots, 10)$: uniform discrete-time distribution on $[1, 10]$.

Results in Figure 6.2 have been obtained with the following initial values for the EM algorithm:

- $q_{0,1}^{(0)} := W_{0.9,1}$: discrete-time Weibull distribution;
- $q_{1,0}^{(0)} := W_{0.9,1.5}$: discrete-time Weibull distribution.

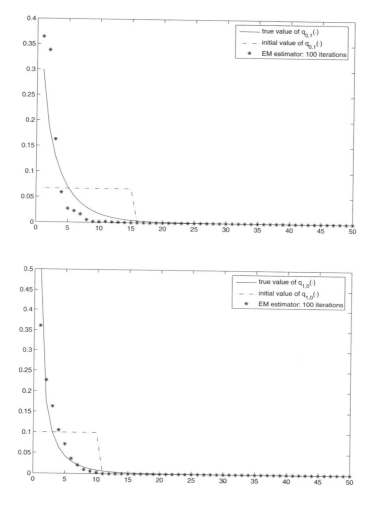

Fig. 6.1. Estimator of SM kernel with uniform initial distribution of q

Note that we have good kernel estimators, even when starting from a distant initial value (uniform distribution when trying to estimate a Weibull distribution).

Obviously, in our example, using a hidden Markov model for detecting the CpG islands does not work since the original semi-Markov process is far from a Markovian one.

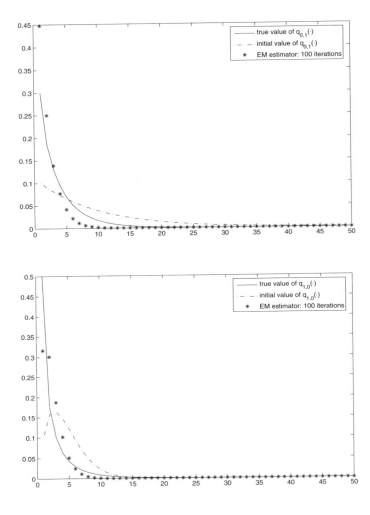

Fig. 6.2. Estimator of SM kernel with Weibull initial distribution of q

Exercises

Exercise 6.1. Let $(Z_n, Y_n)_{n \in \mathbb{N}}$ be a hidden Markov chain of the M1-M0 type, that is:

- $(Z_n)_{n \in \mathbb{N}}$ is a Markov chain with state space $E = \{1, \ldots, s\}$, transition matrix $\mathbf{p} := (p_{ij})_{i,j \in E}$, and initial distribution $\boldsymbol{\alpha}$;
- $(Y_n)_{n \in \mathbb{N}}$ are conditionally independent random variables (in the sense of Equation (6.1)), of state space $A = \{1, \ldots, d\}$, and of conditional distribution $R_{i;j} := \mathbb{P}(Y_n = j \mid Z_n = i), i \in E, j \in A$.

Show that:

- $\mathbb{P}(Y_n = j) = \sum_{i \in E} R_{i;j} (\boldsymbol{\alpha} \mathbf{p}^n)(i)$;
- $\mathbb{P}(Y_0 = j_0, \ldots, Y_n = j_n) = \sum_{i_0 \in E} \cdots \sum_{i_n \in E} \alpha_{i_0} \prod_{k=0}^{n} R_{i_k;j_k} \prod_{l=1}^{n} p_{i_{l-1}, i_l}$.

Exercise 6.2 (hidden semi-Markov chains for detecting an unfair die). Consider the problem of an unfair die detection by means of hidden semi-Markov chains, presented in Example 6.3. Compute the same quantities as we did for a hidden-Markov model in Example 6.2, that is:

1. The probability $\mathbb{P}(Y_n = i \mid Z_0 = 1)$, $1 \leq i \leq 6$;
2. The limit $\lim_{n \to \infty} \mathbb{P}(Y_n = i \mid Z_0 = 1)$, $1 \leq i \leq 6$.

Exercise 6.3 (backward-recurrence times of a SMC). Consider a semi-Markov chain $(Z_n)_{n \in \mathbb{N}}$ and its backward-recurrence times $U_n := n - S_n, n \in \mathbb{N}$.

1. Show that the chain $(Z_n, U_n)_{n \in \mathbb{N}}$ is a Markov chain, with the transition matrix $\widetilde{\mathbf{p}} := (p_{(i,t_1)(j,t_2)})_{i,j \in E, t_1, t_2 \in \mathbb{N}}$ given in Proposition 6.1.
2. If the semi-Markov chain $(Z_n)_{n \in \mathbb{N}}$ is irreducible, show that the Markov chain $(Z_n, U_n)_{n \in \mathbb{N}}$ is irreducible, aperiodic and compute its stationary distribution.

Exercise 6.4 (forward-recurrence times of a SMC). Consider a semi-Markov chain $(Z_n)_{n \in \mathbb{N}}$ and define by $V_n := S_{n+1} - n, n \in \mathbb{N}$ its forward-recurrence times.

1. Show that the chain $(Z_n, V_n)_{n \in \mathbb{N}}$ is a Markov chain and compute its transition matrix in terms of the basic quantities of the SMC.
2. If the semi-Markov chain $(Z_n)_{n \in \mathbb{N}}$ is irreducible, show that the Markov chain $(Z_n, V_n)_{n \in \mathbb{N}}$ is irreducible, aperiodic and compute its stationary distribution.

Exercise 6.5. Let $(Z_n, Y_n)_{n \in \mathbb{N}}$ be a hidden SM-M0 hidden semi-Markov chain, suppose that Assumptions A1, A3, and A4 are fulfilled, and consider $(Y_0, \ldots, Y_M)$ a sample path of observations. In Equation (6.19) we have expressed the transition matrix $\mathbf{p} = (p_{ij})_{i,j \in E, i \neq j}$ of the embedded Markov chain J in terms of the semi-Markov kernel $\mathbf{q} = (q_{ij}(k))_{i,j \in E, i \neq j, k \in D}$. Use this relation and Theorem 6.6 in order to prove the asymptotic normality of the MLE of $\mathbf{p}$, as M tends to infinity.

Exercise 6.6. In the setting of Exercise 6.5, prove that the MLE of conditional sojourn time distributions of the SMC $\mathbf{f} = (f_{ij}(k))_{i,j\in E, i\neq j, k\in D}$ is strongly consistent and asymptotically normal, as M tends to infinity.

A

Lemmas for Semi-Markov Chains

In this appendix, we present some lemmas used for proving the asymptotic normality of estimators obtained for semi-Markov characteristics and for the associated reliability indicators (Chapters 4 and 5). More exactly, these results are used when applying Pyke and Schaufeles' central limit theorem for MRCs (Theorem 3.5). For more details, see the first proof of Theorem 4.2.

We consider (J, S) a MRC with semi-Markov kernel $\mathbf{q}$ and cumulative semi-Markov kernel $\mathbf{Q}$.

Lemma A.1. *Let* $\mathbf{A} \in \mathcal{M}_E(\mathbb{N})$ *be a matrix-valued function. For any fixed* $x, k \in \mathbb{N}$ *and* $i, j \in E$, *we have*

$$(A_{ij} * \mathbf{1}_{\{x=\cdot\}})(k) = \begin{cases} A_{ij}(k-x), & \text{if } x \leq k, \\ 0, & \text{otherwise}, \end{cases}$$

and

$$(A_{ij} * \mathbf{1}_{\{x=\cdot\}})^2(k) = \begin{cases} A_{ij}^2(k-x), & \text{if } x \leq k, \\ 0, & \text{otherwise}. \end{cases}$$

Proof.

$$(A_{ij} * \mathbf{1}_{\{x=\cdot\}})(k) = \sum_{l=0}^{k} A_{ij}(k-l)\mathbf{1}_{\{x=l\}} = \begin{cases} A_{ij}(k-x), & \text{if } x \leq k, \\ 0, & \text{otherwise}. \end{cases}$$

Using the above result, we obtain the second assertion:

$$(A_{ij} * \mathbf{1}_{\{x=\cdot\}})^2(k) = [\sum_{l=0}^{k} A_{ij}(k-l)\mathbf{1}_{\{x=l\}}]^2 = \begin{cases} A_{ij}^2(k-x), & \text{if } x \leq k, \\ 0, & \text{otherwise}. \end{cases}$$

$\square$

Lemma A.2. *Let* $\mathbf{A} \in \mathcal{M}_E(\mathbb{N})$ *be a matrix-valued function. For any fixed* $k \in \mathbb{N}$, i, j, l *and* $r \in E$, *we have*

$$\sum_{x=0}^{\infty}(A_{lr} * \mathbf{1}_{\{x=\cdot\}})(k)q_{ij}(x) = (A_{lr} * q_{ij})(k)$$

and

$$\sum_{x=0}^{\infty}(A_{lr} * \mathbf{1}_{\{x=\cdot\}})^2(k)q_{ij}(x) = (A_{lr}^2 * q_{ij})(k).$$

Proof. Using Lemma A.1 we have

$$\sum_{x=0}^{\infty}(A_{lr} * \mathbf{1}_{\{x=\cdot\}})(k)q_{ij}(x) = \sum_{x=0}^{\infty}A_{lr}(k-x)\mathbf{1}_{\{x\le k\}}q_{ij}(x)$$

$$= \sum_{x=0}^{k}A_{lr}(k-x)q_{ij}(x) = (A_{lr} * q_{ij})(k).$$

In the same way we get the second relation.   □

Lemma A.3. *Let* $\mathbf{A}, \mathbf{B} \in \mathcal{M}_E(\mathbb{N})$ *be two matrix-valued functions. For any fixed* $k, y \in \mathbb{N}$, i, j, l, r, u *and* $v \in E$, *we have*

1. $A_{uv} * \mathbf{1}_{\{y\le\cdot\}}(k) = \begin{cases} \sum_{t=0}^{k-y} A_{uv}(t), & y \le k, \\ 0, & otherwise, \end{cases}$

2. $\displaystyle\sum_{x=0}^{\infty} A_{uv} * \mathbf{1}_{\{x\le\cdot\}}(k)q_{ij}(x) = A_{uv} * Q_{ij}(k),$

3. $\displaystyle\sum_{x=0}^{\infty}\left(A_{uv} * \mathbf{1}_{\{x\le\cdot\}}(k)\right)^2 q_{ij}(x) = \left(\sum_{t=0}^{\cdot} A_{uv}(t)\right)^2 * q_{ij}(k),$

4. $\displaystyle\sum_{x=0}^{\infty} A_{uv} * \mathbf{1}_{\{x=\cdot\}}(k)B_{lr} * \mathbf{1}_{\{x\le\cdot\}}(k)q_{ij}(x) = \left[A_{uv}(\cdot)\sum_{t=0}^{\cdot} B_{lr}(t)\right] * q_{ij}(k).$

Proof.
1.

$$A_{uv} * \mathbf{1}_{\{y\le\cdot\}}(k) = \sum_{t=0}^{k}A_{uv}(k-t)\mathbf{1}_{\{y\le t\}} = \begin{cases} \sum_{t=0}^{k-y} A_{uv}(t), & y \le k, \\ 0, & otherwise. \end{cases}$$

2. We have

$$\sum_{x=0}^{\infty} A_{uv} * \mathbf{1}_{\{x\le\cdot\}}(k)q_{ij}(x) = \sum_{x=0}^{k}\sum_{l=0}^{k}A_{uv}(k-l)\mathbf{1}_{\{x\le l\}}q_{ij}(x)$$

$$= \sum_{l=0}^{k}\left[\sum_{x=0}^{k}\mathbf{1}_{\{x\le l\}}q_{ij}(x)\right]A_{uv}(k-l) = \sum_{l=0}^{k}Q_{ij}(l)A_{uv}(k-l) = A_{uv} * Q_{ij}(k).$$

3. Using Point 1. of the present lemma, we have

$$\sum_{x=0}^{\infty}\Big(A_{uv} * \mathbf{1}_{\{x\le\cdot\}}(k)\Big)^2 q_{ij}(x) = \sum_{x=0}^{k}\Big(\sum_{l=0}^{k-x} A_{uv}(l)\Big)^2 q_{ij}(x)$$

$$= \Big(\sum_{l=0}^{\cdot} A_{uv}(l)\Big)^2 * q_{ij}(k).$$

4. Using Lemma A.1 and Point 1. of the present lemma we have

$$\sum_{x=0}^{\infty} A_{uv} * \mathbf{1}_{\{x=\cdot\}}(k) B_{lr} * \mathbf{1}_{\{x\le\cdot\}}(k) q_{ij}(x)$$

$$= \sum_{x=0}^{k} A_{uv}(k-x)\Big(\sum_{t=0}^{k-x} B_{lr}(t)\Big) q_{ij}(x)$$

$$= \Big[A_{uv}(\cdot)\sum_{t=0}^{\cdot} B_{lr}(t)\Big] * q_{ij}(k).$$

And hence the proof is finished.   □

Recall that, for a matrix-valued function $\mathbf{A} \in \mathcal{M}_E(\mathbb{N})$, we denoted by $\mathbf{A}^+ \in \mathcal{M}_E(\mathbb{N})$ the matrix-valued function defined by $\mathbf{A}^+(k) := \mathbf{A}(k+1)$, $k \in \mathbb{N}$.

Lemma A.4. *Let* $\mathbf{A}, \mathbf{B} \in \mathcal{M}_E(\mathbb{N})$ *be two matrix-valued functions. For any fixed* $k \in \mathbb{N}, i, j, l, r, u$ *and* $v \in E$, *we have*

1. $\displaystyle\sum_{x=0}^{\infty}(A_{uv} * \mathbf{1}_{\{x=\cdot\}})(k)(B_{lr} * \mathbf{1}_{\{x=\cdot\}})(k-1)q_{ij}(x) = (A_{uv}^+ B_{lr}) * q_{ij}(k-1),$

2. $\displaystyle\sum_{x=0}^{\infty}(A_{uv} * \mathbf{1}_{\{x=\cdot\}})(k)(B_{lr} * \mathbf{1}_{\{x\le\cdot\}})(k-1)q_{ij}(x)$

$$= \Big[A_{uv}^+(\cdot)\sum_{t=0}^{\cdot} B_{lr}(t)\Big] * q_{ij}(k-1),$$

3. $\displaystyle\sum_{x=0}^{\infty}(A_{uv} * \mathbf{1}_{\{x=\cdot\}})(k-1)(B_{lr} * \mathbf{1}_{\{x\le\cdot\}})(k)q_{ij}(x)$

$$= \Big[A_{uv}(\cdot)\Big(\sum_{t=0}^{\cdot} B_{lr}(t)\Big)^+\Big] * q_{ij}(k-1),$$

4. $\displaystyle\sum_{x=0}^{\infty}(A_{uv} * \mathbf{1}_{\{x\le\cdot\}})(k-1)(B_{lr} * \mathbf{1}_{\{x\le\cdot\}})(k)q_{ij}(x)$

$$= \Big[\Big(\sum_{t=0}^{\cdot} A_{uv}(t)\Big)\Big(\sum_{t=0}^{\cdot} B_{lr}(t)\Big)^+\Big] * q_{ij}(k-1).$$

Proof.
1. Using Lemmas A.1 and A.2 we have

$$\sum_{x=0}^{\infty}(A_{uv} * \mathbf{1}_{\{x=\cdot\}})(k)(B_{lr} * \mathbf{1}_{\{x=\cdot\}})(k-1)q_{ij}(x)$$

$$= \sum_{x=0}^{k-1} A_{uv}(k-x)B_{lr}(k-1-x)q_{ij}(x)$$

$$= \sum_{x=0}^{k-1}(A_{uv}^{+}B_{lr})(k-1-x)q_{ij}(x) = (A_{uv}^{+}B_{lr}) * q_{ij}(k-1).$$

2. Using Lemmas A.1–A.3 we obtain

$$\sum_{x=0}^{\infty}(A_{uv} * \mathbf{1}_{\{x=\cdot\}})(k)(B_{lr} * \mathbf{1}_{\{x\le\cdot\}})(k-1)q_{ij}(x)$$

$$= \sum_{x=0}^{k-1} A_{uv}(k-x)\Big(\sum_{t=0}^{k-1-x} B_{lr}(t)\Big)q_{ij}(x)$$

$$= \Big[A_{uv}^{+}(\cdot)\sum_{t=0}^{\cdot} B_{lr}(t)\Big] * q_{ij}(k-1).$$

3. In the same way as above, we have

$$\sum_{x=0}^{\infty}(A_{uv} * \mathbf{1}_{\{x=\cdot\}})(k-1)(B_{lr} * \mathbf{1}_{\{x\le\cdot\}})(k)q_{ij}(x)$$

$$= \sum_{x=0}^{k-1} A_{uv}(k-1-x)\Big(\sum_{t=0}^{k-x} B_{lr}(t)\Big)q_{ij}(x)$$

$$= \Big[A_{uv}(\cdot)\Big(\sum_{t=0}^{\cdot} B_{lr}(t)\Big)^{+}\Big] * q_{ij}(k-1).$$

4. We have

$$\sum_{x=0}^{\infty}(A_{uv} * \mathbf{1}_{\{x\le\cdot\}})(k-1)(B_{lr} * \mathbf{1}_{\{x\le\cdot\}})(k)q_{ij}(x)$$

$$= \sum_{x=0}^{k-1}\Big(\sum_{t=0}^{k-1-x} A_{uv}(t)\Big)\Big(\sum_{t=0}^{k-x} B_{lr}(t)\Big)q_{ij}(x)$$

$$= \Big[\Big(\sum_{t=0}^{\cdot} A_{uv}(t)\Big)\Big(\sum_{t=0}^{\cdot} B_{lr}(t)\Big)^{+}\Big] * q_{ij}(k-1),$$

and the proof is finished. $\square$

B

Lemmas for Hidden Semi-Markov Chains

We present here some necessary results on the conditional independence of hidden semi-Markov chains, of the SM-M0 and SM-M1 types. We use these lemmas for deriving the EM algorithm, which practically provides the MLEs of such models. Details can be found in Section 6.4.4.

B.1 Lemmas for Hidden SM-M0 Chains

Let $(Z_n, Y_n)_{n \in \mathbb{N}}$ be a hidden SM-M0 chain and $y_0^M = (y_0, \ldots, y_M)$ a sample of the observed chain Y. The following results deal with conditional independence properties of such a model.

Lemma B.1. *For any $j \in E, 0 \leq n \leq M, 1 \leq l \leq M$, we have*

$$\mathbb{P}(\mathbf{y}_{n+1}^{n+l} \mid Z_{n+1} = j, Z_0^n = \cdot, \mathbf{y}_0^n)$$
$$= \mathbb{P}(\mathbf{y}_{n+1}^{n+l} \mid Z_{S_{N(n+1)}}^{n+1} = j, Z_{S_{N(n+1)}-1} \neq j).$$

Lemma B.2. *For any $j \in E, 0 \leq n \leq M, 1 \leq l \leq M$, we have*

$$\mathbb{P}(\mathbf{y}_{n+1}^{n+l} \mid Z_{n+1-u} = j, Z_0^{n-u} = \cdot, \mathbf{y}_0^n)$$
$$= \mathbb{P}(\mathbf{y}_{n+1}^{n+l} \mid Z_{S_{N(n+1-u)}}^{n+1-u} = j, Z_{S_{N(n+1-u)}-1} \neq j).$$

Lemma B.3. *For any positive integers $0 \leq n \leq M, 0 \leq l \leq M - n, 1 \leq t \leq M - n - l$ and any states $i_n, \ldots, i_{n+l} \in E$, we have*

$$\mathbb{P}(\mathbf{y}_n^{n+l} \mid Z_{n+l+1}^{n+l+t} = \cdot, Z_n^{n+l} = i_n^{n+l}, Z_0^{n-1} = \cdot, \mathbf{y}_0^{n-1})$$
$$= \prod_{k=n}^{n+l} \mathbb{P}(\mathbf{y}_k \mid Z_k = i_k) = \prod_{k=n}^{n+l} R_{i_k; y_k}.$$

B.2 Lemmas for Hidden SM-M1 Chains

Let $(Z_n, Y_n)_{n \in \mathbb{N}}$ be a hidden SM-M1 chain and $y_0^M = (y_0, \ldots, y_M)$ a sample of the observed chain Y. The following results deal with conditional independence properties of such a model.

Lemma B.4. *For any $j \in E, 0 \le n \le M, 1 \le l \le M$, we have*

$$\mathbb{P}(\mathbf{y}_{n+1}^{n+l} \mid Z_{n+1} = j, Z_0^n = \cdot, \mathbf{y}_0^n)$$
$$= \mathbb{P}(\mathbf{y}_{n+1}^{n+l} \mid Z_{S_{N(n+1)}}^{n+1} = j, Z_{S_{N(n+1)}-1} \neq j, \mathbf{y}_n).$$

Lemma B.5. *For all $j \in E, 0 \le n \le M, 1 \le l \le M$, we have*

1. *If $u < k$,*

$$\mathbb{P}(\mathbf{y}_{n+1}^{n+l} \mid Z_{n-k+2}^{n+1-u} = \cdot, Z_{n-k+1} = j, Z_0^{n-k} = \cdot, \mathbf{y}_0^{n-k})$$
$$= \mathbb{P}(\mathbf{y}_{n+1}^{n+l} \mid Z_{n-k+2}^{n+1-u} = \cdot, Z_{S_{N(n-k+1)}}^{n-k+1} = j, Z_{S_{N(n-k+1)}-1} \neq j, \mathbf{y}_{n-k}),$$

2. *If $u \ge k$,*

$$\mathbb{P}(\mathbf{y}_{n+1}^{n+l} \mid Z_{n+1-u} = j, Z_0^{n-u} = \cdot, \mathbf{y}_0^{n-k})$$
$$= \mathbb{P}(\mathbf{y}_{n+1}^{n+l} \mid Z_{S_{N(n+1-u)}}^{n+1-u} = j, Z_{S_{N(n+1-u)}-1} \neq j, \mathbf{y}_{n-k}).$$

Lemma B.6. *For any positive integers $0 \le n \le M, 0 \le l \le M - n, 1 \le t \le M - n - l$ and any states $i_n, \ldots, i_{n+l} \in E$, we have*

$$\mathbb{P}(\mathbf{y}_n^{n+l} \mid Z_{n+l+1}^{n+l+t} = \cdot, Z_n^{n+l} = i_n^{n+l}, Z_0^{n-1} = \cdot, \mathbf{y}_0^{n-1})$$
$$= \mathbb{P}(\mathbf{y}_n^{n+l} \mid Z_{n+1}^{n+l} = i_{n+1}^{n+l}, Z_{S_{N(n)}}^n = i_n, Z_{S_{N(n)}-1} \neq i_n, \mathbf{y}_{n-1})$$
$$= \prod_{k=n}^{n+l} R_{i_k; y_{k-1}, y_k}.$$

C

Some Proofs

Complete proofs of the asymptotic normality of the availability estimator (Theorem 5.2) and of the BMP-failure rate estimator (Theorem 5.4) are provided in this appendix. The proofs are based on what we called the first method, namely, Pyke and Schaufele's CLT for MRCs (Theorem 3.5).

Proof (of Theorem 5.2).

$$\sqrt{M}\left[\widehat{A}(k, M) - A(k)\right]$$

$$= \sqrt{M} \sum_{i=1}^{s} \sum_{j \in U} \alpha_i \left[\widehat{P}_{ij}(k, M) - P_{ij}(k)\right]$$

$$= \sqrt{M} \sum_{i=1}^{s} \sum_{j \in U} \alpha_i \left[\widehat{\psi}_{ij} * \left(\mathbf{I} - diag(\widehat{\mathbf{Q}} \cdot \mathbf{1})\right)_{jj} - \psi_{ij} * \left(\mathbf{I} - diag(\mathbf{Q} \cdot \mathbf{1})\right)_{jj}\right](k)$$

$$= \sqrt{M} \sum_{i=1}^{s} \sum_{j \in U} \alpha_i \left[\widehat{\psi}_{ij} * \left(\mathbf{I} - diag(\widehat{\mathbf{Q}} \cdot \mathbf{1})\right)_{jj} - \psi_{ij} * \left(\mathbf{I} - diag(\mathbf{Q} \cdot \mathbf{1})\right)_{jj}\right.$$

$$+ \widehat{\psi}_{ij} * \left(\mathbf{I} - diag(\mathbf{Q} \cdot \mathbf{1})\right)_{jj} - \widehat{\psi}_{ij} * \left(\mathbf{I} - diag(\mathbf{Q} \cdot \mathbf{1})\right)_{jj}$$

$$+ \psi_{ij} * diag(\widehat{\mathbf{Q}} \cdot \mathbf{1})_{jj} - \psi_{ij} * diag(\widehat{\mathbf{Q}} \cdot \mathbf{1})_{jj}$$

$$\left. + \psi_{ij} * diag(\mathbf{Q} \cdot \mathbf{1})_{jj} - \psi_{ij} * diag(\mathbf{Q} \cdot \mathbf{1})_{jj}\right](k)$$

$$= \sqrt{M} \sum_{i=1}^{s} \sum_{j \in U} \alpha_i \left[(\widehat{\psi}_{ij} - \psi_{ij}) * \left(\mathbf{I} - diag(\mathbf{Q} \cdot \mathbf{1})\right)_{jj}\right.$$

$$\left. - \psi_{ij} * \left(diag(\widehat{\mathbf{Q}} - \mathbf{Q}) \cdot \mathbf{1}\right)_{jj}\right]$$

$$- \sqrt{M} \sum_{i=1}^{s} \sum_{j \in U} \alpha_i \left[(\widehat{\psi}_{ij} - \psi_{ij}) * \left(diag(\widehat{\mathbf{Q}} - \mathbf{Q}) \cdot \mathbf{1}\right)_{jj}\right](k).$$

The last right-hand-side term of the previous equality can be written

$$\sum_{i=1}^{s}\sum_{j\in U}\sum_{r=1}^{s}\sum_{l=0}^{k}\alpha_i\sqrt{M}(\widehat{\psi}_{ij}-\psi_{ij})(l)\Delta Q_{jr}(k-l).$$

For every $l\in\mathbb{N}, l\leq M, \sqrt{M}(\widehat{\psi}_{ij}-\psi_{ij})(l)$ converges in distribution to a normal random variable (Theorem 4.5) when M tends to infinity, and $\Delta Q_{jr}(k-l)$ converges in probability to zero (Theorem 4.3) when M tends to infinity. Thus, using Slutsky's theorem (Theorem E.10), we obtain that $\sqrt{M}\sum_{i=1}^{s}\sum_{j\in U}\alpha_i\left[(\widehat{\psi}_{ij}-\psi_{ij})*\left(diag(\widehat{\mathbf{Q}}-\mathbf{Q})\cdot\mathbf{1}\right)_{jj}\right](k)$ converges in distribution to zero when M tends to infinity. Thus, $\sqrt{M}[\widehat{A}(k,M)-A(k)]$ has the same limit in distribution as

$$\sqrt{M}\sum_{i=1}^{s}\sum_{j\in U}\alpha_i\left[(\widehat{\psi}_{ij}-\psi_{ij})*\left(\mathbf{I}-diag(\mathbf{Q}\cdot\mathbf{1})\right)_{jj}-\psi_{ij}*\left(diag(\widehat{\mathbf{Q}}-\mathbf{Q})\cdot\mathbf{1}\right)_{jj}\right](k).$$

From the proof of Theorem 4.5 we know that $\sqrt{M}[\widehat{\psi}_{ij}(k,M)-\psi_{ij}(k)]$ has the same limit in distribution as $\sqrt{M}[\boldsymbol{\psi}*\Delta\mathbf{q}*\boldsymbol{\psi}]_{ij}(k)$. Writing

$$\Delta q_{ij}(\cdot)=\widehat{q}_{ij}(\cdot,M)-q_{ij}(\cdot)=\frac{1}{N_i(M)}\sum_{l=1}^{N(M)}[\mathbf{1}_{\{J_{l-1}=i,J_l=j,X_l=\cdot\}}-q_{ij}(\cdot)\mathbf{1}_{\{J_{l-1}=i\}}]$$

and

$$\begin{aligned}\Delta Q_{ij}(\cdot)&=\widehat{Q}_{ij}(\cdot,M)-Q_{ij}(\cdot)\\&=\frac{1}{N_i(M)}\sum_{l=1}^{N(M)}[\mathbf{1}_{\{J_{l-1}=i,J_l=j,X_l\leq\cdot\}}-Q_{ij}(\cdot)\mathbf{1}_{\{J_{l-1}=i\}}],\end{aligned}$$

we obtain that $\sqrt{M}[\widehat{A}(k,M)-A(k)]$ has the same limit in distribution as

$$\begin{aligned}&\sqrt{M}\sum_{i=1}^{s}\sum_{j\in U}\alpha_i\sum_{r=1}^{s}\sum_{l=1}^{s}\left[\left(\mathbf{I}-diag(\mathbf{Q}\cdot\mathbf{1})\right)_{jj}*\psi_{ir}*\psi_{lj}*\Delta q_{rl}\right](k)\\&-\sqrt{M}\sum_{i=1}^{s}\sum_{j\in U}\alpha_i\sum_{l=1}^{s}\psi_{ij}*\Delta Q_{jl}(k)\\=\ &\sqrt{M}\sum_{r=1}^{s}\sum_{l=1}^{s}\left[\sum_{i=1}^{s}\sum_{j\in U}\alpha_i\psi_{ir}*\psi_{lj}*\left(\mathbf{I}-diag(\mathbf{Q}\cdot\mathbf{1})\right)_{jj}\right]*\Delta q_{rl}(k)\\&-\sqrt{M}\sum_{i=1}^{s}\sum_{j\in U}\alpha_i\sum_{l=1}^{s}\psi_{ij}*\Delta Q_{jl}(k)\\=\ &\sqrt{M}\sum_{r=1}^{s}\sum_{l=1}^{s}D_{rl}*\Delta q_{rl}(k)-\sqrt{M}\sum_{r\in U}\sum_{l=1}^{s}\left(\sum_{n=1}^{s}\alpha_n\psi_{nr}\right)*\Delta Q_{rl}(k)\end{aligned}$$

$$= \frac{1}{\sqrt{M}} \sum_{n=1}^{N(M)} \sum_{l,r=1}^{s} \frac{M}{N_r(M)} \Big[D_{rl} * \Big(\mathbf{1}_{\{J_{n-1}=r, J_n=l, X_n=\cdot\}} - q_{rl}(\cdot)\mathbf{1}_{\{J_{n-1}=r\}} \Big)(k)$$

$$-\mathbf{1}_{\{r \in U\}} \Big(\sum_{t=1}^{s} \alpha_t \psi_{tr} \Big) * \Big(\mathbf{1}_{\{J_{n-1}=r, J_n=l, X_n \leq \cdot\}} - Q_{rl}(\cdot)\mathbf{1}_{\{J_{n-1}=r\}} \Big)(k) \Big]$$

$$= \frac{1}{\sqrt{M}} \sum_{l=1}^{N(M)} f(J_{l-1}, J_l, X_l),$$

where, for fixed $k \in \mathbb{N}$, we have defined the function $f : E \times E \times \mathbb{N} \to \mathbb{R}$ by

$$f(i,j,x) := \sum_{r=1}^{s} \sum_{l=1}^{s} \frac{M}{N_r(M)} \Big[D_{rl} * \Big(\mathbf{1}_{\{i=r, j=l, x=\cdot\}} - q_{rl}(\cdot)\mathbf{1}_{\{i=r\}} \Big)(k)$$

$$-\mathbf{1}_{\{r \in U\}} \Big(\sum_{t=1}^{s} \alpha_t \psi_{tr} \Big) * \Big(\mathbf{1}_{\{i=r, j=l, x \leq \cdot\}} - Q_{rl}(\cdot)\mathbf{1}_{\{i=r\}} \Big)(k) \Big]$$

$$= \frac{M}{N_i(M)} \Big\{ D_{ij} * \mathbf{1}_{\{x=\cdot\}}(k) - \mathbf{1}_{\{i \in U\}} \Big(\sum_{t=1}^{s} \alpha_t \psi_{ti} \Big) * \mathbf{1}_{\{x \leq \cdot\}}(k)$$

$$-\sum_{l=1}^{s} \Big[D_{il} * q_{il}(k) - \mathbf{1}_{\{i \in U\}} \Big(\sum_{t=1}^{s} \alpha_t \psi_{ti} \Big) * Q_{il}(k) \Big] \Big\}.$$

For $i \in E$ and $k \in \mathbb{N}$, let us set

$$M_i(k) := \sum_{l=1}^{s} \Big[D_{il} * q_{il}(k) - \mathbf{1}_{\{i \in U\}} \Big(\sum_{t=1}^{s} \alpha_t \psi_{ti} \Big) * Q_{il}(k) \Big].$$

In order to apply the Pyke and Schaufeles' CLT, we will now compute A_{ij}, A_i, B_{ij}, B_i, using Lemmas A.1–A.3.

$$A_{ij} = \frac{M}{N_i(M)} \Big\{ \sum_{x=1}^{k} D_{ij} * \mathbf{1}_{\{x=\cdot\}}(k) q_{ij}(x)$$

$$-\sum_{x=1}^{k} \mathbf{1}_{\{i \in U\}} \Big(\sum_{t=1}^{s} \alpha_t \psi_{ti} \Big) * \mathbf{1}_{\{x \leq \cdot\}}(k) q_{ij}(x)$$

$$-\sum_{l=1}^{s} \Big[D_{il} * q_{il}(k) - \mathbf{1}_{\{i \in U\}} \Big(\sum_{t=1}^{s} \alpha_t \psi_{ti} \Big) * Q_{il}(k) \Big] \sum_{x=1}^{\infty} q_{ij}(x) \Big\}$$

$$= \frac{M}{N_i(M)} \Big\{ D_{ij} * q_{ij}(k) - \mathbf{1}_{\{i \in U\}} \Big(\sum_{t=1}^{s} \alpha_t \psi_{ti} \Big) * Q_{ij}(k)$$

$$-\sum_{l=1}^{s} \Big[D_{il} * q_{il}(k) - \mathbf{1}_{\{i \in U\}} \Big(\sum_{t=1}^{s} \alpha_t \psi_{ti} \Big) * Q_{il}(k) \Big] p_{ij} \Big\};$$

$$A_i = \frac{M}{N_i(M)} \Big\{ \sum_{j=1}^{s} D_{ij} * q_{ij}(k) - \sum_{j=1}^{s} \mathbf{1}_{\{i \in U\}} \Big(\sum_{t=1}^{s} \alpha_t \psi_{ti} \Big) * Q_{ij}(k)$$

$$
-\sum_{l=1}^{s}\Big[D_{il}*q_{il}(k)-\mathbf{1}_{\{i\in U\}}\Big(\sum_{t=1}^{s}\alpha_t\psi_{ti}\Big)*Q_{il}(k)\Big]\sum_{j=1}^{s}p_{ij}\Big\}
$$

$$
=0;
$$

$$
B_{ij}=\Big(\frac{M}{N_i(M)}\Big)^2\Big\{\sum_{x=1}^{k}\Big(D_{ij}*\mathbf{1}_{\{x=\cdot\}}(k)\Big)^2 q_{ij}(x)
$$

$$
+\sum_{x=1}^{k}\Big[\mathbf{1}_{\{i\in U\}}\Big(\sum_{t=1}^{s}\alpha_t\psi_{ti}\Big)*\mathbf{1}_{\{x\le\cdot\}}(k)\Big]^2 q_{ij}(x)
$$

$$
+M_i(k)^2\sum_{x=1}^{\infty}q_{ij}(x)-2M_i(k)\sum_{x=1}^{k}D_{ij}*\mathbf{1}_{\{x=\cdot\}}(k)q_{ij}(x)
$$

$$
+2M_i(k)\sum_{x=1}^{k}\mathbf{1}_{\{i\in U\}}\Big(\sum_{t=1}^{s}\alpha_t\psi_{ti}\Big)*\mathbf{1}_{\{x\le\cdot\}}(k)q_{ij}(x)
$$

$$
-2\sum_{x=1}^{k}D_{ij}*\mathbf{1}_{\{x=\cdot\}}(k)\mathbf{1}_{\{i\in U\}}\Big(\sum_{t=1}^{s}\alpha_t\psi_{ti}\Big)*\mathbf{1}_{\{x\le\cdot\}}(k)q_{ij}(x)\Big\}
$$

$$
=\Big(\frac{M}{N_i(M)}\Big)^2\Big\{D_{ij}^2*q_{ij}(k)+\mathbf{1}_{\{i\in U\}}\Big[\sum_{l=0}^{\cdot}\sum_{t=1}^{s}\alpha_t\psi_{ti}(l)\Big]^2*q_{ij}(k)
$$

$$
-2M_i(k)D_{ij}*q_{ij}(k)+2M_i(k)\mathbf{1}_{\{i\in U\}}\Big(\sum_{t=1}^{s}\alpha_t\psi_{ti}\Big)*Q_{ij}(k)
$$

$$
-2\Big[D_{ij}(\cdot)\mathbf{1}_{\{i\in U\}}\Big(\sum_{l=0}^{\cdot}\sum_{t=1}^{s}\alpha_t\psi_{ti}(l)\Big)\Big]*q_{ij}(k)+M_i^2(k)p_{ij}\Big\};
$$

$$
B_i=\Big(\frac{M}{N_i(M)}\Big)^2\Big\{\sum_{j=1}^{s}D_{ij}^2*q_{ij}(k)+\sum_{j=1}^{s}\Big(\mathbf{1}_{\{i\in U\}}\sum_{t=1}^{s}\alpha_t\Psi_{ti}\Big)^2*q_{ij}(k)
$$

$$
+M_i^2(k)\sum_{j=1}^{s}p_{ij}-2M_i(k)\sum_{j=1}^{s}D_{ij}*q_{ij}(k)
$$

$$
+2M_i(k)\sum_{j=1}^{s}\mathbf{1}_{\{i\in U\}}\Big(\sum_{t=1}^{s}\alpha_t\psi_{ti}\Big)*Q_{ij}(k)
$$

$$
-2\sum_{j=1}^{s}\Big[D_{ij}(\cdot)\mathbf{1}_{\{i\in U\}}\Big(\sum_{t=1}^{s}\alpha_t\Psi_{ti}\Big)\Big]*q_{ij}(k)\Big\}
$$

$$
=\Big(\frac{M}{N_i(M)}\Big)^2\Big\{\sum_{j=1}^{s}\Big[D_{ij}-\mathbf{1}_{\{i\in U\}}(\sum_{t=1}^{s}\alpha_t\Psi_{ti})\Big]^2*q_{ij}(k)
$$

$$
-\Big[\sum_{j=1}^{s}\Big(D_{ij}*q_{ij}-\mathbf{1}_{\{i\in U\}}(\sum_{t=1}^{s}\alpha_t\psi_{ti})*Q_{ij}\Big)\Big]^2(k)\Big\};
$$

$$m_f := \frac{1}{\mu_{ii}} r_i = \frac{1}{\mu_{ii}} \sum_{j=1}^{s} A_j \frac{\mu_{ii}^*}{\mu_{jj}^*} = 0;$$

$$B_f := \frac{1}{\mu_{jj}} \sigma_i^2 = \frac{\mu_{jj}^*}{\mu_{jj}} \sum_{i=1}^{s} B_i \frac{1}{\mu_{ii}^*}$$

$$= \frac{\mu_{jj}^*}{\mu_{jj}} \sum_{i=1}^{s} \left(\frac{M}{N_i(M)}\right)^2 \frac{1}{\mu_{ii}^*} \left\{ \sum_{j=1}^{s} \left[D_{ij} - \mathbf{1}_{\{i \in U\}} \left(\sum_{t=1}^{s} \alpha_t \Psi_{ti}\right) \right]^2 * q_{ij}(k)$$

$$- \left[\sum_{j=1}^{s} \left(D_{ij} * q_{ij} - \mathbf{1}_{\{i \in U\}} \left(\sum_{t=1}^{s} \alpha_t \psi_{ti}\right) * Q_{ij} \right) \right]^2 (k) \right\}.$$

Since $N_i(M)/M \xrightarrow[M \to \infty]{a.s.} 1/\mu_{ii}$ (Proposition 3.8), we obtain from the CLT that $\sqrt{M}[\hat{A}(k, M) - A(k)]$ converges in distribution, as M tends to infinity, to a normal random variable with zero mean and the variance $\sigma_A^2(k)$ given by (5.34). And, hence, the proof is finished. □

Proof (of Theorem 5.4).

$$\sqrt{M}[\hat{\lambda}(k, M) - \lambda(k)]$$

$$= \sqrt{M} \frac{R(k)\hat{R}(k-1, M) - \hat{R}(k, M)R(k-1)}{\hat{R}(k-1, M)R(k-1)}$$

$$= \frac{\sqrt{M}[R(k)(\hat{R}(k-1, M) - R(k-1)) - (\hat{R}(k, M) - R(k))R(k-1)]}{\hat{R}(k-1, M)R(k-1)}.$$

From the consistency of the reliability estimator (Theorem 5.1), we have $\hat{R}(k-1, M) \xrightarrow[M \to \infty]{a.s.} R(k-1)$, so in order to obtain the asymptotic normality for the BMP-failure rate, we need only prove that

$$\sqrt{M} \left[R(k)\left(\hat{R}(k-1, M) - R(k-1)\right) - \left(\hat{R}(k, M) - R(k)\right)R(k-1) \right]$$

converges in distribution, as M tends to infinity, to a normal random variable of zero mean and variance $\sigma_1^2(k)$. Using the same argument as in the proof of Theorem 5.1, we obtain that

$$\sqrt{M} \left[R(k)\left(\hat{R}(k-1, M) - R(k-1)\right) - \left(\hat{R}(k, M) - R(k)\right)R(k-1) \right]$$

has the same limit in distribution as

$$\sqrt{M} \left[\sum_{l,r=1}^{s} R(k)D_{rl}^U * \Delta q_{rl}(k-1) - \sum_{r \in U} \sum_{l=1}^{s} R(k)\left(\sum_{n \in U} \alpha_n \psi_{nr}\right) * \Delta Q_{rl}(k-1) \right.$$

$$\left. - \sum_{l,r=1}^{s} R(k-1)D_{rl}^U * \Delta q_{rl}(k) + \sum_{r \in U} \sum_{l=1}^{s} R(k-1)\left(\sum_{n \in U} \alpha_n \psi_{nr}\right) * \Delta Q_{rl}(k) \right]$$

$$
= \frac{1}{\sqrt{M}} \sum_{n=1}^{N(M)} \sum_{l,r=1}^{s} \frac{M}{N_r(M)} \Big[R(k)D_{rl}^U * \Big(\mathbf{1}_{\{J_{n-1}=r, J_n=l, X_n=\cdot\}}
$$

$$
- q_{rl}(\cdot)\mathbf{1}_{\{J_{n-1}=r\}} \Big)(k-1)
$$

$$
- R(k)\mathbf{1}_{\{r\in U\}} \Big(\sum_{t\in U} \alpha_t \psi_{tr} \Big) * \Big(\mathbf{1}_{\{J_{n-1}=r, J_n=l, X_n\le\cdot\}} - Q_{rl}(\cdot)\mathbf{1}_{\{J_{n-1}=r\}} \Big)(k-1)
$$

$$
- R(k-1)D_{rl}^U * \Big(\mathbf{1}_{\{J_{n-1}=r, J_n=l, X_n=\cdot\}} - q_{rl}(\cdot)\mathbf{1}_{\{J_{n-1}=r\}} \Big)(k)
$$

$$
- R(k-1)\mathbf{1}_{\{r\in U\}} \Big(\sum_{t\in U} \alpha_t \psi_{tr} \Big) * \Big(\mathbf{1}_{\{J_{n-1}=r, J_n=l, X_n\le\cdot\}} - Q_{rl}(\cdot)\mathbf{1}_{\{J_{n-1}=r\}} \Big)(k) \Big]
$$

$$
= \frac{1}{\sqrt{M}} \sum_{l=1}^{N(M)} f(J_{l-1}, J_l, X_l),
$$

where we have defined the function $f : E \times E \times \mathbb{N} \to \mathbb{R}$ by

$$
f(i,j,x)
$$

$$
:= \sum_{l,r=1}^{s} \frac{M}{N_r(M)} \Big[R(k)D_{rl}^U * \Big(\mathbf{1}_{\{i=r, j=l, x=\cdot\}} - q_{rl}(\cdot)\mathbf{1}_{\{i=r\}} \Big)(k-1)
$$

$$
- R(k)\mathbf{1}_{\{r\in U\}} \Big(\sum_{t\in U} \alpha_t \psi_{tr} \Big) * \Big(\mathbf{1}_{\{i=r, j=l, x\le\cdot\}} - Q_{rl}(\cdot)\mathbf{1}_{\{i=r\}} \Big)(k-1)
$$

$$
- R(k-1)D_{rl}^U * \Big(\mathbf{1}_{\{i=r, j=l, x=\cdot\}} - q_{rl}(\cdot)\mathbf{1}_{\{i=r\}} \Big)(k)
$$

$$
- R(k-1)\mathbf{1}_{\{r\in U\}} \Big(\sum_{t\in U} \alpha_t \psi_{tr} \Big) * \Big(\mathbf{1}_{\{i=r, j=l, x\le\cdot\}} - Q_{rl}(\cdot)\mathbf{1}_{\{i=r\}} \Big)(k) \Big]
$$

$$
= \frac{M}{N_i(M)} \Big[R(k)D_{ij}^U * \mathbf{1}_{\{x=\cdot\}}(k-1) - R(k-1)D_{ij}^U * \mathbf{1}_{\{x=\cdot\}}(k)
$$

$$
- R(k)\mathbf{1}_{\{i\in U\}} \Big(\sum_{t\in U} \alpha_t \psi_{ti} \Big) * \mathbf{1}_{\{x\le\cdot\}}(k-1)
$$

$$
+ R(k-1)\mathbf{1}_{\{i\in U\}} \Big(\sum_{t\in U} \alpha_t \psi_{ti} \Big) * \mathbf{1}_{\{x\le\cdot\}}(k) - T_i(k) \Big].
$$

Using Lemmas A.2–A.4, we obtain

$$
A_{ij} = \frac{M}{N_i(M)} \Big[\sum_{x=1}^{\infty} R(k)D_{ij}^U * \mathbf{1}_{\{x=\cdot\}}(k-1)q_{ij}(x)
$$

$$
- \sum_{x=1}^{\infty} R(k-1)D_{ij}^U * \mathbf{1}_{\{x=\cdot\}}(k)q_{ij}(x)
$$

$$- \sum_{x=1}^{\infty} R(k) \mathbf{1}_{\{i \in U\}} \Big(\sum_{t \in U} \alpha_t \psi_{ti} \Big) * \mathbf{1}_{\{x \leq \cdot\}} (k-1) q_{ij}(x)$$

$$+ \sum_{x=1}^{\infty} R(k-1) \mathbf{1}_{\{i \in U\}} \Big(\sum_{t \in U} \alpha_t \psi_{ti} \Big) * \mathbf{1}_{\{x \leq \cdot\}} (k) q_{ij}(x) - T_i(k) \sum_{x=1}^{\infty} q_{ij}(x) \Big]$$

$$= \frac{M}{N_i(M)} \Big[R(k) D_{ij}^U * q_{ij}(k-1) - R(k-1) D_{ij}^U * q_{ij}(k)$$

$$- R(k) \mathbf{1}_{\{i \in U\}} \Big(\sum_{t \in U} \alpha_t \psi_{ti} \Big) * q_{ij}(k-1)$$

$$+ R(k-1) \mathbf{1}_{\{i \in U\}} \Big(\sum_{t \in U} \alpha_t \psi_{ti} \Big) * q_{ij}(k) - T_i(k) p_{ij} \Big].$$

$$A_i = \frac{M}{N_i(M)} \Big[T_i(k) - T_i(k) \sum_{j=1}^{s} p_{ij} \Big] = 0.$$

From Lemmas A.3 and A.4 we get

$$B_{ij} = \Big(\frac{M}{N_i(M)} \Big)^2 \Big\{ R^2(k) (D_{ij}^U)^2 * q_{ij}(k-1) + R^2(k-1) (D_{ij}^U)^2 * q_{ij}(k)$$

$$+ R^2(k) \mathbf{1}_{\{i \in U\}} \Big(\sum_{l=0}^{\cdot} \sum_{t \in U} \alpha_t \psi_{ti}(l) \Big)^2 * q_{ij}(k-1)$$

$$+ R^2(k-1) \mathbf{1}_{\{i \in U\}} \Big(\sum_{l=0}^{\cdot} \sum_{t \in U} \alpha_t \psi_{ti}(l) \Big)^2 * q_{ij}(k) + T_i^2(k) p_{ij}$$

$$- 2R(k-1)R(k) \Big(D_{ij}^U (D_{ij}^U)^+ \Big) * q_{ij}(k-1)$$

$$- 2R^2(k) \Big[D_{ij}^U(\cdot) \mathbf{1}_{\{i \in U\}} \Big(\sum_{l=0}^{\cdot} \sum_{t \in U} \alpha_t \psi_{ti}(l) \Big) \Big] * q_{ij}(k-1)$$

$$+ 2R(k-1)R(k) \Big[D_{ij}^U(\cdot) \mathbf{1}_{\{i \in U\}} \Big(\sum_{l=0}^{\cdot} \sum_{t \in U} \alpha_t \psi_{ti}(l) \Big)^+ \Big] * q_{ij}(k-1)$$

$$+ 2R(k-1)R(k) \Big[(D_{ij}^U)^+(\cdot) \mathbf{1}_{\{i \in U\}} \Big(\sum_{l=0}^{\cdot} \sum_{t \in U} \alpha_t \psi_{ti}(l) \Big) \Big] * q_{ij}(k-1)$$

$$- 2R^2(k-1) \Big[D_{ij}^U(\cdot) \mathbf{1}_{\{i \in U\}} \Big(\sum_{l=0}^{\cdot} \sum_{t \in U} \alpha_t \psi_{ti}(l) \Big) \Big] * q_{ij}(k)$$

$$- 2R(k-1)R(k) \mathbf{1}_{\{i \in U\}} \Big[\Big(\sum_{l=0}^{\cdot} \sum_{t \in U} \alpha_t \psi_{ti}(l) \Big) \Big(\sum_{l=0}^{\cdot} \sum_{t \in U} \alpha_t \psi_{ti}(l) \Big)^+ \Big]$$

$$* q_{ij}(k-1)$$

$$- 2T_i(k) \Big[R(k) D_{ij}^U * q_{ij}(k-1) - R(k-1) D_{ij}^U * q_{ij}(k)$$

$$-R(k)\mathbf{1}_{\{i\in U\}}\Big(\sum_{t\in U}\alpha_t\psi_{ti}\Big)*Q_{ij}(k-1)$$

$$+R(k-1)\mathbf{1}_{\{i\in U\}}\Big(\sum_{t\in U}\alpha_t\psi_{ti}\Big)*Q_{ij}(k)\Big]\Big\};$$

$$B_i=\Big(\frac{M}{N_i(M)}\Big)^2\Big\{R^2(k)\sum_{j=1}^{s}(D_{ij}^U)^2*q_{ij}(k-1)$$

$$+R^2(k-1)\sum_{j=1}^{s}(D_{ij}^U)^2*q_{ij}(k)$$

$$+R^2(k)\mathbf{1}_{\{i\in U\}}\sum_{j=1}^{s}\Big(\sum_{t\in U}\alpha_t\Psi_{ti}\Big)^2*q_{ij}(k-1)$$

$$+R^2(k-1)\mathbf{1}_{\{i\in U\}}\sum_{j=1}^{s}\Big(\sum_{t\in U}\alpha_t\Psi_{ti}\Big)^2*q_{ij}(k)+T_i^2(k)\sum_{j=1}^{s}p_{ij}$$

$$-2R(k-1)R(k)\sum_{j=1}^{s}\Big(D_{ij}^U(D_{ij}^U)^+\Big)*q_{ij}(k-1)$$

$$-2R^2(k)\mathbf{1}_{\{i\in U\}}\sum_{j=1}^{s}\Big[D_{ij}^U\Big(\sum_{t\in U}\alpha_t\Psi_{ti}\Big)\Big]*q_{ij}(k-1)$$

$$+2R(k-1)R(k)\mathbf{1}_{\{i\in U\}}\sum_{j=1}^{s}\Big[D_{ij}^U\Big(\sum_{t\in U}\alpha_t\Psi_{ti}^+\Big)\Big]*q_{ij}(k-1)$$

$$+2R(k-1)R(k)\mathbf{1}_{\{i\in U\}}\sum_{j=1}^{s}\Big[(D_{ij}^U)^+\Big(\sum_{t\in U}\alpha_t\Psi_{ti}\Big)\Big]*q_{ij}(k-1)$$

$$-2R^2(k-1)\mathbf{1}_{\{i\in U\}}\sum_{j=1}^{s}\Big[D_{ij}^U\Big(\sum_{t\in U}\alpha_t\Psi_{ti}\Big)\Big]*q_{ij}(k)$$

$$-2R(k-1)R(k)\mathbf{1}_{\{i\in U\}}\sum_{j=1}^{s}\Big[\Big(\sum_{t\in U}\alpha_t\Psi_{ti}\Big)\Big(\sum_{t\in U}\alpha_t\Psi_{ti}^+\Big)\Big]*q_{ij}(k-1)$$

$$-2T_i(k)\sum_{j=1}^{s}\Big[R(k)D_{ij}^U*q_{ij}(k-1)-R(k-1)D_{ij}^U*q_{ij}(k)$$

$$-R(k)\mathbf{1}_{\{i\in U\}}\Big(\sum_{t\in U}\alpha_t\psi_{ti}\Big)*Q_{ij}(k-1)$$

$$+R(k-1)\mathbf{1}_{\{i\in U\}}\Big(\sum_{t\in U}\alpha_t\psi_{ti}\Big)*Q_{ij}(k)\Big]\Big\};$$

Consequently, we obtain

$$B_i=\Big(\frac{M}{N_i(M)}\Big)^2\Big\{R^2(k)\sum_{j=1}^{s}\Big[D_{ij}^U-\mathbf{1}_{\{i\in U\}}\sum_{t\in U}\alpha_t\Psi_{ti}\Big]^2*q_{ij}(k-1)$$

$$+R^2(k-1)\sum_{j=1}^{s}\Big[D_{ij}^{U}-\mathbf{1}_{\{i\in U\}}\sum_{t\in U}\alpha_t\Psi_{ti}\Big]^2 * q_{ij}(k)-T_i^2(k)$$

$$+2R(k-1)R(k)\sum_{j=1}^{s}\Big[\mathbf{1}_{\{i\in U\}}D_{ij}^{U}\sum_{t\in U}\alpha_t\Psi_{ti}^{+}+\mathbf{1}_{\{i\in U\}}(D_{ij}^{U})^{+}\sum_{t\in U}\alpha_t\Psi_{ti}$$

$$-(D_{ij}^{U})^{+}D_{ij}^{U}-\mathbf{1}_{\{i\in U\}}\Big(\sum_{t\in U}\alpha_t\Psi_{ti}\Big)\Big(\sum_{t\in U}\alpha_t\Psi_{ti}^{+}\Big)\Big]*q_{ij}(k-1)\Big\}.$$

Since $N_i(M)/M \xrightarrow[M\to\infty]{a.s.} 1/\mu_{ii}$ (Proposition 3.8), applying the CLT, we obtain the desired result.   □

D

Markov Chains

We give here some basic definitions and results on finite state space Markov chains (discrete-time Markov processes). For references, see, e.g., Feller (1993), Norris (1997), and Girardin and Limnios (2001).

D.1 Definition and Transition Function

Definition D.1. *The random variable sequence* $X = (X_n)_{n \in \mathbb{N}}$ *defined on a probability space* $(\Omega, \mathcal{F}, \mathbb{P})$, *with values in the finite set* $E = \{1, \ldots, s\}$, *is a* Markov chain *if, for any nonnegative integer n and any states* $i, j, i_0, i_1, \ldots, i_{n-1} \in E$, *we have:*

$$\mathbb{P}(X_{n+1} = j \mid X_0 = i_0, X_1 = i_1, \ldots, X_{n-1} = i_{n-1}, X_n = i)$$
$$= \mathbb{P}(X_{n+1} = j \mid X_n = i) = p_{ij;n}. \tag{D.1}$$

Equality (D.1) is called *Markov property*. If $p_{ij;n} = p_{ij}$ does not depend on n, then the Markov chain is called *homogeneous* (with respect to time). In the sequel, we consider only homogeneous Markov chains. The function $(i, j) \to p_{ij}$, defined on $E \times E$, is called the *transition function* of the chain. As we are concerned only with finite state space Markov chains, we can represent the transition function by a squared matrix $\mathbf{p} = (p_{ij})_{i,j \in E} \in \mathcal{M}_E$. For this reason, we will also use the term *transition matrix* when speaking about the transition function.

Example D.1. Let us consider Example 2.2 of the HEV DNA sequence given in Chapter 2. Suppose that, instead of independence assumption, we consider that the sequence is a dependent one. That is, the probability of a basis to appear in place n $(n \geq 1)$, knowing the bases in places 0 to n, depends only on the basis that occupies the place $n - 1$. Thus, the probability of these dependencies are given in the following matrix:

$$\mathbf{p} = \begin{pmatrix} 0.2 & 0.3 & 0.3 & 0.2 \\ 0.3 & 0.2 & 0.2 & 0.3 \\ 0.2 & 0.3 & 0.3 & 0.2 \\ 0.3 & 0.2 & 0.2 & 0.3 \end{pmatrix}.$$

Thus, the sequence $X_n, n \geq 0$, with values in $\{A, C, G, T\}$, is a Markov chain, with transition probability matrix $\mathbf{p}$.

Let us also define the *n-step transition function* by

$$p_{ij}^{(n)} = \mathbb{P}(X_{n+m} = j \mid X_m = i), \quad \text{for any } m \in \mathbb{N}.$$

In matrix form, we write $\mathbf{p}^{(n)} = (p_{ij}^{(n)})_{i,j \in E}$. We also have $\mathbf{p}^{(1)} = \mathbf{p}$ and we set $\mathbf{p}^{(0)} = \mathbf{I}$, with $I(i, j) = \delta_{ij}$ (Kronecker symbol).

The transition function of a Markov chain satisfies the following properties:

1. $p_{ij} \geq 0$,
2. $\sum_{j \in E} p_{ij} = 1$,
3. $\sum_{k \in E} p_{ik}^{(n)} p_{kj}^{(m)} = p_{ij}^{(n+m)}$,

for any $i, j \in E$ and $n, m \geq 0$.

Note that, in matrix notation, $\mathbf{p}^{(n)}$ represents the usual n-fold matrix product of $\mathbf{p}$. For this reason, we write $\mathbf{p}^n$ instead of $\mathbf{p}^{(n)}$.

The last property given above is called the Chapman–Kolmogorov identity (or equation). In matrix form it can be written as

$$\mathbf{p}^m \mathbf{p}^n = \mathbf{p}^n \mathbf{p}^m = \mathbf{p}^{n+m},$$

which shows that the n-step transition matrices $(\mathbf{p}^n)_{n \in \mathbb{N}}$ form a semigroup.

The distribution of X_0 is called the *initial distribution of the chain*.

Proposition D.1. *Let* $(X_n)_{n \in \mathbb{N}}$ *be a Markov chain of transition function* $\mathbf{p}$ *and initial distribution* $\boldsymbol{\alpha}$. *For any* $n \geq 1$, $k \geq 0$, *and any states* $i, j, i_0, i_1, ..., i_n \in E$, *we have:*

1. $\mathbb{P}(X_0 = i_0, X_1 = i_1, \dots, X_{n-1} = i_{n-1}, X_n = i_n) = \alpha_{i_0} p_{i_0 i_1} \cdots p_{i_{n-1} i_n}$,
2. $\mathbb{P}(X_{k+1} = i_1, \dots, X_{k+n-1} = i_{n-1}, X_{k+n} = i_n \mid X_k = i_0) = p_{i_0 i_1} \cdots p_{i_{n-1} i_n}$,
3. $\mathbb{P}(X_{n+m} = j \mid X_m = i) = \mathbb{P}(X_n = j \mid X_0 = i) = p_{ij}^n$.

Proposition D.2. *Let* $(X_n)_{n \in \mathbb{N}}$ *be a Markov chain of transition function* $\mathbf{p}$.

1. *The sojourn time of the chain in state* $i \in E$ *is a geometric random variable on* $\mathbb{N}^*$ *of parameter* $1 - p_{ii}$.
2. *The probability that the chain enters state* j *when it leaves state* i *equals* $\frac{p_{ij}}{1 - p_{ii}}$ *(for* $p_{ii} \neq 1$, *which means that state* i *is non absorbing).*

State Probabilities

Let $E = \{1, \ldots, s\}$ be the state space of the Markov chain $(X_n)_{n \in \mathbb{N}}$. Note, for any state $i \in E$,

$$P_i(n) = \mathbb{P}(X_n = i).$$

This probability is called the probability of state i at time $n \geq 0$. Denote also by

$$P(n) = (P_1(n), \ldots, P_s(n))$$

the state probability vector. For a time $n \geq 0$, $P(n)$ is the distribution of X_n. If $n = 0$, then $P(0)$, denoted by $\boldsymbol{\alpha} := P(0)$, is the initial distribution of the chain.

We can write

$$\mathbb{P}(X_n = j) = \sum_{i \in E} \alpha(i) p_{ij}^n$$

or, in matrix form,

$$P(n) = \boldsymbol{\alpha} \mathbf{p}^n.$$

D.2 Strong Markov Property

Definition D.2 (stopping time or Markov time). *A random variable T, defined on $(\Omega, \mathcal{F}, \mathbb{P})$, with values in $\overline{\mathbb{N}} = \mathbb{N} \cup \{\infty\}$, is called a* stopping time *with respect to the sequence $(X_n)_{n \in \mathbb{N}}$ if the occurrence of the event $\{T = n\}$ is determined by the past of the chain up to time n, $(X_k; k \leq n)$.*
More precisely, let $\mathcal{F}_n = \sigma(X_0, \ldots, X_n), n \geq 0$, be the σ-algebra generated by $X_0, \ldots, X_n$, i.e., the information known at time n. The random variable T is called a stopping time if, for every $n \in \mathbb{N}$, $\{T = n\} \in \mathcal{F}_n$.

Definition D.3 (strong Markov property). *The Markov chain $(X_n)_{n \in \mathbb{N}}$ is said to have the* strong Markov property *if, for any stopping time T, for any integer $m \in \mathbb{N}$ and state $j \in E$ we have*

$$\mathbb{P}(X_{m+T} = j | X_k, k \leq T) = \mathbb{P}_{X_T}(X_m = j) \ a.s.$$

Proposition D.3. *Any Markov chain has the strong Markov property.*

D.3 Recurrent and Transient States

A state of a Markov chain can be characterized as being recurrent or transient. This distinction is fundamental for the study of Markov chains. All the following definitions and results will be given for a Markov chain $(X_n)_{n\in\mathbb{N}}$ of transition matrix $\mathbf{p} = (p_{ij})_{i,j\in E}$.

Let us define :

- $\eta_i = \min\{n \mid n \in \mathbb{N}^*, X_n = i\}$ (with min $\emptyset = \infty$), the first passage time of the chain in state i. If $X_0(\omega) = i$, then η_i is the recurrence time of state i. Note that $\eta_i > 0$.
- $N_i(n) = \sum_{k=0}^{n-1} \mathbf{1}_{\{X_k=i\}}$, the time spent by the chain in state i, during the time interval $[0, n-1]$. If $n = \infty$, then we note $N_i = N_i(\infty)$, with N_i taking values in $\overline{\mathbb{N}}$.
- $N_{ij}(n) = \sum_{k=1}^{n} \mathbf{1}_{\{X_{k-1}=i, X_k=j\}}$, the number of direct transition from i to j, up to time n. If $n = \infty$, then we note $N_{ij} = N_{ij}(\infty)$, with N_{ij} taking values in $\overline{\mathbb{N}}$.

Definition D.4 (recurrent and transient Markov chain). *A state $i \in E$ is called* recurrent *if $\mathbb{P}_i(\eta_i < \infty) = 1$; in the opposite case, when $\mathbb{P}_i(\eta_i < \infty) < 1$, the state i is called* transient. *A recurrent state i is called* positive recurrent *if $\mu_{ii}^* = \mathbb{E}_i[\eta_i] < \infty$ and* null recurrent *if $\mu_{ii}^* = \infty$.*
The Markov chain is said to be (positive/null) recurrent (resp. transient), if all the states are (positive/null) recurrent (resp. transient).

Definition D.5 (irreducible Markov chain). *If for any states i, j there is a positive integer n such that $p_{ij}^{(n)} > 0$, then the Markov chain is said to be* irreducible.

Note that the Markov chain defined in Example D.1 is irreducible and aperiodic.
Let us set $\rho_{ij} := \mathbb{P}_i(\eta_j < +\infty)$.

Lemma D.1. *For any $i, j \in E$ and $m \in \mathbb{N}^*$, we have $\mathbb{P}_i(N_j \geq m) = \rho_{ij}\rho_{jj}^{m-1}$.*

Proposition D.4.   *1. A state $i \in E$ is transient iff $\mathbb{P}(N_i = +\infty) = 0$ or iff $\sum_n p_{ii}^n < \infty$.*
2. A state $i \in E$ is recurrent iff $\mathbb{P}(N_i = +\infty) = 1$ or iff $\sum_n p_{ii}^n = \infty$.

Remark D.1. For a finite-state Markov chain, every recurrent state is positive recurrent. A finite state space irreducible Markov chain is positive recurrent.

Proposition D.5. *For i and j recurrent states, we have*

$$N_i(n)/n \xrightarrow[n\to\infty]{a.s.} \frac{1}{\mu_{ii}^*}, \tag{D.2}$$

$$N_{ij}(n)/n \xrightarrow[n\to\infty]{a.s.} \frac{p_{ij}}{\mu_{ii}^*}. \tag{D.3}$$

The following result is from Karlin and Taylor (1981) (page 73, Theorem 1.1), and is used for proving the renewal theorem (Theorem 2.6).

Let $(X_n)_{n \in \mathbb{N}^*}$ be a sequence of integer-valued, independent, identically distributed random variables and let $S_n := X_1 + \ldots + X_n, n \in \mathbb{N}^*$. Define also $S_0 := 0$. We can show that the sequence $(S_n)_{n \in \mathbb{N}}$ is a Markov chain.

Theorem D.1. *Suppose that the Markov chain $(S_n)_{n \in \mathbb{N}}$ is irreducible. If*

$$\mathbb{E}(| X_k |) < \infty, k \in \mathbb{N}^*$$

and

$$\mu := \mathbb{E}(X_k) = 0, k \in \mathbb{N}^*,$$

then the Markov chain $(S_n)_{n \in \mathbb{N}}$ is recurrent.

D.4 Stationary Distribution

Definition D.6. *A probability distribution ν on E is said to be* stationary *or* invariant *for the Markov chain $(X_n)_{n \in \mathbb{N}}$ if, for any $j \in E$,*

$$\sum_{j \in E} \nu(i) p_{ij} = \nu(j),$$

or, in matrix form,

$$\nu \mathbf{p} = \nu,$$

where $\nu = (\nu(1), \ldots, \nu(s)$ is a row vector.

Proposition D.6. *For a recurrent state i, we have: $\nu(i) = 1/\mu_{ii}^*$.*

Definition D.7. *A state $i \in E$ is said to be* periodic *of period $d > 1$, or d-periodic, if $g.c.d.\{n \mid n > 1, p_{ii}^n > 0\} = d$. If $d = 1$, then the state i is said to be* aperiodic.

Definition D.8. *An aperiodic recurrent state is called* ergodic. *An irreducible Markov chain with one state ergodic (and then all states ergodic) is called* ergodic.

Proposition D.7 (ergodic theorem for Markov chains). *For an ergodic Markov chain we have*

$$p_{ij}^n \xrightarrow[n \to \infty]{} \nu(j)$$

for any $i, j \in E$.

Proposition D.8. *For an ergodic Markov chain, there exists ν the stationary distribution of the chain such that $\mathbf{p}^n$ converges at an exponential rate to $\Pi = \mathbf{1}^\top \nu$, where $\mathbf{1} = (1, \ldots, 1)$.*

Proposition D.9. *For any states $i, j \in E$, we have*

$$\frac{1}{\sqrt{n}}(N_{ij}(n) - p_{ij}N_i(n)) \xrightarrow[n \to \infty]{\mathcal{D}} \mathcal{N}(0, \nu(i)p_{ij}[1 - p_{ij}]).$$

The following two results concern the SLLN and the CLT for a function g : $E \to \mathbb{R}$ of a Markov chain (see, e.g., Dacunha-Castelle and Duflo, 1986). The expression of the asymptotic variance in the CLT was obtained by Trevezas and Limnios (2008a).

Proposition D.10 (strong law of large numbers). *For an ergodic Markov chain $(X_n)_{n \in \mathbb{N}}$, with stationary distribution $\boldsymbol{\nu}$, we have*

$$\frac{1}{n}\sum_{k=0}^{n} g(X_k) \xrightarrow[n \to \infty]{a.s.} \sum_{i \in E} \nu(i)g(i) =: \overline{g}.$$

Proposition D.11 (central limit theorem). *For an ergodic Markov chain $(X_n)_{n \in \mathbb{N}}$, with stationary distribution $\boldsymbol{\nu}$, we have*

$$\sqrt{n}\left(\frac{1}{n}\sum_{k=0}^{n} g(X_k) - \overline{g}\right) \xrightarrow[n \to \infty]{\mathcal{D}} \mathcal{N}(0, \sigma^2),$$

where

$$\sigma^2 := \widetilde{\mathbf{g}}\, diag(\boldsymbol{\nu})[2\mathbf{Z} - \mathbf{I}]\widetilde{\mathbf{g}}^{\top}$$

where $\mathbf{Z} = (\mathbf{I} - \mathbf{p} + \mathbf{\Pi})^{-1}$ is the fundamental matrix of $\mathbf{p}$ and $\widetilde{\mathbf{g}}$ is the row vector with components $\widetilde{g}(i) := g(i) - \overline{g}$, $i \in E$.

D.5 Markov Chains and Reliability

We briefly present here the main reliability measures of a system and we give their explicit forms for a discrete-time finite-state Markov system.

Let the state space $E = \{1, \ldots, s\}$ of the Markov chain $X = (X_n)_{n \in \mathbb{N}}$ be partitioned in the subset $U = \{1, \ldots, s_1\}$ of working states of the system (the up states) and in the subset $D = \{s_1 + 1, \ldots, s\}$ of failure states (the down states), $0 < s_1 < s$. According to the partition of the state space into the up states and the down states, we partition the transition matrix $\mathbf{p}$, its initial distribution $\boldsymbol{\alpha}$ and stationary distribution $\boldsymbol{\nu}$:

$$\mathbf{p} = \begin{pmatrix} \mathbf{p}_{11} & \mathbf{p}_{12} \\ \mathbf{p}_{21} & \mathbf{p}_{22} \end{pmatrix} \begin{matrix} U \\ D \end{matrix}, \quad \boldsymbol{\alpha} = \begin{pmatrix} \boldsymbol{\alpha}_1 & \boldsymbol{\alpha}_2 \end{pmatrix}, \quad \boldsymbol{\nu} = \begin{pmatrix} \boldsymbol{\nu}_1 & \boldsymbol{\nu}_2 \end{pmatrix}.$$

For $m, n \in \mathbb{N}^*$ such that $m > n$, let $\mathbf{1}_{m,n}$ denote the m-dimensional column vector whose n first elements are 1 and last $m - n$ elements are 0; for $m \in \mathbb{N}^*$, let $\mathbf{1}_m$ denote the m-column vector all of whose elements are 1.

For $n \geq 0$, we introduce the following reliability indicators:

- Reliability
$$R(n) := \mathbb{P}(\forall k \in [0, n], X_k \in U);$$

- Availability
$$A(n) := \mathbb{P}(X_n \in U);$$

- Asymptotic or steady-state availability
$$A_\infty := \lim_{n \to \infty} A(n) = \sum_{i \in U} \nu(i);$$

- Maintainability
$$M(n) := 1 - \mathbb{P}(\forall k \in [0, n], X_k \in D);$$

- Mean hitting times
  - MTTF - mean time to failure;
  - MTTR - mean time to repair;
  - MUT - mean up time;
  - MDT - mean down time;
  - MTBF - mean time between failures.

For a discrete-time Markov system, for $n \geq 0$, the above reliability indicators have the following closed forms in terms of transition matrix $\mathbf{p}$, initial distribution $\boldsymbol{\alpha}$ and stationary distribution $\boldsymbol{\nu}$.

- Reliability
$$R(n) = \boldsymbol{\alpha}_1 \mathbf{p}_{11}^n \mathbf{1}_{s_1}.$$

- Availability
$$A(n) := \boldsymbol{\alpha} \mathbf{p}^n \mathbf{1}_{s,s_1}.$$

- Maintainability
$$M(n) := 1 - \boldsymbol{\alpha}_2 \mathbf{p}_{22}^n \mathbf{1}_{s-s_1}.$$

- Mean times
  - $MTTF = \boldsymbol{\alpha}_1 (\mathbf{I} - \mathbf{p}_{11})^{-1} \mathbf{1}_{s_1},$

  - $MTTR = \boldsymbol{\alpha}_2 (\mathbf{I} - \mathbf{p}_{22})^{-1} \mathbf{1}_{s-s_1},$

  - $MUT = \dfrac{\nu_1 \mathbf{1}_{s_1}}{\nu_2 \mathbf{p}_{21} \mathbf{1}_{s_1}},$

  - $MDT = \dfrac{\nu_2 \mathbf{1}_{s-s_1}}{\nu_1 \mathbf{p}_{12} \mathbf{1}_{s-s_1}},$

  - $MTBF = MUT + MDT.$

Example D.2. Let us consider a three-state discrete-time Markov system with

$$E = \{1, 2, 3\}, \quad U = \{1, 2\}, \quad D = \{3\}.$$

Let the transition matrix and the initial distribution be

$$\mathbf{p} = \begin{pmatrix} 0.9 & 0.1 & 0 \\ 0.4 & 0.4 & 0.2 \\ 0.4 & 0 & 0.6 \end{pmatrix}, \quad \alpha = (1, 0, 0).$$

A simple computation gives the stationary distribution

$$\nu = (0.8 \quad 0.1333 \quad 0.0677).$$

The reliability of the system at time n is given by

$$R(n) = (1 \quad 0) \begin{pmatrix} 0.9 & 0.1 \\ 0.4 & 0.4 \end{pmatrix}^n \begin{pmatrix} 1 \\ 1 \end{pmatrix}$$

and the mean time to failure is

$$MTTF = (1 \quad 0) \begin{pmatrix} 0.1 & -0.1 \\ -0.4 & 0.6 \end{pmatrix}^{-1} \begin{pmatrix} 1 \\ 1 \end{pmatrix} = 35.$$

In a similar way, the other reliability quantities can be easily computed.

E

Miscellaneous

results, often used along the book, are presented in this appendix. All the random variables (or vectors) are defined on some probability space $(\Omega, \mathcal{F}, \mathbb{P})$.

Probability

The following result is a direct application of the countable additivity of a probability measure (see, e.g., Billingsley, 1995; Durrett, 1991).

Theorem E.1. *Let* $(\Omega, \mathcal{F}, \mathbb{P})$ *be a probability space and* $A_n \in \mathcal{F}, n \in \mathbb{N}$.

1. **(continuity from below)** *If* $A_n \uparrow A$ *(i.e.,* $A_0 \subset A_1 \subset \dots$ *and* $\cup_{n \in \mathbb{N}} A_n = A$*), then* $\mathbb{P}(A_n) \uparrow \mathbb{P}(A)$ *(i.e.,* $\lim_{n \to \infty} \mathbb{P}(A_n) = \mathbb{P}(A)$ *and the sequence* $(P(A_n))_{n \in \mathbb{N}}$ *is nondecreasing in* n*).*
2. **(continuity from above)** *If* $A_n \downarrow A$ *(i.e.,* $A_0 \supset A_1 \supset \dots$ *and* $\cap_{n \in \mathbb{N}} A_n = A$*), then* $\mathbb{P}(A_n) \downarrow \mathbb{P}(A)$ *(i.e.,* $\lim_{n \to \infty} \mathbb{P}(A_n) = \mathbb{P}(A)$ *and the sequence* $(P(A_n))_{n \in \mathbb{N}}$ *is nonincreasing in* n*).*

The following theorem allows to exchange limit and expected value (see, e.g., Billingsley, 1995).

Theorem E.2 (monotone convergence theorem). *Let* $X_n, n \in \mathbb{N}$, *and* X *be real random variables such that* $X_n \geq 0, n \in \mathbb{N}$, *and* $X_n \uparrow X$ *a.s., as* n *tends to infinity (i.e.,* $X_n \xrightarrow[n \to \infty]{a.s.} X$ *and the sequence* $(X_n(\omega))_{n \in \mathbb{N}}$ *is a nondecreasing sequence in* n, *for any* $\omega \in \Omega$*). Then* $\mathbb{E}[X_n] \uparrow \mathbb{E}[X]$, *as* n *tends to infinity (i.e.,* $\mathbb{E}[X_n] \xrightarrow[n \to \infty]{} \mathbb{E}[X]$ *and the sequence* $(\mathbb{E}[X_n])_{n \in \mathbb{N}}$ *is a nondecreasing sequence in* n*).*

Series

For double series $(a(m,n))_{m,n\in\mathbb{N}}$ we have the following result.

Proposition E.1 (dominated convergence theorem for sequences).
Let $a(m,n)_{m,n\in\mathbb{N}}$ be a double sequence such that $\lim_{m\to\infty} a(m,n)$ exists for each n and that $\mid a(m,n)\mid\leq b(n)$, with $\sum_{n=0}^{\infty} b(n) < \infty$. Then:

1. $\lim_{m\to\infty}\sum_{n=0}^{\infty} a(m,n) = \sum_{n=0}^{\infty}\lim_{m\to\infty} a(m,n) < \infty$;

2. $\lim_{m\to\infty}\sum_{n=0}^{m} a(m,n) = \sum_{n=0}^{\infty}\lim_{m\to\infty} a(m,n) < \infty$.

Remark E.1. The first assertion is proved, e.g., in Port (1994) (Proposition A 1.5., page 840). The second one can be easily obtained from the first one. A slightly less general form of this second assertion can be found in Karlin and Taylor (1975) (proof of Theorem 1.2, pages 83-84).

Note also that this result can be immediately obtained from the continuous-case dominated convergence theorem, by taking as measure the counting measure.

Martingales

Theorem E.3. *Let $(X_n)_{n\in\mathbb{N}}$ be a martingale (or a sub-/supermartingale) with respect to the filtration $\mathcal{F} = (\mathcal{F}_n)_{n\in\mathbb{N}}$, such that $\sup_n \mathbb{E}(\mid X_n\mid) < \infty$. Then, there exists a random variable X_∞, integrable, such that*

$$X_n \xrightarrow[n\to\infty]{a.s.} X_\infty.$$

The following result is the Lindeberg–Lévy Central limit theorem for martingales (see, e.g., Billingsley, 1961b, 1995; Durrett, 1991; Shiryaev, 1996).

Theorem E.4 (central limit theorem for martingales). *Let $(X_n)_{n\in\mathbb{N}^*}$, be a martingale with respect to the filtration $\mathcal{F} = (\mathcal{F}_n)_{n\in\mathbb{N}}$ and define the process $Y_n := X_n - X_{n-1}, n\in\mathbb{N}^*$, (with $Y_1 := X_1$), called a difference martingale. If*

1. $\frac{1}{n}\sum_{k=1}^{n}\mathbb{E}[Y_k^2\mid\mathcal{F}_{k-1}] \xrightarrow[n\to\infty]{P} \sigma^2 > 0$;

2. $\frac{1}{n}\sum_{k=1}^{n}\mathbb{E}[Y_k^2\mathbf{1}_{\{|Y_k|>\epsilon\sqrt{n}\}}] \xrightarrow[n\to\infty]{} 0$, *for all $\epsilon > 0$,*

then

$$\frac{X_n}{n} \xrightarrow[n\to\infty]{a.s.} 0$$

and

$$\frac{1}{\sqrt{n}}X_n = \frac{1}{\sqrt{n}}\sum_{k=1}^{n} Y_k \xrightarrow[n\to\infty]{\mathcal{D}} \mathcal{N}(0,\sigma^2).$$

Stochastic Convergence

The following result is from Gut (1988) (Theorem 2.1., pages 10-11).

Theorem E.5. *Let* $(Y_n)_{n\in\mathbb{N}}$ *be a sequence of random variables and* $(N_n)_{n\in\mathbb{N}}$ *a positive integer-valued stochastic process. Suppose that*

$$Y_n \xrightarrow[n\to\infty]{a.s.} Y \text{ and } N_n \xrightarrow[n\to\infty]{a.s.} \infty.$$

Then, $Y_{N_n} \xrightarrow[n\to\infty]{a.s.} Y.$

The following theorem is from Billingsley (1999) (Theorem 14.4, page 152).

Theorem E.6 (Anscombe's theorem). *Let* $(Y_n)_{n\in\mathbb{N}}$ *be a sequence of random variables and* $(N_n)_{n\in\mathbb{N}}$ *a positive integer-valued stochastic process. Suppose that*

$$\frac{1}{\sqrt{n}} \sum_{m=1}^{n} Y_m \xrightarrow[n\to\infty]{\mathcal{D}} \mathcal{N}(0,\sigma^2) \text{ and } N_n/n \xrightarrow[n\to\infty]{P} \theta,$$

where θ *is a constant,* $0 < \theta < \infty$. *Then,*

$$\frac{1}{N_n} \sum_{m=1}^{N_n} Y_m \xrightarrow[n\to\infty]{\mathcal{D}} \mathcal{N}(0,\sigma^2).$$

Remark E.2. Note that in Anscombe's theorem the sequence of random variables $(Y_n)_{n\in\mathbb{N}}$ does not need to be neither independent, nor identically distributed.

For the following theorem, see, e.g., Van der Vaart (1998) (Theorem 3.1, page 26).

Theorem E.7 (Delta method). *Let* $\phi = (\phi_1, \ldots, \phi_m) : D_\phi \subset \mathbb{R}^k \to \mathbb{R}^m$ *be a differentiable application in* $\theta = (\theta_1, \ldots, \theta_k)$. *Let* θ_n *be random vectors with values in the domain of* ϕ. *If the random vector* $\sqrt{a_n}(\theta_n - \theta)$ *is convergent in distribution, as* a_n *tends to infinity, to the* k-*variate normal distribution* $\mathcal{N}_k(\mu, \Sigma)$, *then the random vector* $\sqrt{a_n}(\phi(\theta_n) - \phi(\theta))$ *is convergent in distribution to the normal distribution* $\mathcal{N}_m(\phi'_\theta \mu, \phi'_\theta \Sigma (\phi'_\theta)^\top)$, *where* ϕ'_θ *is the partial derivative matrix of* ϕ *with respect to* θ:

$$\phi'_\theta = \begin{pmatrix} \frac{\partial\phi_1}{\partial\theta_1} & \cdots & \frac{\partial\phi_1}{\partial\theta_k} \\ \vdots & & \vdots \\ \frac{\partial\phi_m}{\partial\theta_1} & \cdots & \frac{\partial\phi_m}{\partial\theta_k} \end{pmatrix}.$$

For the following result, see, e.g., Shiryaev (1996) (Theorem 3, page 488).

Theorem E.8 (Wald lemma). *Let $(Y_n)_{n \in \mathbb{N}}$ be a sequence of independent random variables with common expected value. Let N be a stopping time for the sequence $(Y_n)_{n \in \mathbb{N}}$ with finite expected value and define $S_n := Y_0 + \ldots + Y_n$, $n \in \mathbb{N}$. Then,*

$$\mathbb{E}[S_N] = \mathbb{E}[Y_0]\mathbb{E}[N].$$

The following theorem (see, e.g., Van der Vaart, 1998, Theorem 2.3, pages 7-8) states that if the sequence of random vectors $(X_n)_{n \in \mathbb{N}}$, converges to X in any of the three modes of stochastic convergence (i.e., in distribution, in probability or almost surely), then the same holds true for the sequence $(g(X_n))_{n \in \mathbb{N}}$, where g is any continuous function.

Theorem E.9 (continuous mapping theorem). *Let X and $X_n, n \in \mathbb{N}$, be random vectors in $\mathbb{R}^k, k \geq 1$, and let $g : \mathbb{R}^k \to \mathbb{R}^m$ be a continuous function at every point of a set C such that $\mathbb{P}(X \in C) = 1$.*

1. *If $X_n \xrightarrow[n \to \infty]{\mathcal{D}} X$, then $g(X_n) \xrightarrow[n \to \infty]{\mathcal{D}} g(X)$.*
2. *If $X_n \xrightarrow[n \to \infty]{P} X$, then $g(X_n) \xrightarrow[n \to \infty]{P} g(X)$.*
3. *If $X_n \xrightarrow[n \to \infty]{a.s.} X$, then $g(X_n) \xrightarrow[n \to \infty]{a.s.} g(X)$.*

The following result is a direct application of continuous mapping theorem (see, e.g., Van der Vaart, 1998, Lemma 2.8, page 11).

Theorem E.10 (Slutsky's theorem). *Let $X, X_n, Y_n, n \in \mathbb{N}$, be random variables or vectors. If $X_n \xrightarrow[n \to \infty]{\mathcal{D}} X$ and $Y_n \xrightarrow[n \to \infty]{\mathcal{D}} c$, with c a constant, then*

1. $X_n + Y_n \xrightarrow[n \to \infty]{\mathcal{D}} X + c,$
2. $Y_n X_n \xrightarrow[n \to \infty]{\mathcal{D}} cX,$
3. $Y_n^{-1} X_n \xrightarrow[n \to \infty]{\mathcal{D}} c^{-1}X,$ *for $c \neq 0$.*

Remark E.3. The constant c in 1. must be a constant vector having the same dimension as X, whereas in 2. and 3. c must be a constant scalar. Nevertheless, one can consider c in 2. and 3. also as being a matrix (see Van der Vaart, 1998, for details).

Notation

$\mathbb{N}$	: set of nonnegative integers, $\mathbb{N}^* = \mathbb{N}\backslash\{0\}$, $\overline{\mathbb{N}} = \mathbb{N} \cup \{\infty\}$
$\mathbb{R}_+$	: set of nonnegative real numbers $[0, \infty)$
$(\Omega, \mathcal{F}, \mathbb{P})$	: probability space
$\mathbb{P}, \mathbb{E}$	: probability, expectation with respect to $\mathbb{P}$
$E = \{1, \ldots, s\}$	: finite state space of the semi-Markov chain Z
$\mathcal{M}_E$	: real matrices on $E \times E$
$\mathcal{M}_E(\mathbb{N})$	: matrix-valued functions defined on $\mathbb{N}$, with values in $\mathcal{M}_E$
$\mathbf{A} * \mathbf{B}$	: discrete-time matrix convolution product of $\mathbf{A}, \mathbf{B} \in \mathcal{M}_E(\mathbb{N})$
$\mathbf{A}^{(n)}$	: n-fold convolution of $\mathbf{A} \in \mathcal{M}_E(\mathbb{N})$
$\mathbf{A}^{(-1)}$	: left inverse in the convolution sense of $\mathbf{A} \in \mathcal{M}_E(\mathbb{N})$
$Z = (Z_n)_{n \in \mathbb{N}}$	: semi-Markov chain (SMC)
$(J, S) = (J_n, S_n)_{n \in \mathbb{N}}$	: Markov renewal chain (MRC)
$S = (S_n)_{n \in \mathbb{N}}$	: jump times
$J = (J_n)_{n \in \mathbb{N}}$	: visited states, embedded Markov chain (EMC)
$X = (X_n)_{n \in \mathbb{N}^*}$	: sojourn times between successive jumps
M	: fixed censoring time
$N(M)$	: number of jumps of Z in the time interval $[1, M]$
$U = (U_n)_{n \in \mathbb{N}}$	: backward-recurrence times of the SMC Z, $U_n := n - S_{N(n)}, n \in \mathbb{N}$
$N_i(M)$	: number of visits to state i of the EMC, up to time M
$N_{ij}(M)$	: number of transitions from state i to state j of the EMC, up to time M
$N_{ij}(k, M)$	: number of transitions from state i to state j of the EMC, up to time M, with sojourn time in state i equal to k
$(Z, Y) = (Z_n, Y_n)_{n \in \mathbb{N}}$	: hidden semi-Markov chain (Z hidden component)
$A = \{1, \ldots, d\}$	: finite state space of chain Y

$N_{i;b}(M)$: number of visits of state b by chain Y, up to time M, when chain Z is visiting state i, for (Z, Y) hidden chain SM-M0

$N_{i;ab}(M)$: number of transitions of Y from a to b, up to time M, with Y in state b, when Z in state i, for (Z, Y) hidden chain SM-M1

$\mathbf{q} = (q_{ij}(k))_{i,j \in E, k \in \mathbb{N}}$: semi-Markov kernel

$\mathbf{p} = (p_{ij})_{i,j \in E}$: transition matrix of the EMC J

$\mathbf{f} = (f_{ij}(k))_{i,j \in E, k \in \mathbb{N}}$: conditional sojourn time distribution in state i, before visiting state j

$\mathbf{F} = (F_{ij}(k))_{i,j \in E, k \in \mathbb{N}}$: conditional cumulative sojourn time distribution in state i, before visiting state j

$\mathbf{h} = (h_i(k))_{i \in E; k \in \mathbb{N}}$: sojourn time distribution in state i

$\mathbf{H} = (H_i(k))_{i \in E; k \in \mathbb{N}}$: cumulative distribution of sojourn time in state i

$\overline{\mathbf{H}} = (\overline{H_i}(k))_{i \in E, k \in \mathbb{N}}$: survival function in state i, $\overline{H_i}(k) = 1 - H_i(k)$

$\mathbf{P} = (P_{ij}(k))_{i,j \in E, k \in \mathbb{N}}$: transition function of the semi-Markov chain Z

$\boldsymbol{\psi} = (\psi_{ij}(k))_{i,j \in E, k \in \mathbb{N}}$: $\psi(k) := \sum_{n=0}^{k} \mathbf{q}^{(n)}(k)$

$\boldsymbol{\Psi} = (\Psi_{ij}(k))_{i,j \in E, k \in \mathbb{N}}$: Markov renewal function

$\mathbf{m} = (m_i)_{i \in E}$: mean sojourn time in state i

μ_{ij} : mean first passage time from state i to state j, for semi-Markov chain Z

μ_{ij}^* : mean first passage time from state i to state j, for embedded Markov chain J

$\boldsymbol{\pi} = (\pi_j)_{j \in E}$: limit distribution of semi-Markov chain Z

$\boldsymbol{\nu} = (\nu(j))_{j \in E}$: stationary distribution of EMC J

$\boldsymbol{\alpha} = (\alpha_j)_{j \in E}$: initial distribution of semi-Markov chain Z

$\mathbf{R} = (R_{i;a})_{i \in E, a \in A}$: conditional distribution of Y, given $\{Z_n = i\}$, for (Z, Y) hidden chain SM1-M0

$\mathbf{R} = (R_{i;a,b})_{i \in E, a, b \in A}$: conditional transition matrix of Y, given $\{Z_n = i\}$, for (Z, Y) hidden chain SM-M1

U : working states of the semi-Markov system

D : failure states of the semi-Markov system

$R(k)$: reliability at time k

$A(k)$: availability at time k

$M(k)$: maintainability at time k

$\lambda(k)$: BMP-failure rate at time k

$r(k)$: RG-failure rate at time k

MTTF : mean time to failure

MTTR : mean time to repair

MUT : mean up time

MDT : mean down time

$\widehat{p}_{ij}(M)$, $\widehat{q}_{ij}(k, M)$, $\widehat{R}(k, M)$, etc. : estimators of p_{ij}, $q_{ij}(k)$, $R(k)$, etc.

$\xrightarrow{a.s.}$	: almost sure convergence (strong consistency)
$\xrightarrow{P}$	: convergence in probability
$\xrightarrow{\mathcal{D}}$	: convergence in distribution
RC	: renewal chain
RP	: renewal process
(DT)MRC	: (discrete-time) Markov renewal chain
EMC	: embedded Markov chain
MC	: Markov chain
SMC	: semi-Markov chain
SMM	: semi-Markov model
HMC/HMM	: hidden Markov chain/hidden Markov model
HSMC/HSMM	: hidden semi-Markov chain/ hidden semi-Markov model
(DT)MRE	: (discrete-time) Markov renewal equation
EM	: EM algorithm (estimation-maximization)
MLE	: maximum-likelihood estimator
SLLN	: strong law of large numbers
CLT	: central limit theorem
r.v.	: random variable
iff	: if and only if

References

J. Aitchinson and S.D. Silvey. Maximum likelihood estimation of parameters subject to restraints. *Ann. Math. Statist.*, 29:813–828, 1958.

M.G. Akritas and G.G. Roussas. Asymptotic inference in continuous-time semi-Markov processes. *Scand. J. Statist.*, 7:73–79, 1980.

A. Aldous, P. Diaconis, J. Spencer, and J. Michael Steele (Eds.). *Discrete Probability and Algorithms*, volume 72 of *The IMA Volumes in Mathematics and Applications*. Springer, 1995.

G. Alsmeyer. The Markov renewal theorem and related results. *Markov Process. Related Fields*, 3:103–127, 1997.

E.E.E. Alvarez. Smothed nonparametric estimation in window censored semi-Markov processes. *J. Statist. Plann. Inference*, 131:209–229, 2005.

P.K. Anderson, O. Borgan, R.D. Gill, and N. Keiding. *Statistical Models Based on Counting Processes*. Springer, New York, 1993.

P.M. Anselone. Ergodic theory for discrete semi-Markov chains. *Duke Math. J.*, 27 (1):33–40, 1960.

G.E.B. Archer and D.M. Titterington. Parameter estimation for hidden Markov chains. *J. Statist. Plann. Inference*, 108(1-2):365–390, 2002.

S. Asmussen. *Applied Probability and Queues*. Wiley, Chichester, 1987.

D. Bakry, X. Milhaud, and P. Vandekerkhove. Statistique de chaînes de Markov cachées à espaces d'états fini. Le cas non stationnaire. *C. R. Acad. Sci. Paris, Ser. I*, 325:203–206, 1997.

N. Balakrishnan, N. Limnios, and C. Papadopoulos. Basic probabilistic models in reliability. In N. Balakrishnan and C.R. Rao, editors, *Handbook of Statistics-Advanced in Reliability*, pages 1–42. Elsevier, North-Holland, 2001.

P. Baldi and S. Brunak. *Bioinformatics–the Machine Learning Approach*. The MIT Press, Cambridge, 2nd edition, 2001.

V. Barbu and N. Limnios. Discrete time semi-Markov processes for reliability and survival analysis–a nonparametric estimation approach. In M. Nikulin, N. Balakrishnan, M. Mesbah, and N. Limnios, editors, *Parametric and Semiparametric Models with Applications to Reliability, Survival Analysis and Quality of Life*, Statistics for Industry and Technology, pages 487–502. Birkhäuser, Boston, 2004.

V. Barbu and N. Limnios. Empirical estimation for discrete time semi-Markov processes with applications in reliability. *J. Nonparametr. Statist.*, 18(7-8):483–498, 2006a.

V. Barbu and N. Limnios. Nonparametric estimation for failure rate functions of discrete time semi-Markov processes. In M. Nikulin, D. Commenges, and C. Hubert, editors, *Probability, Statistics and Modelling in Public Health, Special volume in honor of professor Marvin Zelen*, pages 53–72. Springer, 2006b.

V. Barbu and N. Limnios. Maximum likelihood estimation for hidden semi-Markov models. *C. R. Acad. Sci. Paris, Ser. I*, 342(3):201–205, 2006c.

V. Barbu, M. Boussemart, and N. Limnios. Discrete time semi-Markov model for reliability and survival analysis. *Comm. Statist. Theory Methods*, 33(11):2833–2868, 2004.

R.E. Barlow and F. Prochan. *Statistical Theory of Reliability and Life Testing: Probability Models*. Holt, Rinehart and Winston, New York, 1975.

R.E. Barlow, A.W. Marshall, and F. Prochan. Properties of probability distributions with monotone hazard rate. *Ann. Math. Statist.*, 34:375–389, 1963.

I.V. Basawa and B.L.S. Prakasa Rao. *Statistical Inference for Stochastic Processes*. Academic Press, London, 1980.

L.E. Baum. An inequality and associated maximization technique in statistical estimation for probabilistic functions of Markov processes. In O. Shisha, editor, *Inequalities III: Proceedings of the Third Symposium on Inequalities Held at the University of California, Los Angeles, 1–9 September, 1969*, pages 1–8. Academic Press, New York, 1972.

L.E. Baum and J.A. Eagon. An inequality with applications to statistical estimation for probabilistic functions of Markov processes and to a model for ecology. *Bull. Amer. Math. Soc.*, 73:360–363, 1967.

L.E. Baum and T. Petrie. Statistical inference for probabilistic functions of finite state Markov chains. *Ann. Math. Statist.*, 37:1554–1563, 1966.

L.E. Baum, T. Petrie, G. Soules, and N. Weiss. A maximization technique occurring in the statistical analysis of probabilistic functions of Markov chains. *Ann. Math. Statist.*, 41(1):164–171, 1970.

P.J. Bickel and Y. Ritov. Inference in hidden Markov models i: local asymptotic normality in the stationary case. *Bernoulli*, 2(3):199–228, 1996.

P.J. Bickel, Y. Ritov, and T. Rydén. Asymptotic normality of the maximum-likelihood estimator for general hidden Markov models. *Ann. Statist.*, 26:1614–1635, 1998.

P.J. Bickel, Y. Ritov, and T. Rydén. Hidden Markov model likelihoods and their derivatives behave like i.i.d. ones. *Ann. Inst. H. Poincaré Probab. Statist.*, 38(6):825–846, 2002.

C. Biernacki, G. Celeux, and G. Govaert. Choosing starting values for the EM algorithm for getting the highest likelihood in multivariate Gaussian mixture models. *Comput. Statist. Data Anal.*, 41:561–575, 2003.

P. Billingsley. *Statistical Inference for Markov Processes*. University of Chicago Press, Chicago, 1961a.

P. Billingsley. The Lindeberg–Lévy theorem for martingales. *Proc. Amer. Math. Soc.*, 12:788–792, 1961b.

P. Billingsley. *Probability and Measure*. Wiley, New York, 3rd edition, 1995.

P. Billingsley. *Convergence of Probability Measures*. Wiley, New York, 2nd edition, 1999.

A. Bonafonte, J. Vidal, and A. Nogueiras. Duration modeling with expanded HMM applied to speech recognition. In *Proceedings of ICSLP96*. Philadelphia, 1996.

D. Bosq. *Nonparametric Statistics for Stochastic Processes–Estimation and Prediction*, volume 110 of *Lecture Notes in Statistics*. Springer, New York, 1996.

C. Bracquemond. *Modélisation stochastique du vieillissement en temps discret*. Ph.D. thesis, Laboratoire de Modélisation et Calcul, INPG, Grenoble, 2001.

C. Bracquemond and O. Gaudoin. A survey on discrete lifetime distributions. *Int. J. Reliabil., Qual., Safety Eng.*, 10(1):69–98, 2003.

P. Brémaud. *Markov Chains, Gibbs Fields, Monte Carlo Simulations, and Queues*. Springer, New York, 1999.

J. Bulla and I. Bulla. Stylized facts of financial time series and hidden semi-Markov models. *Comput. Statist. Data Anal.*, 51:2192–2209, 2006.

C. Burge and S. Karlin. Prediction of complete gene structures in human genomic DNA. *J. Mol. Biol.*, 268:78–94, 1997.

O. Cappé. Ten years of HMMs, March 2001. URL: http://www.tsi.enst.fr/~cappe/docs/hmmbib.html.

O. Cappé, E. Moulines, and T. Ryden. *Inference in Hidden Markov Models*. Springer Series in Statistics. Springer, New York, 2005.

G. Celeux and J. Diebolt. The SEM algorithm: a probabilistic teacher algorithm derived from the EM algorithm for the mixture problem. *Comput. Statist. Quart.*, 2:73–82, 1985.

G. Celeux, J. Diebolt, and C. Robert. Bayesian estimation of hidden Markov chains: a stochastic implementation. *Statist. Probab. Lett.*, 16:77–83, 1993.

O. Chryssaphinou, M. Karaliopoulou, and N. Limnios. On discrete time semi-Markov chains and applications in words occurrences. *Comm. Statist. Theory Methods*, 37:1306–1322, 2008.

G. Churchill. Hidden Markov chains and the analysis of genome structure. *Comput. Chem.*, 16:107–115, 1992.

G.A. Churchill. Stochastic models for heterogeneous DNA sequences. *Bull. Math. Biol.*, 51(1):79–94, 1989.

E. Çinlar. *Introduction to Stochastic Processes*. Prentice Hall, New York, 1975.

E. Çinlar. Markov renewal theory. *Adv. in Appl. Probab.*, 1:123–187, 1969.

D.R. Cox. *Renewal Theory*. Methuen, London, 1962.

A. Csenki. Transition analysis of semi-Markov reliability models - a tutorial review with emphasis on discrete-parameter approaches. In S. Osaki, editor, *Stochastic Models in Reliability and Maintenance*, pages 219–251. Springer, Berlin, 2002.

D. Dacunha-Castelle and M. Duflo. *Probability and Statistics*, volume 2. Springer, Berlin, 1986.

H. Daduna. *Queueing Networks with Discrete Time Scale*, volume 2046 of *Lecture Notes in Computer Science*. Springer, 2001.

A.P. Dempster, N.M. Laird, and D.B. Rubin. Maximum likelihood from incomplete data via the EM algorithm. *J. R. Stat. Soc. Ser. B Stat. Methodol.*, 39:1–38, 1977.

P.A. Devijver. Baum's forward-backward algorithm revisited. *Pattern Recog. Lett.*, 3(6):369–373, 1985.

J. Diebolt and E.H.S. Ip. A stochastic EM algorithm for approximating the maximum likelihood estimate. In W.R. Gilks, S.T. Richardson, and D.J. Spiegelhalte, editors, *Markov Chain Monte Carlo in Practice*. Chapman and Hall, London, 1996.

J.L. Doob. *Stochastic Processes*. Wiley, New York, 1954.

R. Douc. Non singularity of the asymptotic Fisher information matrix in hidden Markov models. *École Polytechnique, preprint*, 2005.

R. Douc and C. Matias. Asymptotics of the maximum likelihood estimator for general hidden Markov models. *Bernoulli*, 7(3):381–420, 2001.

R. Douc, E. Moulines, and T. Rydén. Asymptotic properties of the maximum likelihood estimator in autoregressive models with Markov regime. *Ann. Statist.*, 32(5):2254–2304, 2004.

A.P. Dunmur and D.M. Titterington. The influence of initial conditions on maximum likelihood estimation of the parameters of a binary hidden Markov model. *Statist. Probab. Lett.*, 40:67–73, 1998.

J.-B. Durand. *Modèles à structure cachée: inférence, estimation, sélection de modèles et applications*. Ph.D. thesis, Université Grenoble I, Grenoble, 2003.

J.-B. Durand and O. Gaudoin. Software reliability modelling and prediction with hidden Markov chains. *Statistical Modelling–An International J.*, 5(1):75–93, 2005.

R. Durbin, S.R. Eddy, A. Krogh, and G.J. Mitchison. *Biological Sequence Analysis: Probabilistic Models of Proteins and Nucleic Acids*. Cambridge University Press, Cambridge, 1998.

R. Durrett. *Probability: Theory and Examples*. Wadsworth and Brooks/Cole, Pacific Grove, CA., 1991.

B. Efron and D.V. Hinkley. The observed versus expected information. *Biometrika*, 65:457–487, 1978.

D.A. Elkins and M.A. Wortman. On numerical solution of Markov renewal equation: tight upper and lower bounds. *Methodol. Comput. Appl. Probab.*, 3:239–253, 2002.

R.J. Elliott, L. Aggoun, and J.B. Moore. *Hidden Markov Models: Estimation and Control*. Springer, New York, 1995.

Y. Ephraim and N. Merhav. Hidden Markov processes. *IEEE Trans. Inform. Theory*, 48:1518–1569, 2002.

S.N. Ethier and T.G. Kurtz. *Markov Processes: Characterization and Convergence*. Wiley, New York, 1986.

W. Feller. *An Introduction to Probability Theory and Its Applications*, volume 1. Wiley, New York, 3rd edition, 1993.

W. Feller. *An Introduction to Probability Theory and Its Applications*, volume 2. Wiley, New York, 2nd edition, 1971.

J.D. Ferguson. Variable duration models for speech. In *Proc. of the Symposium on the Application of Hidden Markov Models to Text and Speech*, pages 143–179. Princeton, NJ, 1980.

H. Furstenberg and H. Kesten. Products of random matrices. *Ann. Math. Statist.*, 31:457–469, 1960.

I.I. Gerontidis. Semi-Markov replacement chains. *Adv. in Appl. Probab.*, 26:728–755, 1994.

I. Gertsbakh. *Reliability Theory*. Springer, Berlin, 2000.

I.I. Gihman and A.V. Skorohod. *Theory of Stochastic Processes*, volume 1,2,3. Springer, Berlin, 1974.

R.D. Gill. Nonparametric estimation based on censored observations of Markov renewal processes. *Z. Wahrsch. verw. Gebiete.*, 53:97–116, 1980.

V. Girardin and N. Limnios. *Probabilités en Vue des Applications*. Vuibert, Paris, 2001.

V. Girardin and N. Limnios. Entropy for semi-Markov processes with Borel state spaces: asymptotic equirepartition properties and invariance principles. *Bernoulli*, 12(2):1–19, 2006.

V. Girardin and N. Limnios. On the entropy of semi-Markov systems. *J. Appl. Probab.*, 40(4):1060–1068, 2003.

V. Girardin and N. Limnios. Entropy rate and maximum entropy methods for countable semi-Markov chains. *Comm. Statist. Theory Methods*, 33(3):609–622, 2004.

P. Giudici, T. Rydén, and P. Vandekerkhove. Likelihood-ratio tests for hidden Markov models. *Biometrics*, 56:742–747, 2000.

Y. Guédon. Estimating hidden semi-Markov chains from discrete sequences. *J. Comput. Graph. Statist.*, 12(3):604–639, 2003.

Y. Guédon. Computational methods for discrete hidden semi-Markov chains. *Appl. Stochast. Models Business Ind.*, 15:195–224, 1999.

Y. Guédon and C. Cocozza-Thivent. Explicit state occupancy modelling by hidden semi-Markov models: application of Derin's scheme. *Comput. Speech Lang.*, 4: 167–192, 1990.

A. Gut. *Stopped Random Walks. Limit Theorems and Applications*, volume 5 of *Applied Probability. A Series of the Applied Probability Trust*. Springer, New York, 1988.

P. Hall and C. Heyde. *Martingale Limit Theorems and Its Applications*. Academic Press, New York, 1980.

N. Heutte and C. Huber-Carol. Semi-Markov models for quality of life data with censoring. In M. Mesbah, M.-L.T. Lee, and B.F. Cole, editors, *Statistical Methods for Quality of Life Studies*, pages 207–218. Kluwer, Dordrecht, 2002.

C.C. Heyde. *Quasi-Likelihood and Its Application: a General Approach to Optimal Parameter Estimation*, volume 2 of *Springer Series in Statistics*. Springer, New York, 1997.

R. Howard. *Dynamic Probabilistic Systems*, volume 2. Wiley, New York, 1971.

A.V. Huzurbazar. *Flowgraph Models for Multistate Time-to-Event Data*. Wiley Interscience, New York, 2005.

M. Iosifescu. *Finite Markov Processes and Their Applications*. Wiley, Cinchester & Bucharest, 1980.

M. Iosifescu, N. Limnios, and G. Oprişan. *Modèles Stochastiques*. Collection Méthodes Stochastiques Appliquées. Hermes and Lavoisier, Paris, 2nd edition, 2008.

J. Jacod and A.N. Shiryaev. *Limit Theorems for Stochastic Processes*. Springer, Berlin, 1987.

J. Janssen and R. Manca. *Applied Semi-Markov Processes*. Springer, New York, 2006.

J.L. Jensen and N.V. Petersen. Asymptotic normality of the maximum likelihood estimator in state space models. *Ann. Statist.*, 27:514–535, 1999.

N.L. Johnson, A.W. Kemp, and S. Kotz. *Univariate Discrete Distributions*. Wiley, New York, 3rd edition, 2005.

S. Karlin and H.M. Taylor. *A First Course in Stochastic Processes*. Academic Press, New York, 2nd edition, 1975.

S. Karlin and H.M. Taylor. *A Second Course in Stochastic Processes*. Academic Press, New York, 1981.

N.V. Kartashov. *Strong Stable Markov Chains*. VSP Utrecht, TBiMC Kiev, 1996.

J. Keilson. *Markov Chains Models–Rarity and Exponentiality*. Springer, New York, 1979.

J. Kiefer and J. Wolfowitz. Consistency of the maximum likelihood estimator in the presence of infinitely many incidental parameters. *Ann. Math. Statist.*, 27: 887–906, 1956.

M. Kijima. *Stochastic Processes with Applications to Finance*. Chapman and Hall/CRC, London, 2003.

J.F.C. Kingman. Subadditive processes. In *École d'Été de Probabilités de Saint-Flour V–1975*, volume 539 of *Lecture Notes in Mathematics*, pages 167–223. Springer, Berlin, 1976.

T. Koski. *Hidden Markov Models for Bioinformatics*. Kluwer, Dordrecht, 2001.

V. Krisnamurthy and T. Rydén. Consistent estimation of linear and non-linear autoregressive models with Markov regime. *J. Time Ser. Anal.*, 19:291–307, 1998.

A. Krogh, B. Larsson, G. von Heijne, and E.L.L. Sonnhammer. Predicting transmembrane protein topology with a hidden Markov model: application to complete genomes. *J. Mol. Biol.*, 305(3):567–580, 2001.

S.W. Lagakos, C.J. Sommer, and M. Zelen. Semi-Markov models for partially censored data. *Biometrika*, 65(2):311–317, 1978.

C.-D. Lai and M. Xie. *Stochastic Ageing and Dependence for Reliability*. Springer, New York, 2006.

J. Lamperti. *Stochastic Processes*. Springer, New York, 1977.

B. Lapeyre, E. Pardoux, and R. Sentis. *Méthodes de Monte-Carlo pour les Equations de Transport et de Diffusion*. Springer, Paris, 1998.

N.D. Le, B.G. Leroux, and M.L. Puterman. Reader reaction: exact likelihood evaluation in a Markov mixture model for time series of seizure counts. *Biometrics*, 48:317–323, 1992.

L. Le Cam and G.L. Yang. *Asymptotics in Statistics: Some Basic Concepts*. Springer, Berlin, 2nd edition, 2000.

F. Le Gland and L. Mevel. Exponential forgetting and geometric ergodicity in hidden Markov models. *Math. Control Signals Syst.*, 13:63–93, 2000.

M.-L.T. Lee. *Analysis of Microarray Gene Expression Data*. Kluwer, Boston, 2004.

B.G. Leroux. Maximum-likelihood estimation for hidden Markov models. *Stochastic Process Appl.*, 40:127–143, 1992.

B.G. Leroux and M.L. Puterman. Maximum-penalized-likelihood estimation for independent and Markov-dependent mixture models. *Biometrics*, 48:545–558, 1992.

S.E. Levinson. Continuously variable duration hidden Markov models for automatic speech recognition. *Comput. Speech Lang.*, 1:29–45, 1986.

P. Lévy. Processus semi-markoviens. In *Proc. of International Congress of Mathematics, Amsterdam*, 1954.

N. Limnios. A functional central limit theorem for the empirical estimator of a semi-Markov kernel. *J. Nonparametr. Statist.*, 16(1–2):13–18, 2004.

N. Limnios and G. Oprişan. *Semi-Markov Processes and Reliability*. Birkhäuser, Boston, 2001.

N. Limnios and G. Oprişan. A unified approach for reliability and performability evaluation of semi-Markov systems. *Appl. Stoch. Models Bus. Ind.*, 15:353–368, 1999.

N. Limnios and B. Ouhbi. Nonparametric estimation of some important indicators in reliability for semi-Markov processes. *Stat. Methodol.*, 3:341–350, 2006.

N. Limnios and B. Ouhbi. Empirical estimators of reliability and related functions for semi-Markov systems. In *Mathematical and Statistical Methods in Reliability*, volume 7, pages 469–484. World Scientific, Singapore, 2003.

N. Limnios, M. Mesbah, and A. Sadek. A new index for longitudinal quality of life: modelling and estimation. *Environmetrics*, 15:483–490, 2004.

N. Limnios, B. Ouhbi, and A. Sadek. Empirical estimator of stationary distribution for semi-Markov processes. *Comm. Statist. Theory Methods*, 34(4), 2005.

G. Lindgren. Markov regime models for mixed distributions and switching regressions. *Scand. J. Statist.*, 5:81–91, 1978.

T. Lindvall. *Lectures on the Coupling Method.* Wiley, New York, 1992.

L.A. Liporace. Maximum likelihood estimation for multivariate observations of Markov sources. *IEEE Trans. Inform. Theory*, 28:729–734, 1982.

R.Sh. Liptser and A.N. Shiryayev. *Theory of Martingales.* Kluwer, Dordrecht, 1989.

G. Lisnianski and G. Levitin. *Multi-State System Reliability.* World Scientific, NJ, 2003.

R.J.A. Little and D.B. Rubin. *Statistical Analysis with Missing Data.* Wiley, New York, 1987.

T.A. Louis. Finding the observed information matrix when using the EM algorithm. *J. R. Stat. Soc. Ser. B Stat. Methodol.*, 44:226–233, 1982.

I.L. MacDonald and W. Zucchini. *Hidden Markov and Other Models for Discrete-Valued Time Series.* Chapman and Hall, London, 1997.

R.J. MacKay. Estimating the order of a hidden Markov model. *Canad. J. Statist.*, 30(4):573–589, 2002.

M. Maxwell and M. Woodroofe. Central limit theorems for additive functionals of Markov chains. *Ann. Probab.*, 28(2):713–724, 2000.

L. Mazliak, P. Priouret, and P. Baldi. *Martingales et Markov Chains.* Hermann, Paris, 1998.

G.J. McLachlan and T. Krishnan. *The EM Algorithm and Extensions.* Wiley, New York, 1997.

G.J. McLachlan and D. Peel. *Finite Mixture Models.* Wiley, New York, 2000.

S.P. Meyn and R.L. Tweedie. *Markov Chains and Stochastic Stability.* Springer, Berlin, 1993.

C.D. Mitchell and L.H. Jamieson. Modeling duration in a hidden Markov model with the exponential family. In *Proceedings of the 1993 IEEE International Conference on Acoustics, Speech, and Signal Processing*, pages II.331–II.334. Minneapolis, 1993.

C.D. Mitchell, M.P. Harper, and L.H. Jamieson. On the complexity of explicit duration HMMs. *IEEE Trans. Speech Audio Process.*, 3(3):213–217, 1995.

C.J. Mode and G.T. Pickens. Computational methods for renewal theory and semi-Markov processes with illustrative examples. *Amer. Statist.*, 42(2):143–152, 1998.

C.J. Mode and C.K. Sleeman. *Stochastic Processes in Epidemiology.* World Scientific, NJ, 2000.

218   References

E.H. Moore and R. Pyke. Estimation of the transition distributions of a Markov renewal process. *Ann. Inst. Statist. Math.*, 20:411–424, 1968.

F. Muri-Majoube. *Comparaison d'algorithmes d'identification de chaînes de Markov cachées et application à la détection de régions homogènes dans les séquences d'ADN*. Ph.D. thesis, Université Paris 5, Paris, 1997.

K.P. Murphy. Hidden semi-Markov models. Research report, MIT, 2002. URL: http://www.cs.ubc.ca/~murphyk/Papers/segment.pdf.

T. Nakagawa and S. Osaki. The discrete Weibull distribution. *IEEE Trans. Reliabil.*, R-24:300–301, 1975.

M. Neuts. *Matrix-Geometric Solutions in Stochastic Models*. The Johns Hopkins University Press, Baltimore, 1981.

S.F. Nielsen. The stochastic EM algorithm: Estimation and asymptotic results. *Bernoulli*, 6:457–489, 2000.

M. Nikulin, L. Léo Gerville-Réache, and V. Couallier. *Statistique des Essais Accélérés*. Collection Méthodes Stochastiques Appliquées. Hermes, 2007.

J.R. Norris. *Markov Chains*. Cambridge University Press, Cambridge, 1997.

E. Nummelin. *General Irreducible Markov Chains and Non-negative Operators*. Cambridge University Press, Cambridge, 1984.

S. Orey. *Lecture Notes on Limit Theorems for Markov Chain Transition Probabilities*. Van Nostrand Reinhold, London, 1971.

B. Ouhbi and N. Limnios. Non-parametric failure rate estimation of semi-Markov systems. In J. Janssen and N. Limnios, editors, *Semi-Markov models and applications*, pages 207–218. Kluwer, Dordrecht, 1998.

B. Ouhbi and N. Limnios. Nonparametric estimation for semi-Markov processes based on its hazard rate functions. *Stat. Inference Stoch. Process.*, 2(2):151–173, 1999.

B. Ouhbi and N. Limnios. Nonparametric reliability estimation for semi-Markov processes. *J. Statist. Plann. Inference*, 109(1-2):155–165, 2003a.

B. Ouhbi and N. Limnios. The rate of occurrence of failures for semi-Markov processes and estimation. *Statist. Probab. Lett.*, 59(3):245–255, 2003b.

Y.K. Park, C.K. Un, and O.W. Kwon. Modeling acoustic transitions in speech by modified hidden Markov models with state duration and state duration-dependent observation probabilities. *IEEE Trans. Speech, Audio Process.*, 4 (5):389–392, 1996.

E. Parzen. *Stochastic Processes*. SIAM Classics, Philadelphia, 1999.

T. Petrie. Probabilistic functions of finite state Markov chains. *Ann. Math. Statist.*, 40:97–115, 1969.

W. Pieczynski. Chaînes de Markov triplet (triplet Markov chains). *C. R. Acad. Sci. Paris, Ser. I*, 335(3):275–278, 2002.

A. Platis, N. Limnios, and M. Le Du. Hitting times in a finite non-homogeneous Markov chain with applications. *Appl. Stoch. Models Data Anal.*, 14:241–253, 1998.

S.C. Port. *Theoretical Probability for Applications*. Wiley, New York, 1994.

R. Pyke. Markov renewal processes: definitions and preliminary properties. *Ann. Math. Statist.*, 32:1231–1241, 1961a.

R. Pyke. Markov renewal processes with finitely many states. *Ann. Math. Statist.*, 32:1243–1259, 1961b.

R. Pyke and R. Schaufele. Limit theorems for Markov renewal processes. *Ann. Math. Statist.*, 35:1746–1764, 1964.

R. Pyke and R. Schaufele. The existence and uniqueness of stationary measures for Markov renewal processes. *Ann. Math. Statist.*, 37:1439–1462, 1966.

Z. Qingwei, W. Zuoying, and L. Dajin. A study of duration in continuous speech recognition based on DDBHMM. In *Proceedings of EUROSPEECH'99*, pages 1511–1514, 1999.

L.R. Rabiner. A tutorial on hidden Markov models and selected applications in speech recognition. *Proc. IEEE*, 77:257–286, 1989.

S. Resnick. *Adventures in Stochastic Processes*. Birkhäuser, Boston, 4th printing, 2005.

D. Revuz. *Markov Chains*. North-Holland, Amsterdam, 1975.

I. Rezek, M. Gibbs, and S.J. Roberts. Maximum a posteriori estimation of coupled hidden Markov models. *J. VLSI Signal Process.*, 32(1), 2002.

C.P. Robert, G. Celeux, and J. Diebolt. Bayesian estimation of hidden Markov chains: a stochastic implementation. *Statist. Probab. Lett.*, 16(1):77–83, 1993.

C.P. Robert, T. Rydén, and D.M. Titterington. Bayesian inference in hidden Markov models through reversible jump Markov chain Monte Carlo. *J. R. Stat. Soc. Ser. B Stat. Methodol.*, 62:57–75, 2000.

L.C.G. Rogers and D. Williams. *Diffusions, Markov Processes, and Martingales*, volume 1, 2. Wiley, Chichester, 1994.

S.M. Ross. *Stochastic Processes*. Wiley, New York, 1983.

D. Roy and R. Gupta. Classification of discrete lives. *Microelectron. Reliabil.*, 32 (10):1459–1473, 1992.

T. Rydén. Consistent and asymptotically normal parameter estimates for hidden Markov models. *Ann. Statist.*, 22:1884–1895, 1994.

T. Rydén. Estimating the order of hidden Markov models. *Statistics*, 26:345–354, 1995.

T. Rydén. On recursive estimation for hidden Markov models. *Stochastic Process. Appl.*, 66:79–96, 1997.

T. Rydén. Asymptotically efficient recursive estimation for incomplete data models using the observed information. *Metrika*, 44:119–145, 1998.

T. Rydén and D.M. Titterington. Computational bayesian analysis of hidden Markov models. *J. Comput. Graph. Statist.*, 7:194–211, 1998.

A. Sadek and N. Limnios. Nonparametric estimation of reliability and survival function for continuous-time finite Markov processes. *J. Statist. Plann. Inference*, 133(1):1–21, 2005.

A. Sadek and N. Limnios. Asymptotic properties for maximum likelihood estimators for reliability and failure rates of Markov chains. *Comm. Statist. Theory Methods*, 31(10):1837–1861, 2002.

J. Sansom and C.S. Thompson. Mesoscale spatial variation of rainfall through a hidden semi-Markov model of breakpoint data. *J. of Geophysical Research*, 108 (D8), 2003.

J. Sansom and P.J. Thomson. Fitting hidden semi-Markov models to breakpoint rainfall data. *J. Appl. Probab.*, 38A:142–157, 2001.

J. Sansom and P.J. Thomson. On rainfall seasonality using a hidden semi-Markov model. *J. of Geophysical Research*, 112(D15), 2007.

D.N. Shanbhag and C.R. Rao (Eds.). *Stochastic Processes: Theory and Methods*, volume 19 of *Handbook of Statistics*. Elsevier, 2001.

A.N. Shiryaev. *Probability*. Springer, New York, 2nd edition, 1996.

V.M. Shurenkov. *Ergodic Theorems and Related Problems*. VSP, Utrecht, 1998.

D.S. Silvestrov. *Limit Theorems for Randomly Stopped Stochastic Processes.* Probability and Its Applications. Springer, London, 2004.

W.L. Smith. Regenerative stochastic processes. *Proc. R. Soc. Lond. Ser. A Math. Phys. Eng.*, 232:6–31, 1955.

W.L. Smith. Renewal theory and its ramifications. *J. R. Stat. Soc. Ser. B Stat. Methodol.*, 20(2):243–302, 1958.

F. Stenberg. *Semi-Markov models for insurance and option rewards.* Ph.D. thesis no. 38, Mälardalen University, Sweden, December 2006.

L. Takacs. Some investigations concerning recurrent stochastic processes of a certain type. *Magyar Tud. Akad. Mat. Kutato Int. Kzl.*, 3:115–128, 1954.

M.A. Tanner. *Tools for Statistical Inference. Methods for the Exploration of Posterior Distributions and Likelihood Functions.* Springer, New York, 3rd edition, 1996.

S. Trevezas and N. Limnios. Variance estimation in the central limit theorem for Markov chains. *to appear in J. Statist. Plann. Inference*, 2008a.

S. Trevezas and N. Limnios. Maximum likelihood estimation for general hidden semi-Markov processes with backward recurrence time dependence. *to appear in J. of Mathematical Sciences*, 2008b.

A.W. Van der Vaart. *Asymptotic Statistics.* Cambridge Series in Statistical and Probabilistic Mathematics. Cambridge University Press, Cambridge, 1998.

A.W. Van der Vaart and J.A. Wellner. *Weak Convergence and Empirical Processes.* Springer, New York, 1996.

P. Vandekerkhove. Consistent and asymptotically normally distributed estimates for hidden Markov mixtures of Markov models. *Bernoulli*, 11:103–129, 2005.

S.V. Vaseghi. State duration modelling in hidden Markov models. *Signal Process.*, 41(1):31–41, 1995.

P.-C.G. Vassiliou and A.A. Papadopoulou. Non-homogeneous semi-Markov systems and maintainability of the state sizes. *J. Appl. Probab.*, 29:519–534, 1992.

P.-C.G. Vassiliou and A.A. Papadopoulou. Asymptotic behavior of non homogeneous semi-Markov systems. *Linear Algebra Appl.*, 210:153–198, 1994.

Shurenkov V.M. On the theory of Markov renewal. *Theory Probab. Appl.*, 19(2):247–265, 1984.

C.F.J. Wu. On the convergence properties of the EM algorithm. *Ann. Statist.*, 11(1):95–103, 1983.

M Xie, O. Gaudoin, and C. Bracquemond. Redefining failure rate function for discrete distributions. *Int. J. Reliabil., Qual., Safety Eng.*, 9(3):275–285, 2002.

G.G. Yin and Q. Zhang. *Discrete-Time Markov Chains. Two-Time-Scale Methods and Applications.* Springer, New York, 2005.

S.-Z. Yu and H. Kobayashi. A hidden semi-Markov model with missing data and multiple observation sequences for mobility tracking. *Signal Process.*, 83(2):235–250, 2003.

Index

Restricted Parameter Space Estimation Problems

Constance van Eeden

Point process statistics is successfully used in fields such as material science, human epidemiology, social sciences, animal epidemiology, biology, and seismology. Its further application depends greatly on good software and instructive case studies that show the way to successful work. This book satisfies this need by a presentation of the spatstat package and many statistical examples.

2006. 176 pp. (Lecture Notes in Statistics, Vol. 188) Softcover
ISBN 978-0-387-33747-0

The Nature of Statistical Evidence

Bill Thompson

The purpose of this book is to discuss whether statistical methods make sense. That is a fair question, at the heart of the statistician-client relationship, but put so boldly it may arouse anger. The many books entitled something like Foundations of Statistics avoid controversy by merely describing the various methods without explaining why certain conclusions may be drawn from certain data. But we statisticians need a better answer then just shouting a little louder. To avoid a duel, the authors prejudge the issue and ask the narrower question: "In what sense do statistical methods provide scientific evidence?"

2007. 150 pp. (Lecture Notes in Statistics, Vol. 189) Softcover
ISBN 978-0-387-40050-1

Weak Dependence

J. Dedecker, P. Doukhan, G. Lang, J.R. León, S. Louhichi and C. Prieur

This monograph is aimed at developing Doukhan/Louhichi's (1999) idea to measure asymptotic independence of a random process. The authors propose various examples of models fitting such conditions such as stable Markov chains, dynamical systems or more complicated models, nonlinear, non-Markovian, and heteroskedastic models with infinite memory. Most of the commonly used stationary models fit their conditions. The simplicity of the conditions is also their strength.

2007. 322 pp. (Lecture Notes in Statistics, Vol. 190) Softcover
ISBN 978-0-387-69951-6

Easy Ways to Order▶ Call: Toll-Free 1-800-SPRINGER • E-mail: orders-ny@springer.com • Write: Springer, Dept. S8113, PO Box 2485, Secaucus, NJ 07096-2485 • Visit: Your local scientific bookstore or urge your librarian to order.

Printed in the United States